Textile mechanisms in spinning and weaving machines

Textile mechanisms in spinning and weaving machines

Ganapathy Nagarajan

WOODHEAD PUBLISHING INDIA PVT LTD

New Delhi

Published by Woodhead Publishing India Pvt. Ltd.
Woodhead Publishing India Pvt. Ltd.,
303, Vardaan House, 7/28, Ansari Road,
Daryaganj, New Delhi - 110002, India
www.woodheadpublishingindia.com

First published 2015, Woodhead Publishing India Pvt. Ltd.

Woodhead Publishing India Pvt. Ltd. ISBN: 978-93-80308-45-6
Woodhead Publishing India Pvt. Ltd. e-ISBN: 978-93-80308-94-4

Typeset by Mind Box Solutions, New Delhi
Digitally Printed and bound by Replika Press Pvt. Ltd.

Contents

Preface

I am pleased to present a book on "Textile Mechanisms for Spinning and Weaving Machines" for the students of Textile Technology and also for industry professionals. This book gives an exhaustive account on mechanics in textile machinery in spinning, weaving processes in which every textile students must know about the applied mechanics in textile machineries. The book has been written in both SI conventional and still in use F.P.S. systems. The object of this book is to present the subject matter in a most concise, compact and more solved practical problems in a lucid manner.

The aim of the book is to analyze the various mechanisms involved; and many difficult problems are solved in steps which would be of interest to readers. This should help the students to develop an analytical thinking about the machineries. A number of problems involving work energy, impulse-momentum and principle of moments have been taken in to consideration and incorporated in relevant chapters.

The chapters are divided into many sections in order to isolate individual pieces of information as an aid to study. The work is intentionally pitched at a fairly easy level so as to be of the utmost use to as wide range of students not only in India but also for the students studying textile courses in other developed countries.

The subject matter has been amply illustrated by incorporating a good number of solved unsolved and well-graded examples of every variety. At the end of each chapter, numerical questions have been asked so that the students can practice the same which may be asked in their University Examinations.

Although, every care has been taken to check mistakes and misprints, yet it is difficult to attain perfection. Any errors, omissions and suggestions for further improvement will be thankfully acknowledged and the same will be incorporated in the next edition.

I am grateful to my parents, family and my professors who have helped me directly and indirectly to bring out this book. I would like to thank Woodhead Publishing India, who accepted my proposal to publish this book, although other books on this topic are already available. Finally, I hope this book will be of good use to the students of Textile Technology, and I dedicate this book to the great Almighty who has helped and gave me fillip to bring out my maiden book publication.

I look forward in receiving your comments and suggestions to update this book.

G. Nagarajan

Foreword

It gives me immense pleasure to note that the first edition of this book titled "Textile Mechanisms for Spinning and Weaving Machines" is being brought out. I congratulate the author of the book for his dedicated efforts in bringing this book, which has covered almost all critical areas of mechanics involved in textile spinning, weaving and knitting machines thoroughly under 20 different chapters. The chapters discuss types of power drives used in textile machinery, including belt and rope drives, and cover the design of cone drums in speed frames; different types of cams used in ring frames and looms are also discussed, as well as equations of motion and textile applications, energy calculations in textile machines, frictional forces and their calculation, clutches and brakes. The author has worked out sufficient problems in each chapter for better understanding of students and faculty who handle this subject on mechanics. The author has worked in textile industry and also in academics for long years which make this book very practical. This book will be very useful to all undergraduate and postgraduate students of Textile Technology besides students studying textiles in polytechnic colleges.

I wish the author great success in all his future endeavors.

Dr G Ramakrishnan

1

Belt, rope and chain drives

1.1 Introduction

The belt or rope is used to transmit power from one shaft to another by means of pulleys which rotate at the same speed or at different speeds. The amount of power transmitted depends upon the following factors:

(a) The velocity of the belt.
(b) The tension under which the belt is placed on the pulleys.
(c) The arc of contact between the belt and the smaller pulley.
(d) The condition under which the belt is used.

In order to fulfill the above conditions, it may be noted that

1. The shafts should be properly in line to ensure uniform tension across the belt section.
2. The pulleys should not be too close together in order that the arc of contact on the smaller pulley may be as large as possible.
3. The pulleys should not be so far apart which cause the belt to weigh heavily on the shafts and thus increase the friction load on the bearings.
4. A long belt tends to swing from side to side, causing the belt to run out of the pulleys which in turn develops crooked spots.
5. The tight side of the belt should be at the bottom, so that if sag is present on the loose side will increase the arc of contact at the pulleys.
6. In order to obtain good results with the flat belts, the maximum distance between the shafts should not exceed 10 meters and the minimum should not be less than 3.5 times the diameter of the larger pulley.

1.2 Selection of belt drive

The selection of belt drive depends upon various factors which are listed below:

1. Speed of the driving and driven shafts
2. Speed reduction ratio
3. Power to be transmitted

4. Centre distance between the shafts
5. Positive drive requirements
6. Layout of the shafts
7. Space available
8. Service conditions

1.3 Types of belt drives

(a) *Light drives* – These are used to transmit small powers at belt speed of up to 10 m/s. It is commonly used in agricultural machines and small machine tools.

(b) *Medium drives* – These are used to transmit medium power at belt speeds over 10 m/s up to maximum 22 m/s as in machine tools.

(c) *Heavy drives* – These are used to transmit large powers at belt speeds above 22 m/s. It is commonly found in compressors and generators.

1.4 Types of belts

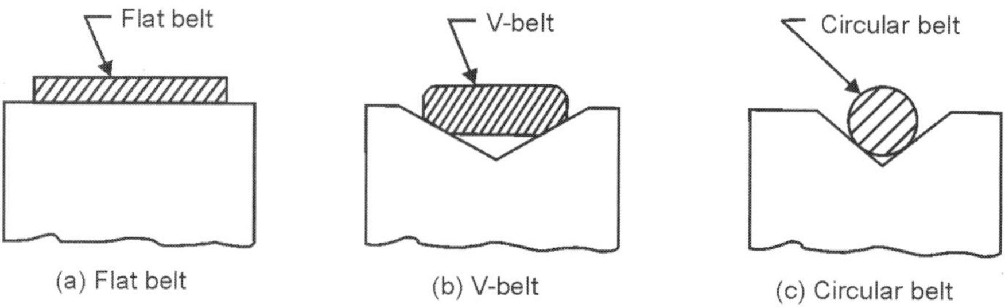

Figure 1.1 Types of belts

1. *Flat belt* – The flat belt as shown in Fig. 1.1(a) is mostly used in factories and workshops, where a moderate amount of power is to be transmitted from one pulley to another. The two pulleys should not be more than 8 meters apart.
2. *V-belt* – The V-belt as shown in Fig. 1.1(b) is mostly used in workshops and factories where a moderate amount of power is to be transmitted from one pulley to another, when the two pulleys are very near to each other.
3. *Circular belt or rope* – The circular belt or rope as shown in Fig. 1.1(c) are commonly used in industries and workshops, where large amount of power is to be transmitted from one pulley to another and also when the two pulleys are more than 8 m apart.

1.5 Flat belt drives

The power from one pulley to another may be transmitted by any of the following types of belt drives.

1. Open belt drive

The open belt drive as shown in Fig. 1.2 is used with shafts arranged in parallel and rotating in the same direction. In this case, the driver A pulls the belt from one side (i.e. lower side RQ and delivers it to the other side, i.e. upper side LM). Thus the tension in the lower side belt will be more than that in the upper side belt. The lower side belt (because of more tension) is known as tight side whereas the upper belt side (because of less tension) is known as slack side as shown in Fig. 1.2.

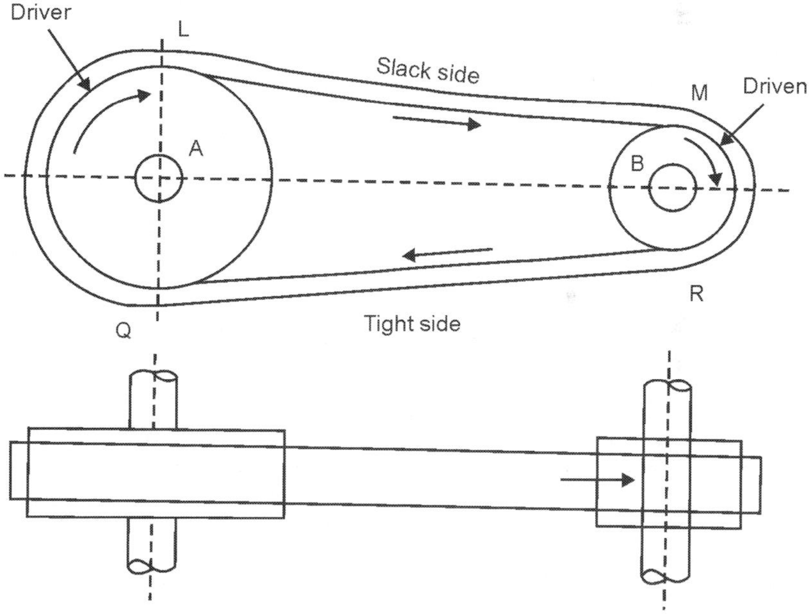

Figure1.2 Open belt drive

2. Crossed belt drive

The crossed belt drive as shown in Fig. 1.3 is used with shafts arranged parallel and rotating in opposite direction.

In this case, the driver pulls the belt from one side (i.e. RQ) and delivers it to the other side (i.e. LM). Thus the tension in the belt RQ will be more than that in the belt LM. The belt RQ (because of more tension) is known as tight side, whereas the belt LM is known as slack side because of less tension (as shown in Fig. 1.3).

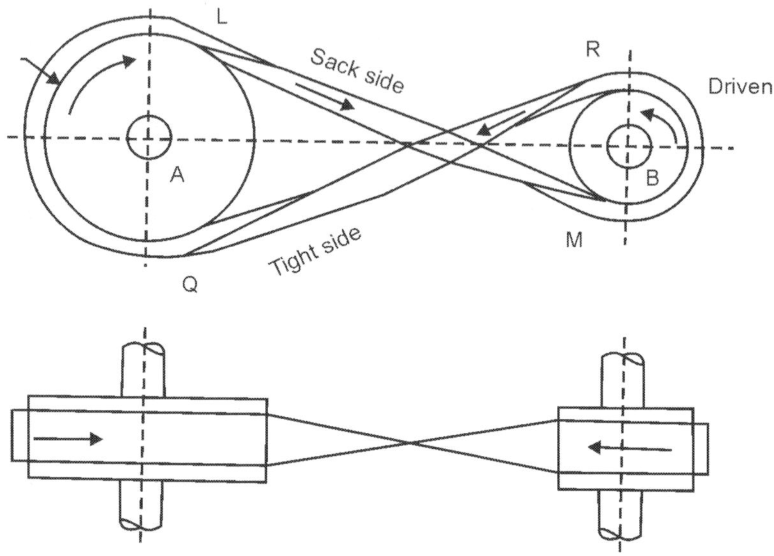

Figure 1.3 Crossed belt drive

At the point where the belt crosses, it rubs against each other and there will be excessive wear and tear. In order to avoid this, the shafts should be placed at a maximum distance of 20 b, where b is the width of the belt and the speed of the belt should be less than 15 m/s.

3. Quarter turn belt drive

The quarter turn belt drive (Fig. 1.4a), also known as right angle belt drive, is used with shafts arranged at right angles and rotating in one definite direction. In order to prevent the belt from leaving the pulley, the width of the face of the pulley should be greater or equal to 1.4 b, where b is the width of the belt.

In some cases if the pulleys cannot be arranged as shown in Fig. 1.4(a), or when the reversible motion is desired, then a quarter turn belt drive with guide pulley as shown in Fig. 1.4(b) may be used.

4. Belt drive with idler pulleys

A belt drive with idler pulley, as shown in Fig. 1.5(a), is used with shafts arranged parallel and when an open belt drive cannot be used due to small angle of contact on the smaller pulley. This type of drive is provided to obtain high velocity ratio and when the required belt tension cannot be obtained by other means.

When it is desired to transmit motion from one shaft to several shafts, all arranged in parallel, a belt drive with many idler pulleys, as shown in Fig. 1.5(b) may be utilized.

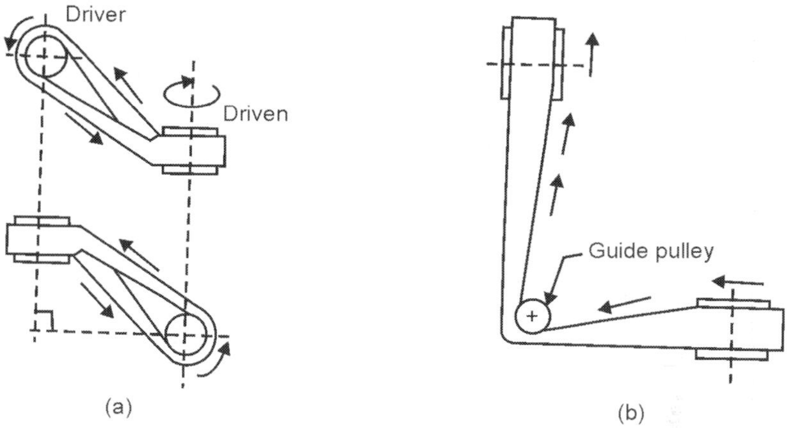

Figure 1.4 (a) Quarter turn belt drive; (b) Quarter turn belt drive with guide pulley

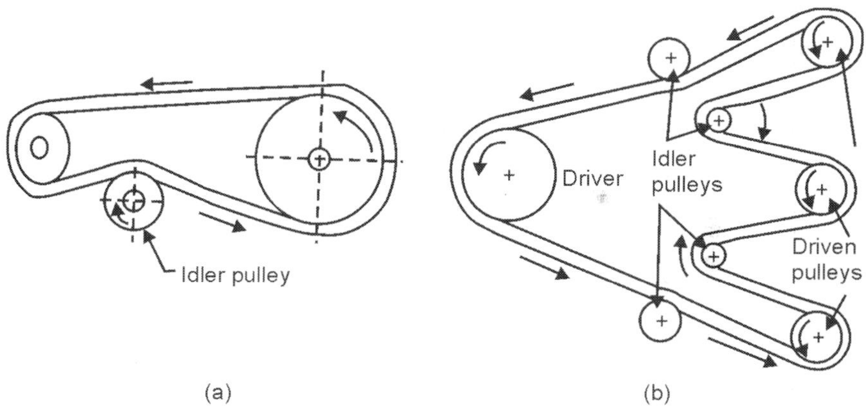

Figure 1.5 (a) Belt drive with single idler pulley; (b) Belt drive with many idler pulleys

5. Compound belt drive

A compound belt drive as shown in Fig. 1.6 is used when power is transmitted from one shaft to another through a number of pulleys.

6. Stepped or cone drive pulley drive

A stepped or cone drive pulley as shown in Fig. 1.7 is used for changing the speed of the driven shaft while the main or driving shaft runs at constant speed. This is accomplished by shifting the belt from one part of the steps to the other (Fig. 1.7).

7. Fast and loose pulley drive

A fast and loose pulley drive as shown in Fig. 1.8 is used, when the driven or machine shaft is to be started or stopped whenever desired without interfering with the driving shaft. A pulley which is keyed to the machine shaft is called fast pulley and runs at the same speed as that of the machine shaft. A loose pulley runs freely over the machine shaft and is incapable of transmitting any power. When the driven shaft is required to be stopped, the belt is shifted on to the loose pulley by means of sliding bar having belt forks.

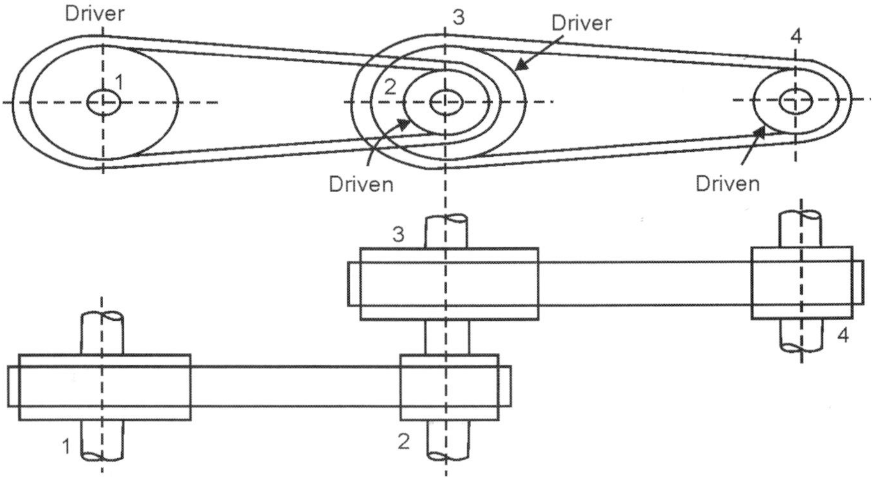

Figure 1.6 Compound belt drive

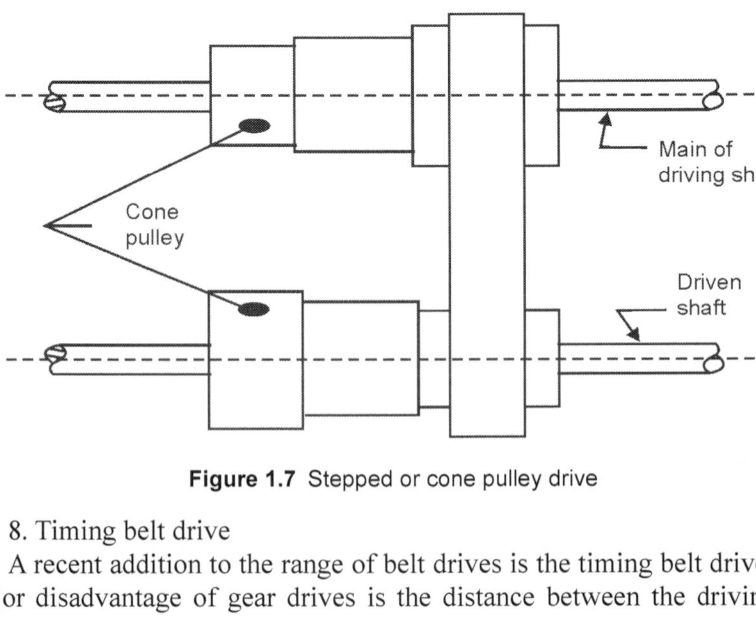

Figure 1.7 Stepped or cone pulley drive

8. Timing belt drive

A recent addition to the range of belt drives is the timing belt drive. One major disadvantage of gear drives is the distance between the driving and driven shafts. Large gear wheels may be necessary or there may be smaller wheels plus a series of idler wheels. One disadvantage of gear drives is negative character and slippage of gears. The timing belt overcomes both disadvantages. Timing belt is a flat belt built with a tough flexible structure and minimum stretch with neoprene teeth molded on its inner surface. The driving and driven pulleys are cut with a matching teeth and thus a positive drive is achieved. Calculation of speed with such a drive is similar to the normal gears. This timing belt drive is found in spinning preparatory machines like carding and draw frame machines.

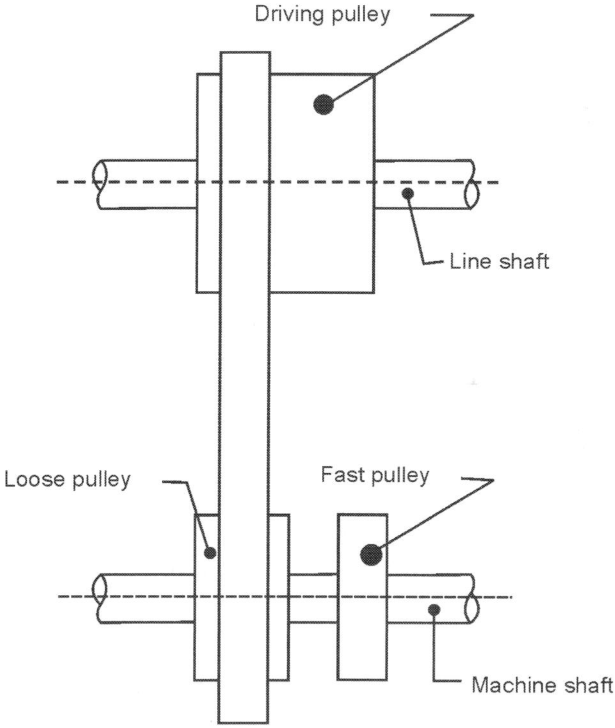

Figure 1.8 Fast and loose pulley drive

9. Cone pulley drive

Cone pulley drive is another form of drive common in textile machines. Such type of drive is used in sizing machines to control the warp stretch in the wet section as shown in Fig. 1.9.

A side shaft driven by the main motor at headstock is employed to drive the drag roller at the head stock and the sizing roller in the size box through the intermediate cone pulleys and a differential. The position of the belt on the cone pulleys is adjusted automatically by a spring-mounted tension feeler which is mounted on the front of the drying chamber. The drive to the drag roller is taken from the side shaft through the manually adjustable expanding pulleys with a view to controlling the dry stretch between the drag roller and the take-up roller at the head stock. This system of controlling stretch gives good results, if properly set.

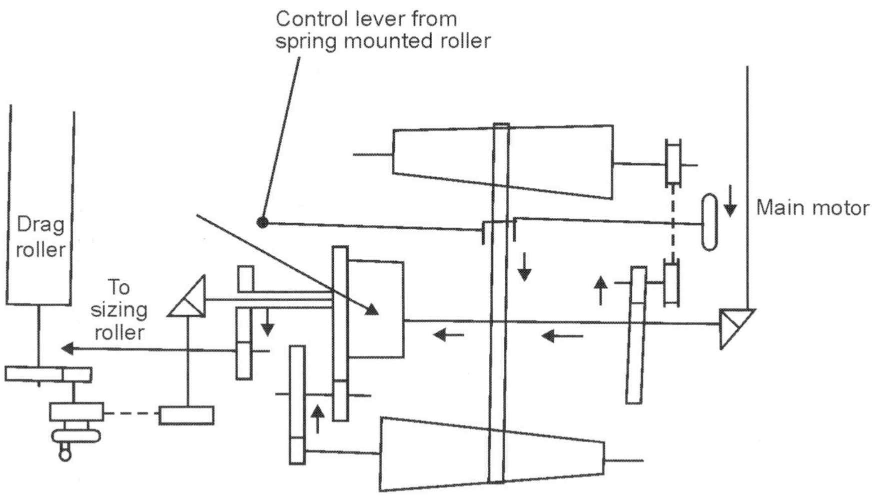

Figure 1.9 Cone pulley drive

10. Positively Infinite Variable (PIV) gear drive

PIV gear comprises two sets of expanding pulleys with serrations on their inner face, which are linked by a chain belt, consisting of thin plates, engaging the serrations. The relative speed of the driving and driven parts is changed by turning a knob, which shifts the belt towards the centre of one set and away from the centre of other set of expanding pulleys as shown in Fig. 1.10.

The PIV drive consists of only one PIV gear for controlling the relative speeds of the sizing rollers in size box and the take-up roller at the head stock. There is no separate control of stretch in dry and wet sections. A tension feeler roller is provided at the front of the drying chamber. The movement of the feeler alters the position of a pointer on a dial between the two limit switches, one each on either side. The contact of the pointer with one of the switches is utilized in sending an impulse to an auxiliary motor capable of running in either direction.

The auxiliary motor changes the setting of the PIV gear in such a manner which changes the speed of the sizing rollers so as to correct the rise or fall in tensions in the warp sheet.

This mechanism prevents the creel tension from reaching the wet sheet between the immersion roller and the rear squeezing nip. Thus the importance of correct setting of PIV gear need not be emphasized.

It should also be ensured that optimum setting of the drive mechanism and proper synchronization among different motors of the multi-motor drives is a pre-requisite for obtaining the lowest possible stretch in warp during sizing process.

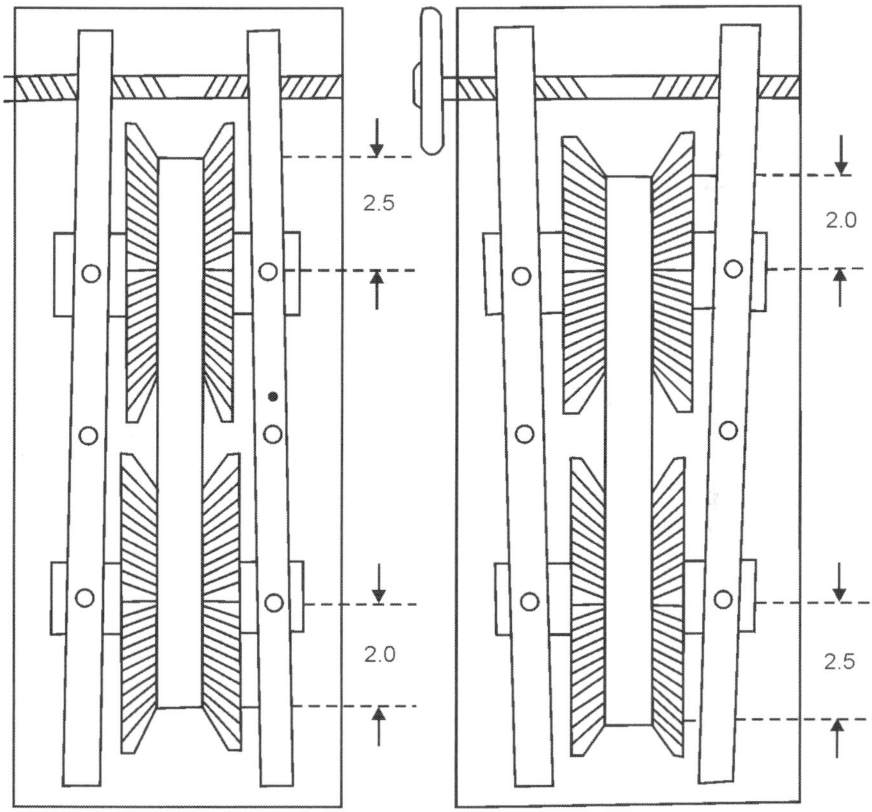

Figure 1.10 PIV gear assembly

1.6 Velocity ratio of belt drive

Velocity ratio of the belt drive is expressed as the ratio between the velocities of the driver and the driven or follower. It may be expressed mathematically, as mentioned below:

D_1 = Diameter of the driver

D_2 = Diameter of the driven or follower

N_1 = Speed of the driver in rpm

N_2 = Speed of the driven in rpm

Therefore, the length of the belt that passes over the driver in 1 minute $= \pi \times D_1 \times N_1$

Similarly, the length of the belt passes over the driven in 1 minute $= \pi \times D_2 \times N_2$

Since, the length of the belt passes over the driven in 1 minute is equal to the length of the belt that passes over the follower in 1 minute, therefore

$$\pi \times D_2 \times N_1 = \pi \times D_2 \times N_2$$

Therefore, Velocity ratio $= \dfrac{N_2}{N_1} = \dfrac{D_1}{D_2}$

If the thickness of the belt (t) is to be considered, then velocity ratio, $= \dfrac{N_2}{N_1} = \dfrac{D_1 + t}{D_2 + t}$

The velocity ratio of a belt drive may also be obtained as discussed below:

$$= V_1 = \frac{\pi \times D_1 \times N_1}{60} \text{ m/s}$$

And the peripheral velocity of the driven or follower pulley is $V_1 = \dfrac{\pi \times D_1 \times N_1}{60}$ m/s

When there is no slip, then $V_1 = V_2$

$$\frac{\pi \times D_1 \times N_1}{60} = \frac{\pi \times D_2 \times N_2}{60} \text{ or}$$

$$\frac{N_2}{N_1} = \frac{D_1}{D_2} \quad \ldots(1.1)$$

1.6.1 Velocity ratio of a compound belt drive

In some occasions, the power is transmitted from one shaft to another through a number of pulleys as shown in Fig. 1.6. Consider a pulley 1 driving the pulleys 2 and 3 keyed to the same shaft; therefore the pulley 1 also drives the pulley 3 which in turn drives the pulley 4.

Let D_1= Diameter of the pulley 1

D_2 = Speed of the pulley 1 in rpm

D_2, D_3, D_3 and N_2, N_3, N_4, = corresponding values for pulleys 2, 3, and 4.

We know that velocity ratio of pulleys 1 and 2,

$$\frac{N_2}{N_1} = \frac{D_1}{D_2} \qquad \text{...(1.2)}$$

Similarly, velocity ratio of pulleys 3 and 4

$$\frac{N_4}{N_3} = \frac{D_3}{D_4} \qquad \text{...(1.3)}$$

Multiplying equations (1.2) and (1.3)

$$\frac{N_2}{N_1} \times \frac{N_4}{N_3} = \frac{D_1}{D_2} \times \frac{D_3}{D_4}$$

or $\frac{N_4}{N_3} = \frac{D_1 \times D_3}{D_2 \times D_4}$ (since $N_2 = N_3$ being keyed to the same shaft)

or $\frac{\text{Speed of the last driven}}{\text{Speed of the first driver}} = \frac{\text{Product of diameters of drivers}}{\text{Product of diameters of driven}}$

1.6.2 Slip of belt

In the previous discussions, we have always considered that there is a firm grip between the belts and shafts. Sometimes, the frictional grip becomes insufficient. This may cause some forward motion of the driver without carrying the belt with it and also the driven pulley. This is called as slip of the belt and is normally expressed as percentage.

The result of the belt slippage is to reduce the velocity ratio of the system. As the slipping of the belt is a common phenomenon, belt should never be used where a definite velocity ratio is required.

Let $S_1\%$ = slip between driver and belt

$S_2\%$ = slip between driven and belt

Therefore, velocity ratio of the belt passing over the driver per second,

$$v = \frac{\pi \times D_1 \times N_1}{60} - \left(\frac{\pi \times D_1 \times N_1}{60} \times \frac{S_1}{100}\right) \qquad \text{...(1.4)}$$

$$= \frac{\pi \times D_1 \times N_1}{60} - \left(1 - \frac{S_1}{100}\right)$$

and the velocity of the belt passing over the driven per second

$$\frac{\pi \times D_2 \times N_2}{60} = v - \left(v \times \frac{S_2}{100}\right) = v\left(1 - \frac{S_2}{100}\right)$$

Substituting the value of v from Eq. (1.4),

$$\frac{\pi \times D_2 \times N_2}{60} = \frac{\pi \times D_1 \times N_1}{60}\left(1 - \frac{S_1}{100}\right)\left(1 - \frac{S_2}{100}\right)$$

$$\frac{N_2}{N_1} = \frac{D_1}{D_2}\left(1 - \frac{S_1}{100} - \frac{S_2}{100}\right)$$

$$= \frac{D_1}{D_2}\left(1 - \frac{S_1 + S_2}{100}\right) = \frac{D_1}{D_2}\left(1 - \frac{S}{100}\right) \text{Neglecting } \frac{S_1 \times S_2}{100 \times 100}$$

Where $S - S_1 + S_2$ = total percentage of slip.

If thickness t of the belt is considered, then,

$$\frac{N_2}{N_1} = \left(\frac{D_1 + t}{D_2 + t}\right)\left(1 - \frac{S}{100}\right) \qquad \ldots(1.5)$$

Example 1.1: (In SI units) An engine running at 150 rpm drives a line shaft by means of a belt. The engine pulley is 750 mm diameter and the pulley on the line shaft being 450 mm. A 900 mm diameter pulley on the line shaft drives a 150 mm diameter pulley keyed to a dynamo shaft. Find the speed of the dynamo shaft when

1. There is no slip
2. There is a slip of 2% at each drive (Fig. 1.11)

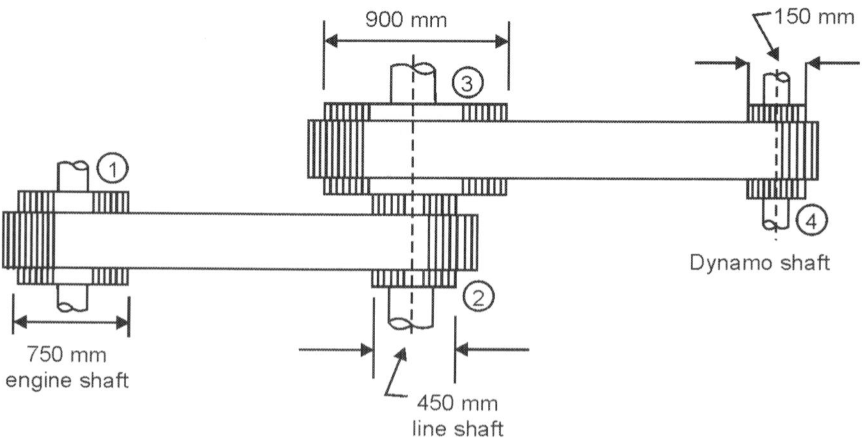

Figure 1.11 Speed of dynamo shaft at: (a) no slip; (b) slip of 2% at each drive

Solution:

Given $N_1 = 150$ rpm; $d_1 = 750$ mm; $d_2 = 450$ mm; $d_3 = 900$ mm; $d_4 = 150$ mm

1. When there is no slip

$$\frac{N_4}{N_1} = \frac{d_1 \times d_3}{d_2 \times d_4}$$

$$\frac{N_4}{150} = \frac{750 \times 900}{450 \times 150} = 10$$

$N_4 = 150 \times 10 = 1500$ rpm

2. When there is a slip of 2%

$$\frac{N_4}{N_1} = \frac{d_1 \times d_3}{d_2 \times d_4}\left(1 - \frac{S_1}{100}\right)\left(1 - \frac{S_2}{100}\right)$$

$$\frac{N_4}{150} = \frac{750 \times 900}{450 \times 150}\left(1 - \frac{2}{100}\right)\left(1 - \frac{2}{100}\right) = 9.6$$

$N_4 = 15089.6 = 1440$ rpm

Example 1.2: Find the speed of a shaft which is driven with the help of a belt by an engine running at 200 rpm. The diameter of the engine pulley is 51 cm and that of the shaft is 30 cm.

Solution:

Speed of the driver (engine) $N_1 = 200$ rpm

Diameter of the driver (engine pulley) $d_1 = 51$ cm

Diameter of driven shaft $d_2 = 50$ cm

Let N_2 be the speed of the driven shaft,

Therefore,

$$\frac{N_2}{N_1} - \frac{d_1}{d_2}$$

$$N_2 = \frac{d_2}{d_1} \times N_1 = \frac{51}{30} \times 200 = 340 \text{ rpm}$$

Example 1.3: If in the above problem, if the thickness of the belt is 10 mm, find the speed of the shaft.

Solution:

Thickness (t) = 10 mm = 1 cm

$N_1 = 200$ rpm

$N_2 = ?$

$d_1 = 51$ cm

$d_2 = 30$ cm

We know that,

$$\frac{N_2}{N_1} = \frac{d_2 + t}{d_2 + t}$$

$$= N_2 = \left(\frac{d_1 + t}{d_2 + t}\right) \times N_1$$

$$= \left(\frac{50 + 1}{30 + 1}\right) \times 200 = \frac{52}{31} \times 200 = 335.4 \text{ rpm}$$

Example 1.4: A shaft running at 200 rpm is to drive a parallel shaft at 300 rpm. The pulley on the driving shaft is 60 cm in diameter. Calculate the diameter of the pulley on the driven shaft.

(i) Neglecting belt thickness

(ii) Belt thickness of 5 mm

(iii) Assuming in the latter case a total slip of 4%

Solution:

$N_1 = 200$ rpm

$N_2 = 300$ rpm

$d_1 = 60$ cm

Thickness ($t = 5$ mm $= 0.5$ cm)

Slip = 4%

$d_2 = ?$

(i) Neglecting belt thickness

$$\frac{N_2}{N_1} = \frac{d_1}{d_2}$$

$$d_2 = \frac{N_1}{N_2} \times d_1$$

$$= \frac{200}{300} \times 60 = 40 \text{ cm}$$

(ii) Taking belt thickness in to account.

$$\frac{N_2}{N_1} = \frac{d_1 + t}{d_2 + t}$$

$$d_2 + t = (d_2 + t) \times \frac{N_1}{N_2}$$

$$= (60 + 0.5) \times \frac{200}{300}$$

$$= 40.33$$

$$d_2 = 40.33 - 0.5 = 39.83 \text{ cm}$$

(iii) Considering belt thickness and total slip.

$$\frac{N_2}{N_1} = \left(\frac{d_1 + t}{d_2 + t}\right)\left(1 - \frac{S}{100}\right)$$

$$\frac{300}{200} = \left(\frac{60.5}{d_2 + 0.5}\right)\left(1 - \frac{4}{100}\right)$$

$$d_2 \times 0.5 = 60.5 \times 0.96 \times \frac{200}{300} = 38.72 \text{ cm}$$

$$d_2 = 38.72 - 0.5 = 38.22 \text{ cm}$$

1.6.3 Creep of belt

When the belt passes from the slack side to the tight side, a certain portion of the belt extends and contracts again when the belt passes from the tight side to slack side. Due to these changes of length, there is relative motion between the belt and the pulley surfaces. This relative motion is called as creep. The net effect of creep is to reduce slightly the speed of the driven pulley or follower. Considering the creep, the velocity ratio is given by

$$\frac{N_2}{N_1} = \frac{d_1}{d_2} \times \frac{E + \sqrt{f_2}}{E + \sqrt{f_1}} \qquad \text{...(1.6)}$$

Example 1.5: The power is transmitted from a pulley of 1 m diameter running at 200 rpm. to a pulley of 2.5 m diameter by means of a belt. Find the speed lost by the driven pulley as a result of the creep, if the stress on the tight and slack side of the belt is 1.44 N/mm^2 and 0.49 N/mm^2. Respectively, the Young's Modulus of the material of the belt is 100 N/mm^2.

Solution:

Diameter of the driver pulley d_1 = 1 m

Diameter of driven pulley d_2 = 2.5 m

Stress on the tight side (f_1) = 1.44 N/mm^2

Stress on the slack side (f_2) = 0.40 N/mm^2

Young's Modulus E = 100 N/mm^2

Let be the speed of the driven pulley,

Using Eq. (1.6), we get

$$\frac{N_2}{N_1} = \frac{d_1}{d_2} \times \frac{E + \sqrt{f_2}}{E + \sqrt{f_1}}$$

$$\frac{N_2}{200} = \frac{1}{2.5} \times \frac{100 + \sqrt{0.49}}{100 + \sqrt{1.44}}$$

$$= \frac{1}{2.5} \times \frac{100.7}{101.12}$$

$$= N_2 = \frac{200}{2.5} \times \frac{100.7}{101.12}$$

$$= 79.67 \text{ rpm}$$

If we neglect the creep, then

$$\frac{N_2}{N_1} = \frac{d_1}{d_2}$$

$$N_2 = \frac{d_1}{d_2} \times N_1$$

$$N_2 = \frac{2.5}{200} \times 200 = 80 \text{ rpm}$$

Speed lost by driven pulley due to creep = 80 – 79.67 = 0.33 rpm

Example 1.6: If in the above problem, the belt is 50 cm wide and 10 mm thick, then find the speed in rpm lost by the driven pulley as a result of creep, if the tensions on the tight side and slack sides of the belt are 10 N/mm width and 4 N/mm width, respectively. E value of the belt is 100 N/mm^2.

Solution:

From Example 1.5, the data are

d_1 = 1 m; = 200 rpm; d_2 = 2.5 m

Width of the belt (b) = 50 cm = 500 mm

Thickness of the belt (t) = 10 mm

Area of cross-section = $b \times t$

= 500 × 10 = 5000 mm^2

Tension in the tight side = 10 N/mm width

= 10 × Width = 10 × 500 = 5000 N

Tension on slack side = 4 N/mm width

$= 4 \times 500 = 2000$ N

Stress on tight side $f_1 = \dfrac{\text{Tension on tight side}}{\text{Area of cross-section}}$

$= \dfrac{5000}{5000} = 1$ N/mm^2

Stress on slack side $f_2 = \dfrac{\text{Tension on slack side}}{\text{Area of cross – section}}$

$= \dfrac{2000}{5000} = 0.4$ N/mm^2

Value of $E = 100$ N/mm^2

Let N_2 be the speed of the driven pulley.

Therefore $\dfrac{N_2}{N_1} = \dfrac{d_1}{d_2} \times \dfrac{E + \sqrt{f_2}}{E + \sqrt{f_1}}$

$$\frac{N_2}{200} = \frac{1}{2.5} \times \frac{100 \times \sqrt{0.4}}{100 + \sqrt{1}}$$

$$N_2 = \frac{200}{2.5} \times \frac{100 + \sqrt{0.4}}{101} = 79.09 \text{ rpm}$$

If creep is neglected, then

$$\frac{N_2}{N_1} = \frac{d_1}{d_2}$$

$$N_2 = \frac{d_1}{d_2} \times N_1$$

$$\frac{1}{2.5} \times 200 = 80 \text{ rpm}$$

Speed lost due to creep $= 80 - 79.09 = 0.01$ rpm

Example 1.7: With the help of a belt, an engine running at 200 rpm drives a line shaft. The diameter of the pulley on the engine is 80 cm and the diameter of the pulley on the line shaft is 40 cm. A 100 cm pulley on the line shaft drives a 20 cm diameter pulley keyed to a dynamo shaft . Find the speed of the dynamo shaft when there is:

(a) No slip

(b) A slip of 2.5% at each drive

Solution:

Engine speed $N_1 = 200$ rpm

Diameter of driver pulley on engine $d_1 = 80$ cm

Diameter of follower pulley $d_2 = 40$ cm

Diameter of driven pulley on line shaft $d_3 = 100$ cm

Diameter of follower pulley on dynamo shaft $d_4 = 20$ cm

Slip at each drive $S_1 = S_2 = 2.5\%$

Let N_4 be the speed of dynamo shaft,

(i) When there is no slip.

$$\frac{N_4}{N_1} = \frac{d_1 \times d_3}{d_2 \times d_4}$$

$$N_4 = \frac{d_1 \times d_3}{d_2 \times d_4} \times N_1$$

$$= \frac{80 \times 100}{40 \times 20} \times 200 = 2000 \text{ rpm}$$

(ii) When there is a slip of 2.5%

$$\frac{N_4}{N_1} = \frac{d_1 \times d_3}{d_2 \times d_4}\left(1 - \frac{S_1}{100}\right)\left(1 - \frac{S_2}{100}\right)$$

$$N_4 = N_1 \times \frac{d_1 \times d_3}{d_2 \times d_4}\left(1 - \frac{S_1}{100}\right)\left(1 - \frac{S_2}{100}\right)$$

$$= \frac{200 \times 80 \times 100}{40 \times 20} \times \left(1 - \frac{2.5}{100}\right)\left(1 - \frac{2.5}{100}\right)$$

$$= 2000 \times \frac{97.5}{100} \times \frac{97.5}{100} = 1901.25 \text{ rpm}$$

1.7 Power transmitted by a belt

Figure 1.12 shows the driving pulley or driver A and the driven pulley or follower B.

As discussed that the driving pulley pulls the belt from one side and delivers the same to the other side, it is thus obvious that the tension on the tight side will be greater than the slack side, as shown in Fig. 1.12.

Let T_1 and T_2 = Tensions at tight side and slack side of the belt in Newton's.

r_1 and r_2 = Radii of the driver and driven, respectively

v = velocity of the belt in m/s

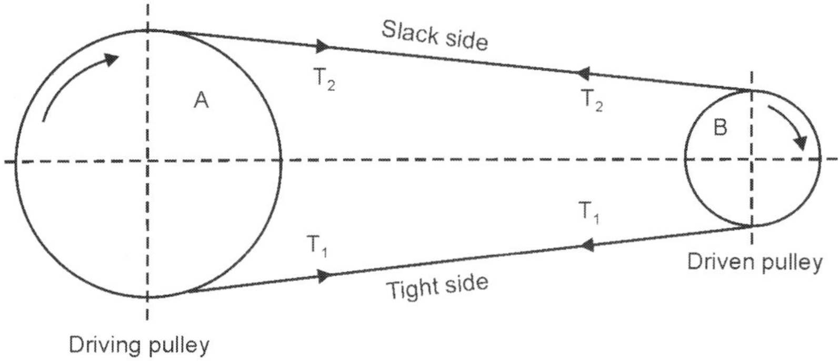

Figure 1.12 Power transmitted by belt

The effective turning (driving) force at the circumference of the follower is the difference between the two tensions($T_1 - T_2$)

Therefore, work done per second = Force × Velocity

$= (T_1 - T_2) \times v$ Nm.

and the power transmitted $P = (T_1 - T_2)v$ Watts.

$$= \frac{(T_1 - T_2) \times v}{1000} \text{ kW}$$

Torque exerted on driving pulley $= (T_1 - T_2) \times r_1$

Torque exerted on driven pulley $= (T_1 - T_2) \times r_2$

Example 1.8: (SI units) A belt is running over a pulley of diameter 120 cm at 200 rpm; the angle of contact is 165 degrees and coefficient of friction between the belt and the pulley is 0.3. If the maximum tension in the belt is 3000 N, find the power transmitted by the belt.

Solution:

Diameter of pulley (d) = 120 cm = 1.2 m

Speed of the pulley (N) = 200 rpm

Angle of lap or contact = 165°

$= \frac{165 \times \pi}{180}$ radians (since 1 degree $= \frac{\pi}{180}$ radian)

Coefficient of friction (μ) = 0.3

Maximum tension in the belt (T_1) = 3000 N

Velocity of the belt $(v) = \frac{\pi \times D \times N}{60}$

$$\frac{\pi \times 1.2 \times 200}{60} = 12.56 \text{ m/s}$$

Let T_2 bet the tension on the slack side of the belt.

Therefore
$$\frac{T_1}{T_2} = e^{\mu\theta}$$
$$= e^{0.3} \times \frac{165 \times \pi}{180}$$
$$= 2.3714$$
$$\frac{T_1}{T_2} = 2.3714$$

or
$$\frac{3000}{T_2} = 2.3714$$
$$T_2 = 1265 \text{ N}$$

Power transmitted by the belt $= (T_1 - T_2) \times \frac{v}{1000}$

$= (3000 - 1265) \times \frac{12.56}{1000} = 21.79$ kW

Example 1.9: (SI units) An open belt drive connects two pulleys of 120 cm and 50 cm diameters, on parallel shafts 4 m apart. The maximum tension in the belt is 1855.3 N. The coefficient of friction is 0.3. The driver pulley of diameter is 120 cm runs at 200 rpm. Calculate:

(a) The power transmitted
(b) Torque on each of the two shafts

Solution:

Diameter of larger pulley (d_1) = 120 cm = 1.2 m
Radius of the larger pulley (r_1) = 1.2/2 = 0.6 m
Diameter of smaller pulley (d_2) = 50 cm = 0.5 m
Radius of smaller pulley (r_2) = 0.5/2 = 0.25 m
Distance between shafts (x) = 4 m
Max. tension (T_1) = 1855.3 N
Coefficient of friction (μ) = 0.3
Speed of the driver pulley (N_1) = 200 rpm
We know that the velocity of belt is given by
$$v = \frac{\pi \times d_1 \times N_1}{60}$$
$$= \frac{\pi \times 1.2 \times 200}{60}$$
$$= 12.56 \text{ m/s}$$

Let us now find the angle of contact (θ) for an open belt drive. For an open belt drive, the angle of contact is given by

$$\theta = 180 - 2\alpha \qquad \ldots(1.7)$$

$$\text{Sin }\alpha = \frac{r_1 \times r_2}{x}$$

$$= \frac{0.6 \times 0.25}{4}$$

$$= 0.0875$$

$$\alpha = \sin^{-1} 0.0875$$

$$= 5.02°$$

Substituting this value of α in Eq. (1.7)

$$\theta = 180 - 2 \times 5.02$$

$$= 169.96°$$

$$= 169.96 \times \frac{\pi}{180} \text{ radians} = 2.967$$

We know that

$$\frac{T_1}{T_2} = e^{\mu\theta}.$$

$$\frac{1855.3}{T_2} = 2.718^{0.3} \times 2.967 = 2.435$$

$$\frac{1855.3}{T_2} = 2.435$$

$$T_2 = \frac{1855.3}{2.435} = 761.8 \text{ N}$$

(i)Power transmitted = $(T_1 - T_2) \times \frac{v}{1000}$

= (1855.3 – 761.8) × 12.56

= 13.73 kW

(ii)Torque on the driver shaft = $(T_1 \times T_2) \times r_1$

= (1855.3 – 761.8) × 0.6

= 656.1 N

Torque on driven shaft

= $(T_1 - T_2) \times r_1$

= (1855.3 – 761.8) × 0.25

= 273.4 N

1.8 Ratio of driving tensions for flat belt drive

Consider a driven pulley rotating in the clockwise direction as shown in Fig. 1.13.

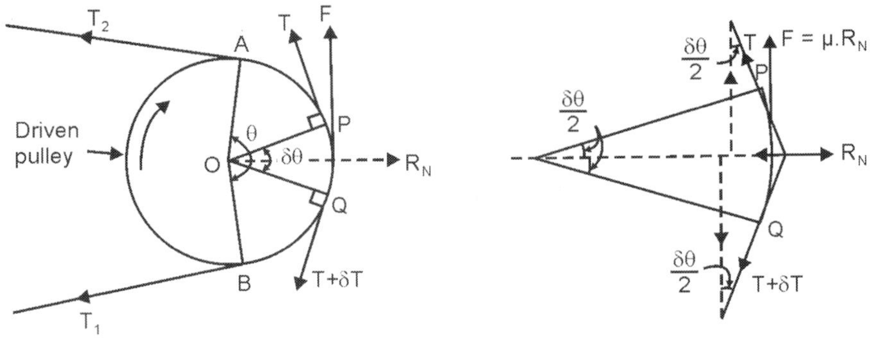

Figure 1.13 Ratio of driving tensions for flat belt

Let

T_1 = Tension in the tight side of the belt

T_2 = Tension in the slack side of the belt

θ = Angle of contact in radians (i.e. angle subtended by the arc AB, along which the belt touches the pulley at the centre)

Now consider a small portion of the belt PQ subtending at an angle dθ at the centre of the pulley as shown in Fig. 1.13. The belt PQ is in equilibrium under the following forces.

1. Tension T in the belt at P
2. Tension $(T + dT)$ in the belt Q
3. Normal reaction R_N and
4. Frictional force, $F = \mu \times R_N$, where μ is the coefficient of friction between the belt and the pulley.

Resolving the forces horizontally and equating the same,

$$R_N = (T + dT)\sin\frac{d\theta}{2} + T\sin\frac{d\theta}{2} \quad \ldots(1.8)$$

Since the angle dθ is very small, therefore putting $\sin\frac{d\theta}{2} = \frac{d\theta}{2}$ = in Eq. (1.8)

$$R_N = (T + dT)\,(d\theta/2) + (T \times d\theta/2) = T.d\theta/2 = dt.d\theta/2 + Td\theta/2$$

$$= T.d\theta \quad \ldots(1.9)$$

(Neglecting $dT.\ d\theta/2$)

Now resolving the forces vertically, we have

$\mu \times R_N = (\mathrm{T} + dT_{\square}) \cos (d\theta/2_{\square} - \mathrm{T} \cos d\theta/2_{\square}$ ...(1.10)

Since the angle $dT_{\square}$ is very small, therefore putting $\cos d\theta/2_{\square} = 1$ in Eq. (1.10),

$\mu \times R_N = \mathrm{T}\, dT_{\square} + = \mathrm{T} = dT_{\square}$ or $R_N = dT/\mu_{\square}$ ...(1.11)

equating the values of R_N from Eqs. (1.9) and (1.11),

$\mathrm{T}.\, d\theta_{\square} = dT/\mu_{\square}$ or $dT/T_{\square} = \mu.\, d\theta_{\square}$.

Integrating both sides between the limits T_1 and T_2 and from θ to θ respectively,

$\log_e (T_1/T_2) = \mu\theta$ or $e^{\mu\theta}_{\square}$ ...(1.12)

Equation (1.12) can be expressed in terms of corresponding logarithm to the base 10,

i.e. $2.3 \log (T_1/T_2) = e^{\mu\theta}$

The above expression gives the relation between the tight side and slack side tensions, in terms of co-efficient of friction and angle of contact.

1.9 Centrifugal tension

Since the belt continuously runs over the pulleys, therefore some centrifugal force is caused whose effect is to increase the tension on both, tight as well as the slack sides. The tension caused by the centrifugal force is called *centrifugal tension.* At lower speeds (less than 10 m/s), the centrifugal tension is very small, but at higher belt speeds (more than 10 m/s), its effect is considerable and must be taken in to account.

Consider a small portion PQ of the belt subtending an angle dθ at the centre of the pulley as shown in Fig. 1.14.

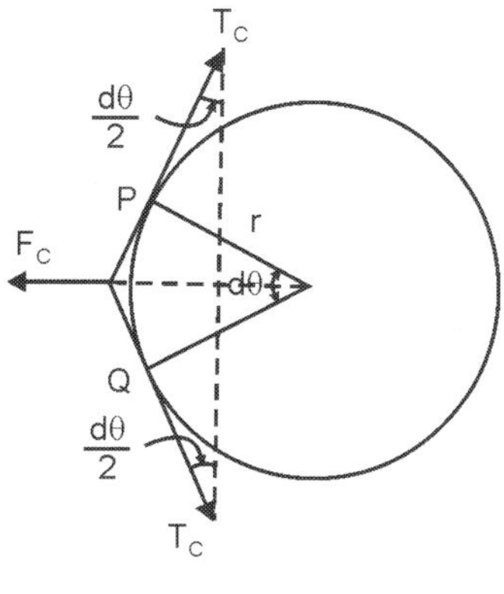

Figure 1.14 Centrifugal tension

Let m = Mass of the belt per unit length in kg

v = Linear velocity of the belt in m/s

r = Radius of the pulley over which the belt runs in meters, and

T_c = Centrifugal tension acting tangentially at P and Q in Newton's.

We know the length of the belt PQ

$= r.d\theta$.

and mass of the belt PQ = m.r. $d\theta$.

Therefore centrifugal force acting on the belt PQ

$F_C = (m.r.d\theta)\, v^2/r = m.d\theta.v^2$

The centrifugal tension acting tangentially at P and Q keep the belt in equilibrium.

Now resolving the forces (i.e. centrifugal force and centrifugal tension) horizontally and equating the same,

$T_C \sin (d\theta/2) + T_C \sin (d\theta/2) = F_C = m.d\theta.\ v^2$.

As the angle $d\theta$ is very small, putting $\sin (d\theta/2 = d\theta/2$ in the above expression

$= 2\ T_C\, (d\theta/2) = m.d\theta.v^2$

or $T_C = m.v^2$

1. When the centrifugal tension is taken in to account, then total tension in the tight side,

   $T_{r1} = T_1 + T_C$

   And total tension in the slack side

   $T_{r2} = T_2 + T_C$

2. Power transmitted

   $\mathrm{P} = (T_{r1} = T_{r2}) \times v$

   $= (T_1 + T_C) - (T_2 + T_C)v = (T_1 + T_2)v$

   From the above expression, the centrifugal tension has no effect on power transmitted.

3. The ratio of driving tensions may be written as

   $2.3 \log (T_{r1} - T_C)/(T_{r2} - T_C)$

   $= \mu.\theta$

Example 10.1: A belt embraces the smaller pulley at an angle of lap of 185° and runs at a speed of 1700 m/min. The belt dimensions are width = 20 cm and thickness = 8 mm. Its density is 1 g/cm^3. Determine the maximum power that can be transmitted at the above speed, if the maximum permissible stress in the belt is not to exceed 250 N/cm^2 and $\mu = 0.25$.

Solution:

Angle of contact or lap = 165° = 165 × π/180 radians = 2.88 rad.

Speed of the belt (v) = 1700 m/min

1700 / 60 = 28.33 m/s

Width of the belt (b) = 20 cm

Thickness of the belt (t) = 8 mm = 0.8 cm

Density of the belt = 1 g/cm^3

= 1/1000 kg/cm^3

Max. permissible stress (f) = 250 N/cm^2

Coefficient of friction $\mu = 0.25$

Let us first find the maximum tension and centrifugal tension in the belt

The maximum tension in the belt (T_m) = f × (Area of the belt) = f × (b × t)

= 250 × 20 × 0.8 = 4000 N.

Centrifugal tension (T_C) is given by

$$T_C = mv^2$$

Where m = Mass of the belt per meter length

= Density × Volume of belt of 1 m length

= Density × (b × t × L)

= 1/1000 × (20 × 0.8 × 100) kg

[Length of 1 m = 100 cm]

= 1.6 kg

Substituting the value of m = 1.6 kg in

$$T_C = mv^2$$

= 1.6 × 28.33 *28.33

= 1284 N

We know that

$$T_m = T_1 + T_C$$

$$T_1 = T_m + T_C$$

= 4000 – 1284

= 2716 N

Let T_2 be the tension on the slack side of the belt

$$\frac{T_1}{T_2} = e^{\mu\theta}$$

= e0.25 + 2.88 = 2.056

$$T_2 = T_1/2.056$$

$$T_2 = 2716/2.056$$
$$= 1321\text{N}$$

Maximum power transmitted by the belt

$$= (T_1 - T_2) \times v/1000$$
$$= 2716 - 1321 \times 28.33 / 1000$$
$$= 39.52 \text{ kW}$$

Example 1.10: A belt of density 1 g/cm³ has a permissible stress of 250 N/cm². Determine the maximum power that can be transmitted by a belt of 20 cm × 1.2 cm if the ratio of the tight side to the slack side tension is 2.

Solution:

Density of the belt = 1 g/cm³ = 1/1000 kg/cm³

Max. permissible stress f = 250 N/cm²

Width of the belt b = 20 cm

Thickness of the belt t = 1.2 cm

Ratio of tensions = $T_1/T_2 = 2.0$

Let us find the mass of 1 m length of the belt and also the maximum tension in the belt

m = Mass of 1 m length of belt

= Density × Volume of 1 m length

= $1/\ 1000 \times (b \times t \times 1000)$

= $1/1000 \times (20 \times 1.2 \times 100)$ kg

= 2.40 kg

Let T_m = Maximum tension

= (Max. stress) × Area of cross-section of belt

= $250 \times b \times t = 250 \times 20 \times 1.2$

= 6000 N

Now, for maximum power transmitted, velocity of the belt is given by

$$v = \sqrt{}T_m/3m$$
$$= \sqrt{}6000/3 \times 2.4$$
$$= 28.86 \text{ m/s}$$

Now, the power transmitted is

$$P = (T_1 \times T_2) \times v/1000$$

Let us find the values of T_1 and T_2.

We know that $T_m = T_1 + T_C$

T_C = Centrifugal tension

T_1 = Tension on tight side

But for max. power transmission,

$T_C = 1/3\ T_m$

= Substituting the value of T_C in

$T_m = T_1 = T_C$

$T_m = T_1 + 1/3\ T_m$

$T_1 = T_m - 1/3\ T_m = 2/3\ T_m$

$T_1 = 2/3 \times 6000 = 4000$ N

But $\frac{T_1}{T_2} = 2.0 : T_2 = \frac{T_1}{2}$

= 4000/2 = 2000 N

Therefore, power transmitted =

$P = (T_1 - T_2) \times v / 1000$

$= (4000 - 2000) \times 28.86 / 1000$

= 57.18 kW

1.10 Initial tension in the belt

The tension in the belt which is passing over the two pulleys (i.e. driver and follower) when the pulleys are stationary is known as initial tension in the belt.

When power is transmitted from one shaft to another shaft with the help of the belt passing over the two pulleys which are keyed to the driver and driven shafts, there should be firm grip between the pulleys and belt. When the pulleys are stationary, the firm grip is increased by tightening the two ends of the belt. Hence the belt is subjected to some tension. This tension is known as *initial tension in the belt.*

Let T_0 = Initial tension in the belt

T_1 = Tight side tension of the belt

T_2 = Slack side tension of the belt

α = Co-efficient of increase of belt length per unit force

When power is supplied to one of the pulleys and transmitted to the other, the tension in the two free length of the belts will be changed. The tight side of the belt stretches until the pull is increased from T_0 to T_1 and slack side shortens until the pull is decreased from T_0 to T_2.

Therefore, increase of tension in tight side = ($T_1 - T_0$)
and increase in tension in slack side = ($T_0 - T_2$)
Now increase in length of belt on tight side = α × increase in tension
= α × ($T_1 - T_0$)
Similarly, decrease in length of belt on slack side = α × ($T_1 \times T_0$)

For a perfectly elastic material of the belt, the length of the belt remains constant when it is at rest or in motion.

Hence increase in length on tight side should be equal to decrease of length on the slack side.

$$\alpha.(T_1 = T_2) = \alpha.\ (T_0 - T_2)$$
$$T_1 - T_0 = T_0 - T_2$$
$$T_0 = T_1 + T_2/2 \quad ...(1.13)$$

If centrifugal tension is considered, then

$$T_0 = T_1 + T_2 + 2T_C / 2 \quad ...(1.14)$$

Example 1.11: An open belt running over two pulleys of 24 cm and 60 cm diameters connects two parallel shafts 3 m apart and transmits 3.75 kW from the smaller pulley which rotates at 300 rpm. Co-efficient of friction between the belt and the pulleys is 0.3 and the safe working tension is 100 N/cm width. Determine the following:

(i) Minimum width of the belt
(ii) Initial tension in the belt
(iii) Length of the belt required

Solution:

Diameter of larger pulley d_1 = 60 cm; r_1 = 30 cm
Diameter of smaller pulley d_2 = 24 cm; r_2 = 12 cm
Distance between two shafts x = 3 m = 300 cm
Power transmitted = 3.75 kW
Speed of the smaller pulley (N_2) = 300 rpm
Co-efficient of friction (μ) = 0.3
Safe working stress = 100 N/cm width
Let b = width of belt
Then T_{max} = 100 × b = 100 b N
For an open belt drive, angle of contact is given by

$$\theta = 180 - 2\alpha$$

A is given by

$$\text{Sin } \alpha = (r_1 - r_2)/x$$
$$= 30 - 12 / 300$$
$$= 0.06$$
$$\alpha = \sin^{-1} 0.06$$
$$= 3.45°$$

Substituting the value of $\alpha = 3.45°$ in

$$\theta = 180 - 2\,\alpha$$
$$\theta = 180 - 2 \times 3.45$$
$$\theta = 173.1°$$
$$= 173.1 \times 3.13/180 = 3.02 \text{ radians}$$

Now we know that

$$T_1/T_2 = e^{\mu\theta}$$
$$= e^{0.3 \times 3.02}$$
$$= 2.474$$
$$T_1 = 2.474\,T_2$$
$$\text{Power transmitted} = (T_1 - T_2) \times v / 1000$$
$$3.75 = (T_1 - T_2) \times \text{v}/1000$$
$$v = \pi \times d_2 N_2 / 60$$
$$= \pi \times 24 \times 300/60$$
$$= 376.9 \text{ cm/s or } 3.769 \text{ m/s}$$

Substituting this value in power equation,

$$3.75 = (T_1 - T_2) \times v/1000$$
$$(T_1 - T_2) = 3.75 \times 1000/3.769$$
$$= 994.7 \text{ N}$$
$$(T_1 - T_2) = 994.7 \text{ N}$$

Substituting the value of $T_1 = 2.474$ in (T_1)

$$(2.474T_2 - T_2) = 994.7 \text{ N}$$
$$1.474T_2 = 994.7 \text{ N}$$
$$T_2 = 994.7/1.474 = 674.8 \text{ N}$$
$$T_1 = 2.474T_2$$
$$T_1 = 2.474 \times 674.8 = 1669.5 \text{ N}$$

(i) Minimum width of the belt required, neglecting centrifugal tension, we have $T_{max} = T_1$

$100\,b = 1669.5\,N$

$b = 1669.5/100 = 16.7$ cm

(ii) Initial belt tension $= T_0 = T_1 + T_2/2$
$= 1669.5 + 674.8/2 = 1172.15\ N$

(iii) Length of the belt required.
For an open belt drive, length of the belt required is
$L = \pi\,(r_1 + r_2) = (r_1 - r_2)^2\,/x + 2x$
$= \pi\,(30 + 12) = (30 - 12)^2/300 + 2 * 300$
$= 733$ cm or 7.33 m

1.11 V-belt drive

V-belt drives are most commonly used in factories and workshops where a great amount of power is to be transmitted from one pulley to another when the two pulleys are very near to each other.

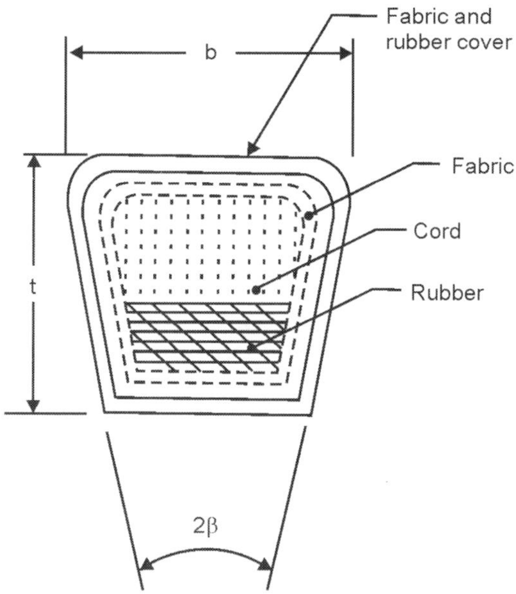

Figure 1.15 (a) Cross-section of V-belt

The V-belts are made of fabric and cords molded in rubber and covered with fabric and rubber, as shown in Fig. 1.15(a). These belts are molded to a trapezoidal shape and are made endless. These are suited for short drives, i.e. when the shafts are at short distance apart. The included angle for the V-belt is usually 30 to 40 degrees. In case of flat belt drive, the belt runs over the pulleys whereas in V-belt drives, the rim of the pulley is grooved in which the V-belt runs. The effect of the groove is to increase the frictional grip of the

V-belt on the pulley and thus reduce the tendency to slipping. The power is transmitted by the wedging action between the belt and the V-groove in the pulley.

A clearance must be provided at the bottom of the groove as shown in Fig. 1.15(b) in order to prevent touching to the bottom as it becomes narrower from wear. The V-belt drive may be inclined at any angle with tight side either at top or bottom. In order to increase the power output, several V-belts are operated side by side. In case of multiple v-belt drives, all the belts should stretch at the same rate so that the load is equally divided between them. If one of the set of belts break, the entire set should be replaced at the same time. If only one belt is replaced, the new belt will be tightly stretched and will move with different velocity.

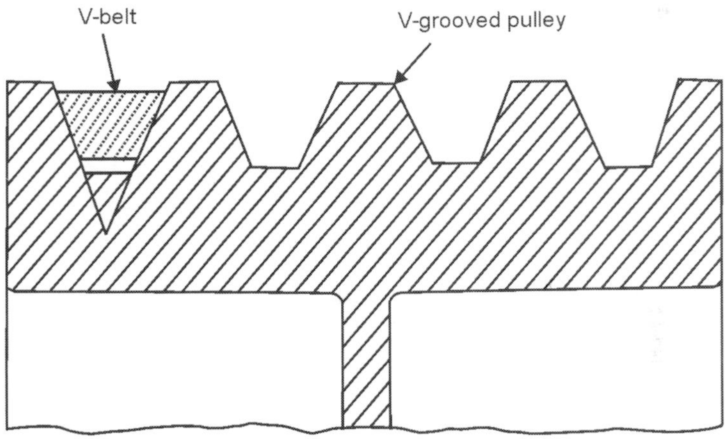

Figure 1.15 (b) Cross-section of a V-grooved pulley

Advantages of V-belt drives:

1. The V-belt gives compactness due to the small distance between the centres of the pulleys.
2. The drive is positive, since the slip between the belt and the pulley groove is negligible.
3. The drive is smooth since the belt are endless and there is no joint problem.
4. The life of the v-belt is longer up to 3–5 years.
5. It can be easily installed and removed.
6. The belts have the cushioning effect to bear the shock when the machine is started.
7. The high velocity ratio (maximum 10 may be obtained).

8. The wedging action of the belt in the groove gives high value of limiting ratio of tensions. Therefore the power transmitted by V-belts is more than flat belt for the same coefficient of friction, arc of contact and allowable tension in the belts.
9. The V-belt may be operated in either direction with tight side of the belt at the top or bottom. The centre line may be horizontal, vertical or inclined.

Disadvantages of V-belt drives

1. The V-belt drive cannot be used with large centre distances.
2. V-belts are not as durable as flat belts.
3. The construction of pulleys for V-belts is more complicated than pulleys for flat belts.
4. The belt life is greatly influenced with temperature changes, improper belt tension and mismatching of belt lengths.
5. The centrifugal tension prevents the use of V-belts at speeds below 5m/s and above 50 m/s.

Example 1.12: The maximum allowable tension in a V-belt of groove angle of 45° is 1500 N. The angle of lap is 170° and the co-efficient of friction between the belt and the pulley is 0.27. If the belt is running at 2 m/s, determine:

(i) Net driving tension

(ii) Power transmitted by the pulley. Ignore the effect of centrifugal tension.

Solution:

Angle of groove $2\alpha = 45°$; $\alpha = 22.5°$

Maximum allowable tension on the belt (T_{max}) = 1500 N

Angle of lap (θ) = 170° = $170 \times \pi/180$ = 2.96 radians

Co-efficient of friction (μ) = 0.27

Belt speed (v) = 2 m/s

We know that

$$T_{max} = T_1 + T_C$$

Where T_C is centrifugal tension. If centrifugal tension is ignored, then

$$T_{max} = T_1 = 1500\ NT_{max}$$

Now for V-belt drive, the ratio of tensions is given by

$$T_1/T_2 = e^{\mu\theta} = e^{2.96 \times \text{cosec } 22.5°}$$

$$= 8.109$$

$$T_2 = T_1/8.109$$
$$T_2 = 1500/8.109 = 185 \text{ N}$$

(i) Net driving tension = $(T_1 - T_2)$
= (1500 – 185) = 1315 N

(ii) Power transmitted
= $(T_1 - T_2) \times v$ Watts
= 1315 × 2 = 2630 W

1.11.1 Ratio of driving tensions for V-belt

A V-belt with grooved pulley is shown in Fig. 1.16.

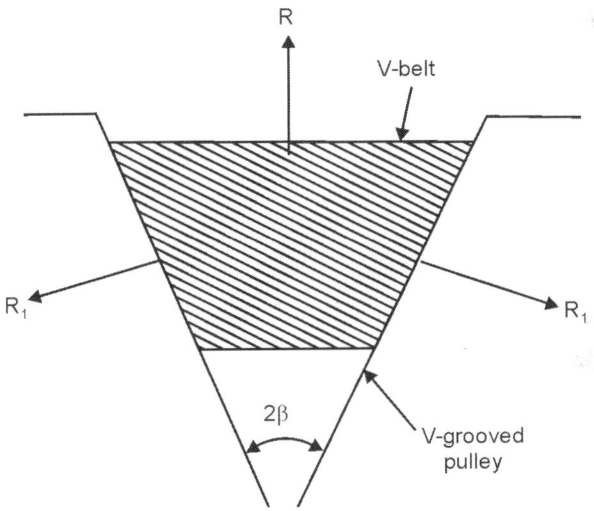

Figure 1.16 V-belt with grooved pulley

Let R_1 = Normal reaction between the belt and sides of the groove
R = Total reaction in the plane of the groove
2β = Angle of the groove
μ = Co-efficient of friction between the belt and sides of the groove
Resolving the reactions vertically to the groove,
$R = R_1 \sin\beta + R_2 \sin\beta = 2R_1 \sin\beta$
or $R_1 = R/2 \sin\beta$
We know that the frictional force
$= 2\,\mu\, R_1 = 2\,\mu \times R/\sin\beta = \mu\, R/\sin\beta = \mu\, R \text{ Cosec } \beta$.

Thus the relationship between T_1 and T_2 for the V-Belt drive will be

$$2.3 \log \left(\frac{T1}{T2}\right) = \mu\ \theta\ \text{Cosec}\ \beta$$

1.12 Rope drive

The rope drives are widely used where a large amount of power is to be transmitted from one pulley to another over a considerable distance. It may be noted that the use of flat belts is limited for the transmission of moderate power from one pulley to another when the two pulleys are not more than 8 m apart. The flat belts require excessive cross-section to transmit large amount of power. The frictional grip in the rope drive is more than that of the V-belt drive.

One of the main advantage of rope drive is that a number of separate drives may be taken from one driving pulley. For example, this type of drives can be seen in many conventional spinning mills, the line shaft on each floor is driven by ropes passing directly from the main engine pulley on the ground floor.

The rope drives are classified into two types:

1. Fibre ropes
2. Wire ropes

1.12.1 Fibre ropes

The ropes for transmitting power are usually made from fibrous materials such as hemp, manila and cotton. The hemp and manila ropes are rough and hence the ropes made out of them are not very flexible and possess poor mechanical properties. Hemp ropes have less strength as compared to manila ropes. When the hemp and manila ropes are bent over the sheave or pulley, there is some sliding of fibers, causing the ropes to wear and chafe internally. In order to minimize the defect, the rope fibres are lubricated with a tar, tallow or graphite. The lubrication also makes the rope moisture proof. The hemp ropes are suitable only for hand-operated hoisting machinery and as the ropes for lifting tackle, hooks, etc.

The cotton ropes are very soft and smooth. The lubrication of the cotton ropes is not required. It may be noted that manila ropes are stronger than cotton ropes.

Advantages of fiber rope drives:

1. The operation is smooth, quiet and steady.
2. They give high mechanical efficiency.
3. They are less affected by outdoor conditions.

Sheave for fiber ropes

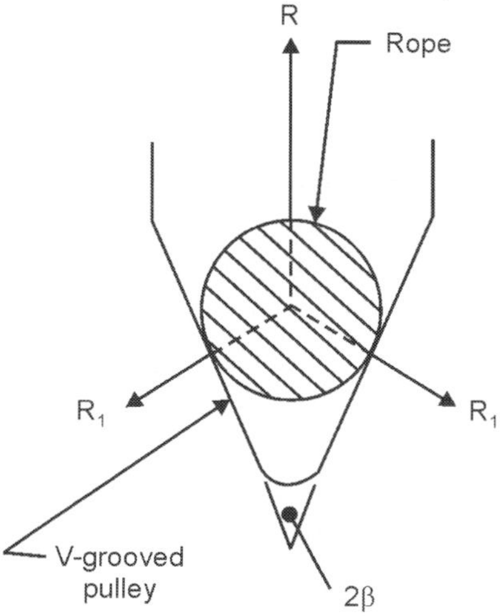

(a) Cross-section of a rope

The fiber ropes are circular in cross-section as shown in Fig. 1.17(a). The sheave for the fiber ropes is shown in Fig. 1.17(b). The groove angle of the pulley for rope drives is usually 45°. The groove in the pulleys are made narrow at the bottom and the rope is pinched between the edges of the V-groove to increase the holding power of the rope on the pulley.

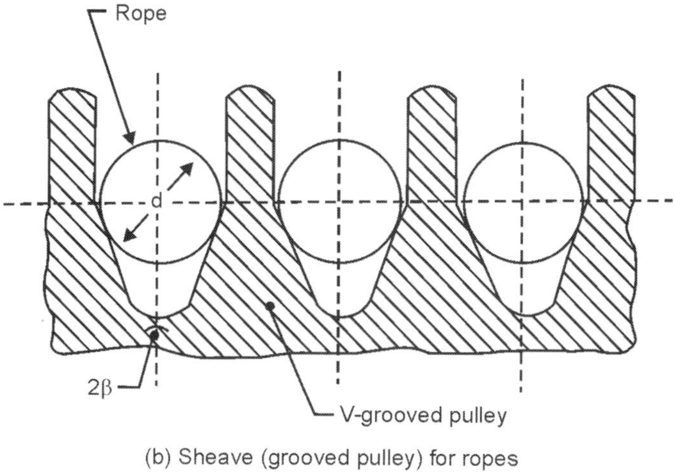

(b) Sheave (grooved pulley) for ropes

Figure 1.17 (a) Cross-section of a rope; (b) sheave for fibre ropes

1.12.2 Wire ropes

When a large amount of power is to be transmitted over long distances from one pulley to another (i.e. when the pulleys are up to 150 m apart), then wire ropes are used. The applications of wire ropes are in elevators, cranes, mine hoists and suspension bridges. The wire ropes run on grooved pulleys but they rest on the bottom of the grooves and are not wedged between the sides of the grooves.

The wire ropes have the following advantages over cotton ropes.

(a) Lighter in weight
(b) Smooth operation and withstand shock loads
(c) They are more reliable
(d) They do not fail suddenly
(e) They are more durable, have high efficiency and low cost

Example 1.13: A rope drive transmits 75 kW through a 150 cm diameter 45° grooved pulley rotating at 200 rpm. Co-efficient of friction between the ropes and pulley grooves is 0.3. and angle of lap is 160°. Each rope has a mass of 0.6 kg/m and can safely run with a pull of 800 N. Allow for centrifugal tension. Determine

(i) Number of ropes required for the drive
(ii) Initial rope tension

Solution:

Power = 75 kW

Diameter of driver pulley (d_1) = 150 cm = 1.5 m

Speed of driver pulley (N_1) = 200 rpm

Angle of groove (2α) = 45°; α = 22.5°

Angle of lap = 160 × π/180 = 2.79 radians

Mass of one rope per meter

length (m) = 0.6 kg

Max. safe tension in one rope = 800 N

Now the velocity (v) of the rope = $\pi \times d_1 \times N_1 / 60$

$= \pi \times 1.5 \times 200 / 60 = 15.7$ m/s

(i) Let us find the power transmitted by one rope.

Centrifugal tension is given by

$T_C = m \times v^2$

$0.6 \times 15.7 \times 15.7 = 147.894$ N

But $T_{max} = T_1 + T_C$

$T_1 = T_{max} - T_C$

$= 800 - 147.894$

$= 652.106$ N

Now the tension in the rope is given by

$$\frac{T_1}{T_2} = e^{0.3 \times 2.70 \times \text{Cosec } 22.5°}$$

$$e = 8.95$$

$$T_2 = T_1 / 8.95$$

$$T_2 = 652.106 / 8.95$$

$$= 72.86 \text{ N}$$

Power transmitted by one rope

$$= (T_1 - T_2) \times v / 1000 \text{ W}$$

$$= (652.106 - 72.86) \times 15.7 / 1000$$

$$= 9.094 \text{ kW}$$

Total power transmitted = 75 kW

Number of ropes required = Total power transmitted / Power transmitted by one rope

$$75/9.094 = 8.24$$

(ii) Initial tension (T_0), if centrifugal tension is considered.

$$T_0 = T_1 + T_2 + 2T_C / 2$$

$$= 652.106 + 72.86 + 2 * 147.894 / 2$$

$$= 510.377 \text{ N}$$

Example 1.14: A pulley is used to transmit power by means of ropes, has a diameter of 3.6 m and has 15 ropes of 45° angle. The angle of contact is 170° and μ between the ropes and the groove side is 0.28. The maximum possible tension in the ropes is 960 N. and the mass of the rope is 1.5 kg per m length. What is the speed of the pulley in rpm and the power transmitted if the conditions of maximum power prevail?

Solution:

Diameter of pulley $D = 3.6$ m

No. of ropes = 15

Angle of groove = $2\alpha = 45°$; $\alpha = 22.5°$

Angle of contact = $170° \times \pi / 180 = 2.965$

Co-efficient of friction (μ) = 0.28

Max. tension (T_{max}) = 960 N

Mass of rope (m) = 1.5 kg per m length

For maximum power

(i) Centrifugal tension

$$T_C = 1/3 \times 960 = 320 \text{ N}$$

Velocity of the rope is given by

$$\begin{aligned} v &= \sqrt{T_m/3_m} \\ &= \sqrt{960 / 3 * 1.5} \\ &= 14.60 \text{ m/s} \end{aligned}$$

Velocity of the rope is given by

$$\begin{aligned} v &= \pi \times DN/60 \\ N &= 14.60 \times 60 / \pi \times 14.60 \\ &= 77.45 \text{ rpm} \end{aligned}$$

Now the ratio of tensions in the rope is given by

$$\begin{aligned} T_1/T_2 &= e^{\mu \text{Cosec } \theta} \\ &= e^{0.28 \times 2.965 \times \text{Cosec } 22.5°} \\ &= 8.756 \\ T_2 &= T_1/8.756 \end{aligned}$$

But
$$\begin{aligned} T_m &= T_1 + T_C \\ 960 &= T_1 + 320 \\ T_1 &= 960 - 320 = 640 \, N \\ T_1 &= 640 / 8.756 \\ &= 73.08 \text{ N} \end{aligned}$$

Maximum power transmitted by one rope =

$(T_1 - T_2) \times v / 1000$

$(640 - 73.68) \times 14.60 / 1000$

= 8.277 kW

Total power transmitted

= Power of one rope × no. of ropes

= 8.277 × 15

= 124.16 kW

1.13 Chain drives

We have seen in belt and rope drives, slipping may occur. In order to avoid slipping, steel chains are used. The chains are made up of rigid links which are hinged together to provide the necessary flexibility for warping, around the driving and driven wheels. The wheels have projecting teeth and fit in to the corresponding recesses in the links of the chain as shown in Fig. 1.18. The wheels and the chains are constrained together to move together without slipping and ensures perfect velocity ratios. The toothed wheels are known as sprocket wheels or sprockets. These wheels resemble to spur gears.

The chains are mostly used to transmit motion and power from one shaft to another, when the distance between the centers of the shafts is short such as in bicycles, motor cycles, agricultural machinery and road rollers, etc.

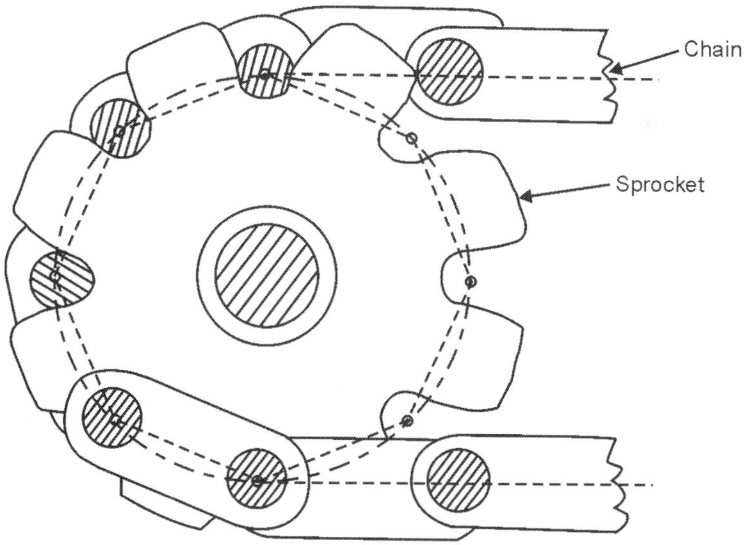

Figure 1.18 Sprocket and chain

Advantages and disadvantages of chain drives over belt or rope drives

(a) Advantages

1. As no slip takes place in chain drives, perfect velocity ratio can be obtained.
2. They occupy less space in width, since they are made of metal than a belt or rope drive.
3. The chain drives may be used when the distance between the shafts is less.
4. The chain drive gives high transmission efficiency up to 98%.
5. The chain drive gives less load on the shafts.

(b) Disadvantages

1. The production cost of the chains is high.
2. The chain drive requires accurate mounting and careful maintenance.

Numerical questions

1. Find the speed of a shaft which is driven by a belt on an engine running at 300 rpm. The diameter of the engine pulley is 60 cm and that of the shaft is 40 cm.

   Ans. 450 rpm

2. A shaft running at 120 rpm has to drive a parallel shaft at 180 rpm. The pulley on the driving shaft is 75 cm in diameter. Calculate the diameter of the pulley on the driven shaft: (i) neglecting belt thickness; (ii) taking belt thickness in to account which is 15 mm; (iii) assuming in the latter case a total slip of 4.5%.

   Ans. 50 cm, 49.7 cm, 47.8 cm

3. The power is transmitted from a pulley of 1 m diameter running at 200 rpm to a pulley of 2.25 m diameter by means of a belt. Find the speed lost by the pulley as a result of the creep, if the stress on the tight and slack side of the belt is 1.4 N/mm^2 and 0.5 N/mm^2. The Young's Modulus of the material of the belt is 100 N/mm^2.

   Ans. 0.2 rpm

4. Two parallel shafts 120 cm apart are to have a velocity ratio of 2.5 and are to be connected by a flat belt. The pulley on the high speed shaft is 24 cm diameter. Determine the length of the belt required: (i) if the belt is open, (ii) if the belt is crossed.

   Ans. (i) 373.4 cm (ii) 386 cm

5. Two parallel shafts 12 m apart are to be connected by a belt running over the pulleys of diameters 480 cm and 80 cm, respectively. Determine the length of the belt: (i) if the belt is open, (ii) if the belt is crossed.

   Ans. (i) 33.13 m, (ii) 33.45 m

6. A shaft running at 200 rpm drives another shaft at 300 rpm, and transmits 6 kW through a belt. The belt is 10 cm wide and 1 cm thick. The distance between the shafts is 4 m. The smaller pulley is 50 cm in diameter. Calculate the stress in (i) open belt, (ii) crossed belt , if $\mu = 0.3$.

   Ans. (i) 126.8 N/cm^2, (ii) 118.47 N/cm^2

7. The maximum allowable tension in a flat belt is 1500 N. The angle of lap is 170° and coefficient of friction between the belt and the material of the pulley is 0.27. Neglecting the effect of centrifugal

tension, calculate the net driving tension and the power transmitted if the belt speed is 2 m/s.

Ans. (i) 826.7 N, (ii) 1.6534 kW.

8. A V-belt having an angle of lap of 180° has a cross-sectional area of 6.2 cm^2 and runs in a groove of included angle of 40°. The density of the belt is 0.0013 kg/cm^3 and maximum stress is limited to 400 N/ cm^2, the coefficient of friction is 0.2. Find the maximum power that can be transmitted, if the wheel has mean diameter of 30 cm and runs at 1000 rpm.

   Ans. 30 kW

9. Find the number of ropes required in a rope drive which is used to transmit 230 kW from a pulley of 1 m diameter running at 450 rpm. The safe pull in each rope is 800 N and mass of the rope is 0.46 kg per m length. The angle of lap is 160° where the angle of groove is 45°. The co-efficient of friction between the rope and the pulley is 0.3.

   Ans. 21

10. A V-belt of mass 0.9 kb/m length is used to transmit power between two shafts. The angle of lap is 170° and angle of groove is 45°. The co-efficient of friction between the belt and the pulley is 0.17 and maximum allowable tension in the belt is 2.2 kN. Find (i) velocity of belt for maximum power, (ii) power transmitted at this velocity.
    Ans. (1) 28.54 m/s, (ii) 30.67 kW

2
Gear trains

2.1 Introduction

A combination of two or more gears is made to mesh with each other to transmit power from one shaft to another. Such a combination is called gear train or train of toothed wheels. The nature of the train wheel used depends upon the velocity ratio required and the relative position of the axes of shafts. A gear train may consist of spur, helical or bevel gears.

2.2 Nomenclature of gears

The following terms are illustrated in Fig. 2.1.

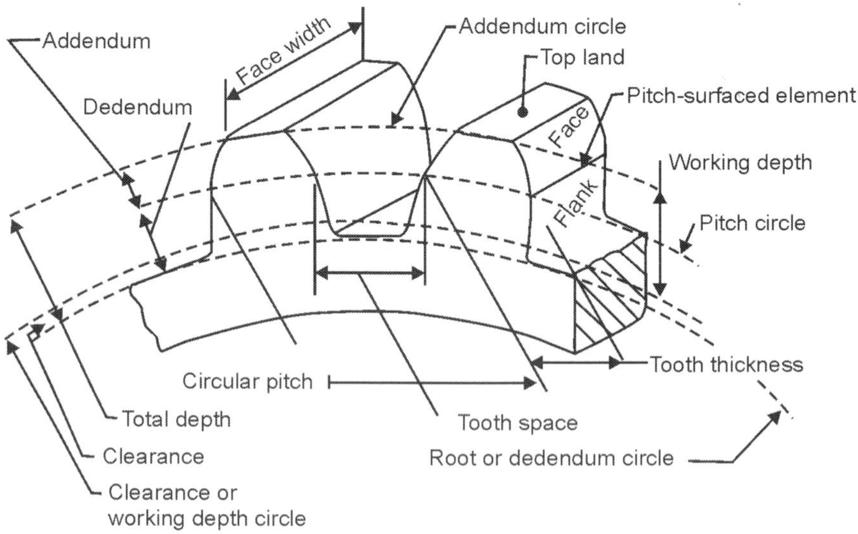

Figure 2.1 Terms used in gears

1. *Pitch circle:* It is an imaginary circle which by rolling action would give the same motion as the actual gear.

2. *Pitch circle diameter:* It is the diameter of the pitch circle. The size of the gear is usually specified by the pitch circle diameter. It is also known as pitch diameter.
3. *Pitch point:* It is a common point of contact between two pitch circles.
4. *Pitch surface:* It is the surface of the rolling discs which the meshing gears have replaced at the pitch circle.
5. *Pressure angle or angle of obliquity:* It is the angle between the common normal to two gear teeth at the point of contact and the common tangent at the pitch point. It is usually denoted by α. The standard pressure angles are 14.5° and 20°.
6. *Addendum:* It is the radial distance of a tooth from the pitch circle to the top of the tooth.
7. *Dedendum:* It is the radial distance of a tooth from the pitch circle to the bottom of the tooth.
8. *Addendum circle:* It is the circle drawn through the top of the teeth and is concentric with the pitch circle.
9. *Dedendum circle:* It is the circle drawn through the bottom of the teeth. It is also called root circle.
10. *Circular pitch:* It is the distance measured on the circumference of the pitch circle from a point of one tooth to the corresponding point on the next tooth. It is usually denoted by pc.

2.2.1 Types of gear trains

Following are the different types of gear trains:

1. Ordinary gear trains are further classified as
   (a) Simple gear train
   (b) Compound gear train (The compound gear train may be a reverted gear train or non-reverted gear train)
2. Epicyclic gear train

2.2.2 Simple gear train

When there is only one gear on each shaft as shown in Fig. 2.2 (a, b, c), it is known as simple gear train. The gears are represented by pitch circles.

When the distance between the two gears is small, the two gears 1 and 2 are made to mesh with each other to transmit motion from one shaft to the other as shown in Fig. 2.2(a). The gear 1 drives gear 2, therefore gear 1 is called driver and gear 2 is called driven or follower. It may be noted that the rotation of the driven gear is opposite to the rotation of the gear 1.

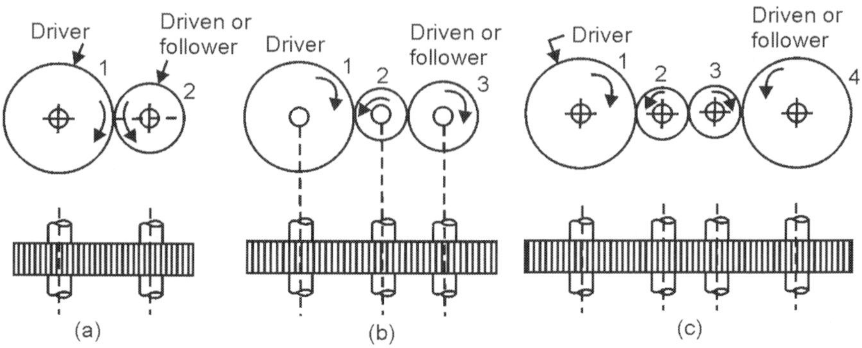

Figure 2.2 Simple gear train

Let N_1 = Speed of gear 1 (or driver) in rpm

N_2 = Speed of the gear 2 (or driven) in rpm

T_1 = Number of teeth in gear 1

T_2 = Number of teeth in gear 2

The speed ratio or velocity ratio of gear trains is the ratio of the speed of the driver to the speed of the driven or follower and ratio of speeds of any pair of gears in mesh is the inverse of their number of teeth.

Therefore,

$$\text{Speed ratio} = N_1/N_2 = T_1/T_2$$

It may be noted that ratio of the speed of the driven or follower to the speed of the driver is known as train value of the gear. Mathematically,

$$\text{Train value} = N_2/N_1 = T_1/T_2$$

Hence, the train value is the reciprocal of speed ratio.

In some occasions, the distance between two gears is large. The motion from one gear to another in such a case may be transmitted by either of the following two methods.

1. By providing large sized gears
2. By providing two or more intermediate gears

In the first method, providing large sized gears is inconvenient and uneconomical.

In the second method, providing one or more intermediate gears is convenient and economical.

It may be noted that when the number of intermediate gears is odd, the rotation of both the gears (i.e. driver and driven) will be in the same direction as shown in Fig. 2.2(b).

If the number of intermediate gears is even, the rotation of driver and driven will be in opposite direction as shown in Fig. 2.2(c).

Velocity ratio of simple gear train

Consider a simple gear train of gears with one intermediate gear as shown in Fig. 13.1(b).

Let N_1 = Speed of driver in rpm

N_2 = Speed of intermediate gear in rpm

N_3 = Speed of the driven or follower in rpm

T_1 = Number of teeth in driver

T_2 = Number of teeth in intermediate gear

T_3 = Number of teeth in driven or follower

Since, the driving gear 1 is in mesh with intermediate gear 2; therefore, the speed ratios of these two gears is

$$N_1/N_2 = T_2/T_1 \quad \text{... (2.1)}$$

Similarly, the intermediate gear 3 is in mesh with driven or follower gear 2; therefore the speed ratios of these two gears is

$$N_2/N_3 = T_3/T_2 \quad \text{... (2.2)}$$

The speed ratio of the gear train shown in Fig. 2.2(b) is obtained by multiplying Eqs. (2.1) and (2.2).

$$N_1/N_2 \times N_2/N_3 = T_2/T_1 \times T_3/T_2$$

$$N_1/N_2 = T_2/T_1$$

or

Speed ratio of driver / speed ratio of driven or follower = Number of teeth on driven / number of teeth on driver

Train value = Speed of driven / speed of driver = Number of teeth in driver / number of teeth in driven

The above equation holds good even if there are any number of intermediate gears. From the above, it may be noted that the speed ratio and train value is independent of the size and number of intermediate gears. These intermediate gears are called as idle gears. These idle gears do not affect the speed ratio or train value of the system. The idle gears are used for the following two purposes.

(a) To connect gears where a large distance is required.

(b) To get required direction of rotation of the driven or follower (clockwise or anti-clockwise).

2.2.3 Compound gear train

When there is more than one gear on a shaft as shown in Fig. 2.3, it is called a compound train of gear.

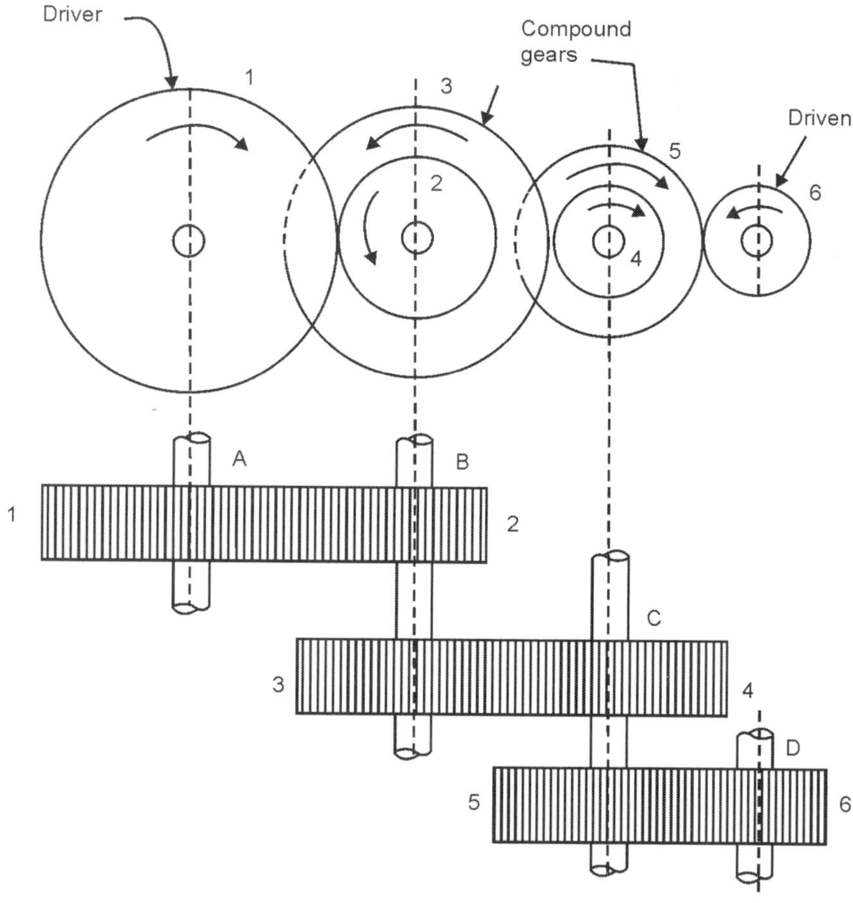

Figure 2.3 Compound gear train

We have seen in the simple train of gears, the idle gears do not affect the speed ratio of the system. These gears are useful in bridging the space between the driver and the follower.

But whenever the distance between the driver and the driven or follower has to be bridged over by intermediate gears and at the same time a great (or much less) speed ratio is required, then the advantage of intermediate gears is intensified by providing compound gears on intermediate shafts. In such cases, each intermediate shaft has two gears rigidly fixed to it so that they have the same speed. One of these two gears meshes with the driver and the other with the driven or follower attached to the next shaft as shown in Fig. 2.3.

As shown in Fig. 2.3, the gear 1 is the driving gear fixed on the shaft A, gears 2 and 3 are compound gears which are fixed on shaft B. The gears 4 and 5 are also compound gears which are fixed on shaft C and the gear 6 is the driven gear fixed on shaft D.

Let N_1 = Speed of driving gear 1

T_1 = Number of teeth on driving gear 1

$N_2, N_3, \ldots N_6$ = Speed of the respective gears in rpm

$T_2, T_3, \ldots T_6$ = Number of teeth on respective gears

Since gear 1 is in mesh with gear 2, therefore its speed ratio is

$$N_1/N_2 = T_2/T_1 \quad \ldots(2.3)$$

Similarly, the gears 3 and 4, speed ratio is

$$N_3/N_4 = T_4/T_3 \quad \ldots(2.4)$$

And for gears 5 and 6, the speed ratio is

$$N_5/N_6 = T_6/T_5 \quad \ldots(2.5)$$

The speed ratio of the compound gear trains is obtained by multiplying Eqs. (2.3), (2.4) and (2.5)

$$N_1/N_2 \times N_3/N_4 \times N_5/N_6 = T_2/T_1 \times T_4/T_3 \times T_6/T_5$$

or

$$*N_1/N_6 = T_2/T_1 \times T_4/T_3 \times T_6/T_5$$

*Since the gears 2 and 3 are fixed on tone shaft B, therefore = . Similarly gears 4 and 5 are fixed on shaft C, therefore $N_4 = N_5$

i.e. Speed ratio of the first driver / speed ratio of the last driven or follower

= Product of number of teeth on the driven or followers / product of number of teeth on drivers

and

Train value = Speed of the last driven or follower / speed of the first driver

= Product of number of teeth on the drivers / product of number of teeth on driven or followers

The advantage of compound gear train over a simple gear train is that a much larger reduction from the first shaft to the last shaft can be obtained with small gears. Usually, for a speed reduction in excess of 7 to 1, simple train is not used, a compound or worm gearing is used.

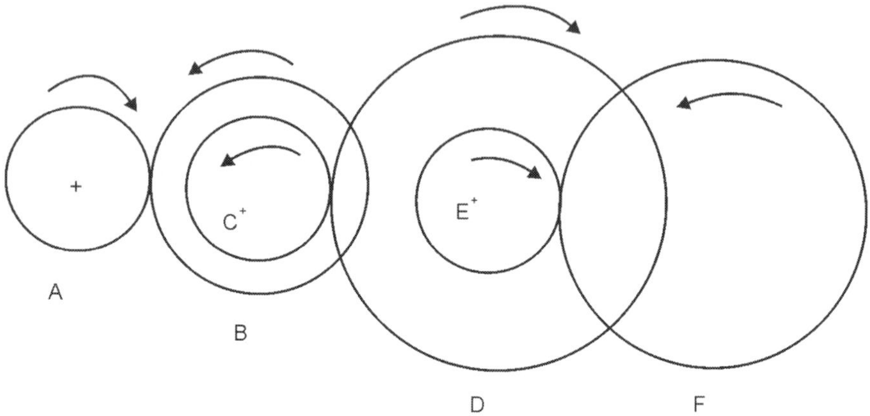

Figure 2.4 Gearing of a textile machine

Example 2.1: The gearing diagram of a textile machine is shown in Fig. 2.4. The motor shaft is connected to shaft A and rotates at 975 rpm. The gear wheels B, C, D and E are fixed to parallel shafts rotating together. The final gear F is fixed to the output shaft. What is the speed of the gear F? The number of teeth on each gear is as follows.

Gear A B C D E F

No. of teeth 20 50 25 75 26 65

Solution:

$N_A = 975$ rpm: $T_A = 20$: $T_B = 50$: $T_C = 25$: $T_D = 75$: $T_E = 26$: $T_F = 65$

From Fig. 2.4, the gears *A, C* and *E* are drivers, while the gears *B, D* and *F* are driven or followers. Let the gear A rotates in clockwise direction. Since the gears *B* and *C* are fixed on the same shaft, therefore it is a compound gear and the direction of rotation of both these gears is in anti-clockwise direction. Similarly the gears *D* and *E* are fixed on the same shaft, which is also a compound gear; the rotation of these gears is in clockwise direction. The gear *F* will rotate in anticlockwise direction.

Let N_F = Speed of the gear *F*, i.e. last driven or follower.

We know that

Speed of the first driver / Speed of the last driven or follower = Product of no. of teeth on driven / Product of no. of teeth on drivers

$$
\begin{aligned}
N_A/N_F &= T_B \times T_D \times T_C/T_A \times T_C \times T_E \\
&= 50 \times 75 \times 65/20 \times 25 \times 26 \\
&= 18.75
\end{aligned}
$$

$$N_F = N_A/18.75$$
$$= 975/18.75$$
$$= 52 \text{ rpm}$$

2.2.4 Reverted gear train (Fig. 2.5)

When the axes of the first gear (i.e. first driver) and the last gear (i.e. driven or follower) are co-axial, then the gear train is known as reverted gear train as shown in Fig. 2.5.

From Fig. 2.5., gear 1 (first driver) is driving the gear 2 (last driven or follower) in the opposite direction. Since the gears 2 and 3 are fixed on the same shaft, therefore they form a compound gear and the gear 3 will rotate in the same direction as that of gear 2. The gear 2 (which is now the second driver) drives the gear 4 (the last driven or follower) in the same direction as gear 1.

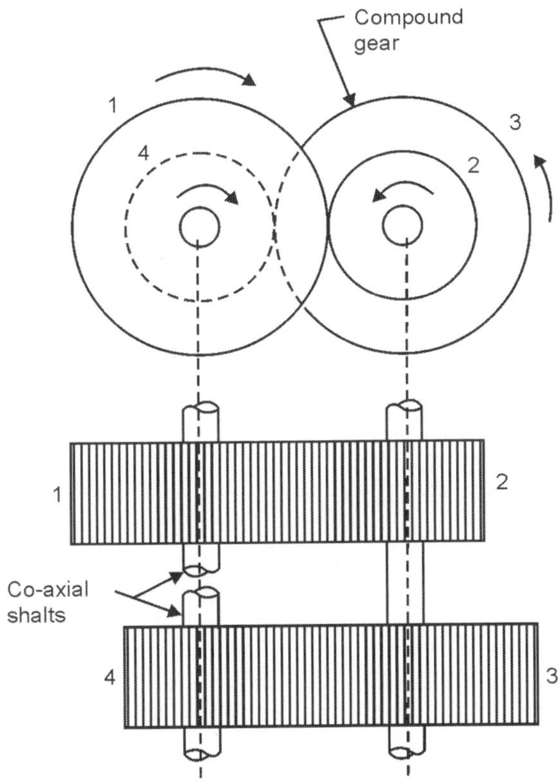

Figure 2.5 Reverted gear train

Thus in reverted gear train, the rotation of the first and the last gear is same.

Let T_1 = Number of teeth on gear 1

r_1 = Pitch circle radius of gear 1

N_1 = Speed of gear1

Likewise,

T_2, T_3, T_4 = Number of teeth on respective gears

r_2, r_3, r_4 = Pitch circle radius of respective gears

N_2, N_3, N_4 = Speed of respective gears

Since the distance between the centers of the shafts of gears 1 and 2 as well as gears 3 and 4 is same, therefore

$$r_1 + r_2 = r_3 + r_4 \qquad \ldots(2.6)$$

Also the circular pitch or module of all the gears is assumed to be same. Therefore, number of teeth on each gear is directly proportional to its circumference or radius.

$$*T_1 + T_2 = T_3 + T_4 \qquad \ldots(2.7)$$

Speed ratio = Product of number of teeth on driven / Product of number of teeth on drivers

or

$$N_1/N_4 = T_2 \times T_4/T_1 \times T_3 \qquad \ldots(2.8)$$

From Eqs. (2.6), (2.7) and (2.8), we can determine the number of teeth on each gear for the given centre distance, speed ratio and module only when the number of teeth on one gear is chosen arbitrarily. The reverted gear trains are used in automotive transmission, lathe back gears, industrial speed reducers and in clocks where the minute and hour hand shafts are co-axial.

- We know the circular pitch, p

$$p = 2 * \pi * r/T = \pi * m$$

or $$r = mT/2 \text{ where } m \text{ is the module.}$$

$$r_1 = mT/2;\ r_2$$
$$= mT_2/2;\ r_3$$
$$= mT_3/3;\ r_4 = mT_4/4$$

Applying in Eq. (2.6),

$$mT_1/2 + mT_2/2 = mT_3/3 + mT_4/4$$
$$T_1 = T_2 = T_3 + T_4$$

2.2.5 Epicyclic gear train (Fig. 2.6)

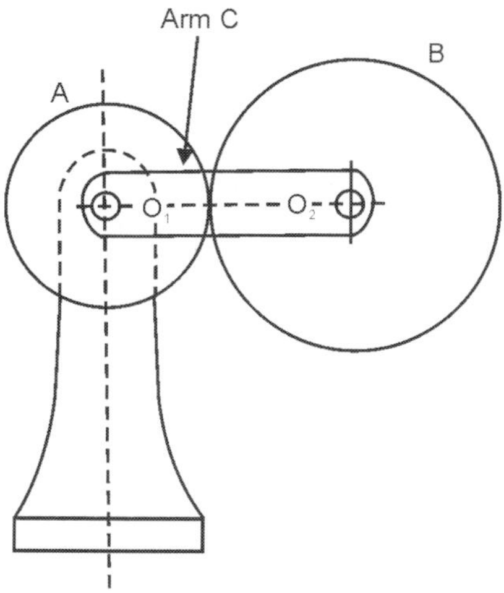

Figure 2.6 Epicyclic gear train

In an epicyclic gear train, the axes of the shafts, over which the gears are mounted, may move relative to affixed axis. A simple epicyclic gear train is shown in Fig. 2.6. In Fig. 2.6, the gear A and the arm C have a common axis at O_1 about which they can rotate. The gear B meshes with gear A and has its axis on the arm O_2, about which the gear can rotate. If the arm is fixed, the gear train is simple and the gear A can drive gear B or vice-versa, but if gear A is fixed and the arm is rotated about the axis of the gear A (O_1), then the gear B is forced to rotate *upon* and *around* gear A. Such a motion is called *Epicyclic* and the gear trains arranged in a manner that one or more of their members move *upon* and *around* another member are known as epicyclic gear train type (Epi means upon and cyclic means around). Epicyclic gear train may be *simple* or *compound* type.

Epicyclic gear trains find their applications in textile machines especially in differential gear box of speed frames and in comber machines. It is useful for transmitting high velocity ratios with gears of moderate sizes in comparatively lesser space.

2.2.6 Velocity ratio of epicyclic gear train

Two methods are in practice to find out the velocity ratio of epicycle gear train:

1. Tabular method

2. Algebraic method

These methods are discussed in detail as follows.

1. Tabular method

Consider an epicyclic gear train as shown in Fig. 2.6.

Let T_A = Number of teeth on gear A

T_B = Number of teeth on gear B

Let us suppose that the arm is fixed. Therefore the axes of both the gears are also fixed relative to each other. When the gear makes one revolution anti-clockwise, the gear B will make T_A/T_B revolutions. Assuming anti-clockwise rotation as positive and clockwise rotation negative, we can say that gear A makes +1 revolution anti-clockwise, gear B will make $(-T_A/T_B)$ revolutions. This statement of relative motion is entered in the first row of Table 2.1.

Secondly, if the gear A makes $+x$ revolutions, then the gear B will make $(-x * T_A/T_B)$ revolutions. This statement is entered in the second row of Table 2.1.

In other words, multiply the each motion entered in the first row by x.

Thirdly, each element of an epicyclic train is given $+y$ revolutions and entered in the third row. Finally, the motion of each element of the gear train is added up and entered in the fourth row.

Table 2.1 Table of motions

Step no.	Conditions of motions	Arm C	Gear A	Gear B
1	Arm fixed,. Gear A rotates through +1 revolution. i.e. anti-clockwise	0	+1	$-T_A/T_B$
2	Arm fixed. Gear A makes +x revolutions	0	+x	$-x * T_A/T_B$
3	Add +y revolutions to all elements	+y	+y	+y
4	Total motion(adding 2 and 30	+y	+x +y	$y - x * T_A/T_B$

When two conditions about the motion of rotation of any two elements are known, then the unknown speed of the third element may be obtained by substituting the given data in the third column of the fourth row.

2. Algebraic method

In the algebraic method, the motion of each element of the epicyclic gear train relative to the arm is obtained in the form of equation. The number of equations depends upon the number of elements in the gear train. But the two conditions are usually applied in any epicyclic gear train, i.e. some element is fixed and the other has specified motion. These two conditions are sufficient to solve all the equations and hence to determine the motion of any element in the epicyclic gear train.

Consider the epicyclic gear train shown in Fig. 2.6.

The arm C is fixed. Therefore the speed of the gear a relative to the arm C is $= N_A - N_C$

and the speed of the gear B relative to the arm C is $= N_B - N_C$

Since the gears A and B are meshing directly, they revolve in opposite direction.

$= N_B - N_C/N_A - N_C = -T_A/T_B$

Since the arm C is fixed, its speed $N_C = 0$

$$N_B/N_A = -T_A/T_B$$

If the gear A is fixed, the speed of gear $N_A = 0$

$$N_B - N_C/0 - N_C = -T_A/T_B$$

or
$$N_B/N_C = 1 + T_A/T_B$$

Among the two methods, tabular method is easier and hence mostly used in solving problems on epicyclic gear train.

Example 2.2: In an epicyclic gear train, shown in Fig. 2.7, an arm carries two gears A and B, each having 36 and 45 teeth, respectively. If the arm rotates at 150 rpm in the anti-clockwise direction about the centre of the gear A which is fixed, determine the speed of the gear B. If the gear A instead of being fixed makes 300 rpm in the clockwise direction, what will be the speed of gear B?

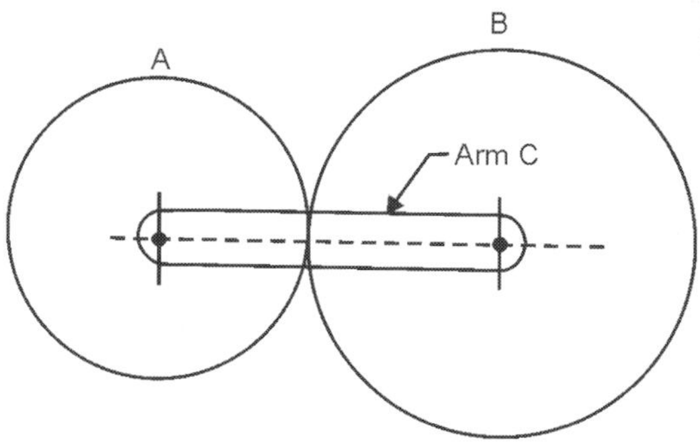

Figure 2.7

Solution:

$T_A = 36$; $T_B = 45$; $N_C = 150$ rpm (anti-clockwise)

We shall solve the problem first by tabular method and then by algebraic method.

1. Tabular method

Table 2.2 Table of motions

Step no.	Conditions of motions	Arm C	Gear A	Gear B
1	Arm fixed. Gear A rotates through +1 revolution, i.e. anti-clockwise	0	+1	$-T_A/T_B$
2	Arm fixed. Gear A makes +x revolutions	0	+x	$-x * T_A/T_B$
3	Add +y revolutions to all elements	+y	+y	+y
4	Total motion (adding 2 and 3)	+y	+x+ y	$y - x * T_A/T_B$

Speed of gear B when A is fixed

Speed of the arm *C* is 150 rpm anti-clockwise, therefore from the fourth row of the table

$$y = +150 \text{ rpm}$$

also, the gear *A* is fixed, therefore

$$x + y = 0 \text{ or } x = -y = -150 \text{ rpm}$$

Speed of gear *B*, $NB = y - x * T_A/T_B$

$$= 150 + 150\ (36/45 = +\ 270) \text{ rpm}$$

Speed of gear $B = +270$ rpm anti-clockwise

Speed of gear *B* when *A* makes 300 rpm clockwise.

Since the gear A makes 300 rpm clockwise, therefore from the fourth row of table

$$x + y = 300$$

Or $$-300 + y = -300 - 150 = -450 \text{ rpm}$$

Speed of gear $B =$

$$N_B = y - x * T_A/T_B$$
$$= 150 + 450 * 36/45$$
$$= 510 \text{ rpm anti-clockwise}$$

2. Algebraic method

Let N_A = Speed of gear *A*

N_B = Speed of gear *B*

N_C = Speed of arm *C*

Assuming the arm *C* is fixed, speed of gear *A* relative to arm *C*

$$= N_A - N_C$$

and speed of gear *B* relative to arm *C* is

$$= N_B - N_C$$

Since the gears A and B revolve in opposite direction,

$$N_B - N_C/N_A - N_C = -T_A/T_B$$

Speed of gear *B*, when gear *A* is fixed

When gear A is fixed, the arm C rotates at 150 rpm in anti-clockwise direction.

$$N_A = 0; N_C = +150 \text{ rpm}$$

$$N_B - 150/0 - 150 = -36/45$$

$$N_B = -150 \times -0.8 + 150 = 120 + 150 = 270 \text{ rpm}$$

Speed of gear B when gear A makes 300 rpm clockwise.

Since the gear A makes 300 rpm clockwise, therefore

$$N_A = -300 \text{ rpm}$$

$$N_B - 150/300 - 150 = -36/45$$

$$N_B = -450 \times -0.8 = 150 = 360 + 150 = 510 \text{ rpm}$$

2.2.7 Compound epicyclic gear train (Sun and Planet gear)

A compound epicyclic gear train is shown in Fig. 2.8. It consists of two co-axial shafts, S_1 and S_2, an annulus gear A which is fixed, the compound gear or planet gear B, C, the sun gear D and the arm H. The annulus gear has internal teeth and the compound gear is carried by the arm and revolves freely on a pin of the arm H. The sun gear is co-axial with the arm and annulus gear but independent of them.

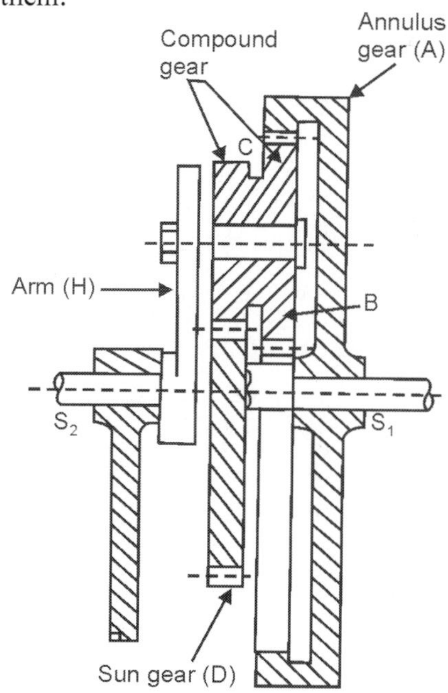

Figure 2.8

The annulus gear *A* meshes with gear *B* and the sun gear *D* meshes with the gear *C*. It is to be noted that when the annulus gear is fixed, the sun gear provides the drive and when the sun gear is fixed, the annulus gear provides the drive. In both cases, the arm acts as a follower.

Note: The gear at the centre is called the sun gear and the gears whose axes move are called planet gears.

Let T_A, T_B, T_C and T_D be the teeth and N_A, N_B, N_C and N_D be the speeds for the gears *A, B, C* and *D*, respectively. When the arm is fixed and the sun gear *D* is turned anti-clockwise, then the compound gear *B-C* and the annulus gear *A* will rotate in the clockwise direction.

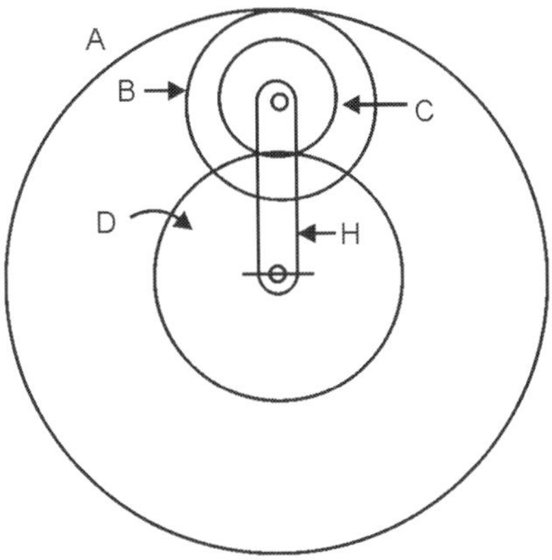

Let T_A, T_B, T_C and T_D be the teeth and N_A, N_B, N_C and N_D be the speeds for the gears *A, B, C* and *D*, respectively. When the arm is fixed and the sun gear *D* is turned anti-clockwise, then the compound gear *B-C* and the annulus gear A will rotate in the clockwise direction.

The table of motions of the various elements is shown below.

Table 2.3 Table of motions

Step no.	Conditions of motions	Arm	Gear D	Compound gear B-C	Gear A
1	Arm fixed. Gear D rotates through +1 revolution	0	+1	$-T_D/T_C$	$-T_D/T_C \times T_B/T_A$
2	Arm fixed. Gear D rotates through +x revolutions	0	+x	$-x\ {*}T_D/T_C$	$-x\ {*}T_D/T_C \times {*}T_B/T_C$
3	Add +y revolutions to all elements	+y	+y	+y	+y
4	Total motion	+y	x + y	$y + x{*}T_D/T_C$	$y - x\ {*}T_D/T_C \times T_B/T_C$

If the annulus gear A is rotated through one revolution anti-clockwise with the arm fixed, then the compound gear rotates through T_A/T_B revolutions in the same sense and the sun gear D rotates through $T_A/T_B \times T_C/T_D$ revolutions in clockwise direction.

Example 2.3: An epicyclic train of gears is arranged as shown in Fig. 2.9. How many revolutions does the arm, to which the pinions B and C are attached make?

1. When A makes one revolution clockwise and D makes half-a revolution anti-clockwise and
2. When A makes one revolution clockwise and D is stationary. The number of teeth on the gears A and D are 40 and 90 teeth, respectively.

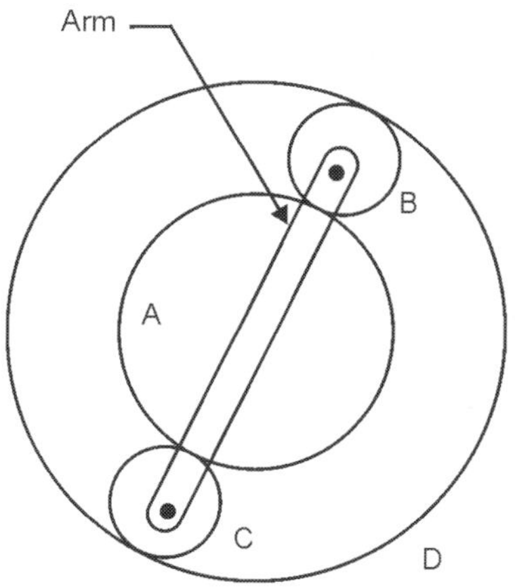

Figure 2.9

Solution:

$T_A = 40$; $T_D = 90$

First we have to find number of teeth on gears B and C (T_B and T_C). Let d_A, d_B, d_C and d_D be the pitch circle diameters of gears A, B, C and D, respectively. Therefore from the geometry of figures, we have

$$d_A + d_B + d_C = d_D \text{ or } d_A + 2d_B = d_D \; (d_B = d_C)$$

Since the number of teeth are proportional to the pitch circle diameters, therefore

$$T_A + 2T_B = T_D \text{ or } 40 + 2T_B = 90$$

From which $T_B = 25$ and $T_C = 25$ ($T_B = T_C$)

The table of motions is shown below.

Table 2.4 Table of motions

Step no.	Conditions of motions	Arm	Gear A	Compound gear B-C	Gear D
1	Arm fixed. Gear A rotates through −1 revolution (i.e. 1 rev. clockwise)	0	−1	$+T_A/T_B$	$+ T_A/T_B \times T_B/T_D$ $= + T_A/T_D$
2	Arm fixed. Gear A rotates through -x revolution	0	−0x	$+x *T_A/T_D$	$+x *T_A/T_B$
3	Add −y revolutions	−y	−y	−y	−y
4	Total motion	−y	−x−y	$x^* T_A/T_B - y$	$x^*T_A/T_B - y$

1. Speed of .arm when A makes 1 revolution clockwise and *D* makes half revolution anti-clockwise.

Since the gear *A* makes 1 revolution clockwise, therefore from the fourth row of the table

$$-x - y = -1 \text{ or } x - y = 1 \quad \text{...(2.9)}$$

And also, the gear *D* makes half revolution anti-clockwise, therefore

$$x *T_A/T_B - y = ½ \text{ or } x * 40/90 - y = ½$$

$$40x - 90y = 45 \text{ or } x - 2.25\,y = 1.125 \quad \text{...(2.10)}$$

From Eqs. (2.9) and (2.10), $x = 1.04$ and $y = -0.04$

$$\text{Speed of arm} = -y = -(-0.04) = +0.04$$

$$= 0.04 \text{ revolution anti-clockwise}$$

1. Speed of arm when *A* makes 1 revolution clockwise and *D* is stationary.

Since the gear *A* makes 1 revolution clockwise, therefore from the fourth row of table

$$-x - y = -1 \text{ or } x + y = 1 \quad \text{...(2.11)}$$

Also the gear *D* is stationary, therefore,

$$x^* T_A TD/T_D - y = 0 \text{ or } x * 40/90 - y = 0$$

$$40x - 90y = 0 \text{ or } x - 2.25y = 0 \quad \text{...(2.12)}$$

From Eqs. 2.11 and 2.12, we get

$$x = 0.692 \text{ and } y = 0.308$$

Therefore, speed of arm = $-y = 0.308 = 0.308$ revolution clockwise.

Example 2.4: In an epicyclic gear train shown in Fig. 2.10. The internal wheels *A* and *B* and compound wheels *C* and *D* rotate independently about axis *O*.The wheels *E* and *F* rotate on pins fixed to the arms *G*. *E* gears with *A* and C and F gears with B and D. All the wheels are having same module and the number of teeth are: $T_A = 28$: $T_D = 26$: $TE = TF = 18$.

1. Find the number of teeth on A and B.
2. If the arm G makes 100 rpm clockwise and A is fixed, find the speed of B.
3. If the arm G makes 100 rpm clockwise and the wheel A makes 10 rpm anti-clockwise, find the speed of wheel B.

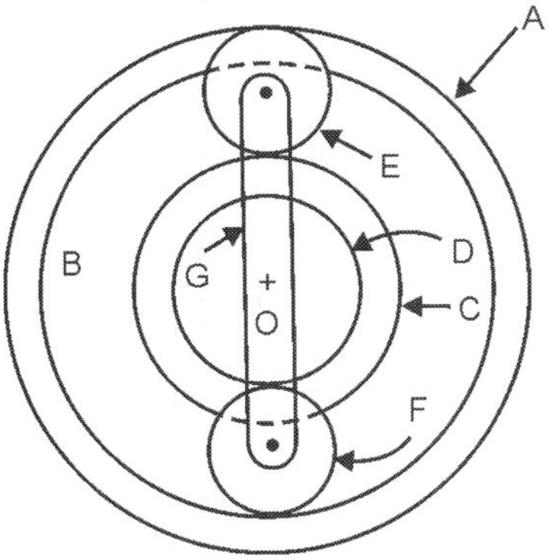

Figure 2.10

Solution:

$T_C = 28$: $T_D = 26$: $T_E = T_F = 18$.

1. Number of teeth on wheels B and C

   Let T_A = Number of teeth on wheel A

   T_B = Number of teeth on wheel B

   If d_A, d_B, d_C, d_D, d_E and d_F are the pitch circle diameters of wheels A, B, C, D, E and F respectively, then from the geometry of figures

   $$d_A = d_C + 2d_E$$

   and $$d_E = d_D + 2d_F$$

   Since the number of teeth are proportional to their pitch circle diameters for the same module,

   $$T_A = T_C + 2T_E = 28 + 2 \times 18 = 64$$

   $$T_B = T_D = 2T_F = 26 + 2 \times 18 = 62$$

2. Speed of wheel B when arm G makes 100 rpm clockwise and wheel A is fixed.

   This can be calculated by constructing the table of motions which is shown in table below.

Table 2.5 Table of motions

Step no.	Conditions of motions	Arm G	Wheel A	Wheel E	Compound wheel C-D	Wheel F	Wheel B
1	Arm fixed. Wheel A makes +1 rev. anticlockwise	0	+1	$+T_A/T_B$	$-T_A/T_B \times T_E/T_C$	$= T_A/T_C \times T_D/T_F$	$+T_A/T_C \times T_D/T_F \times T_F/T_B = +TA/T_C \times T_D/T_B$
2	Arm fixed. Wheel A rotates through +x revolutions	0	+x	$+x * T_A/T_E$	$-x * T_A/T_C$	$+x * T_A/T_C \times T_D/T_F$	$+x * T_A/T_C \times T_D/T_B$
3	Add +y revolutions to all elements	+y	+y	+y	+y	+y	+y
4	Total motions	+y	x + y	$y + x * T_A/T_E$	$y - x * T_A/T_C$	$y + x * T_A/T_C * T_D/T_F$	$y + x * T_A/T_C * T_D/T_B$

Since the arm G makes 100 rpm clockwise, from the fourth row of table,

$$y = -100$$

Also, the wheel A is fixed, therefore,

$$x + y = 0 \text{ or } x = -y = 100$$

Speed of wheel $B = y + x \times T_A/T_C \times T_D/T_B$

$$= -100 + 100 \times 64/28 \times 26/62$$
$$= -100 + 95.8 = -4.2 \text{ rpm clockwise}$$

3. Speed of wheel B when arm G makes 100 rpm clockwise and wheel A makes 10 rpm counter clockwise

   Since the arm G makes 100 rpm clockwise, therefore

   $$y = -100$$

   Also the wheel A makes 10 rpm counter clockwise, therefore

   $$x + y = 10$$

   or $x = 10 - y = 10 + 100 = 110$

   Speed of wheel $B = y + x \times T_A/T_C * T_D/T_B$

   $$= -100 + 110 \times 64/28 \times 26/62$$
   $$= -100 + 105.4 \text{ rpm}$$
   $$= +5.4 \text{ rpm counter or anti-clockwise}$$

2.2.8 Epicyclic gear train with bevel gears (Fig. 2.11)

Epicyclic bevel gears are used to make a compact epicyclic system, and they permit a very high speed reduction with few gears. The useful application of bevel gears is found in differential mechanism in speed frame machines. In the speed frame differential mechanism, the speed of rotation of the bobbin has to be reduced when the diameter of the bobbin builds up. Since the delivery rate is constant from the front roller nip, when the bobbin diameter increases it tends to pull more material from the front roller. But the front roller delivery is always constant throughout the build of the bobbin, the only possibility is to reduce the bobbin speed so as to continue the winding process. This requirement is fulfilled by the differential gears.

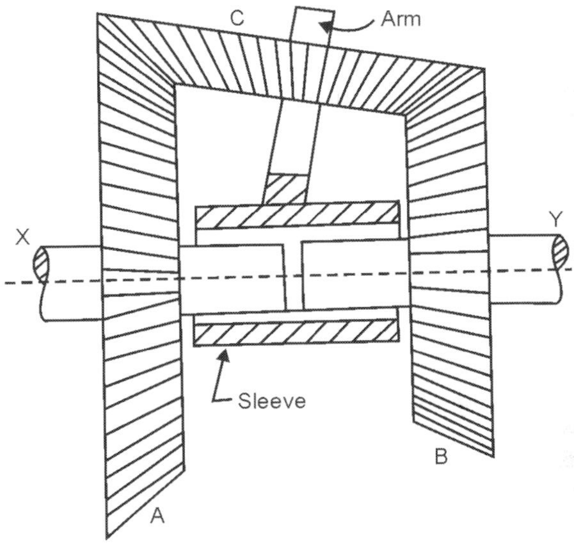

Figure 2.11

Example 2.5: Two bevel gears *A* and *B* having 40 teeth and 30 teeth are rigidly mounted on two co-axial shafts *X* and *Y*. A bevel gear *C* having 50 teeth meshes with *A* and *B* and rotates freely on one end of arm. At the other end of the arm is welded a sleeve and the sleeve is riding freely loose on the axes of the shafts *X* and *Y*.

If the shaft *X* rotates at 100 rpm clockwise and the arm rotates 100 rpm anti-clockwise, find the speed of shaft Y.

Solution:

$T_A = 40$; $T_B = 30$; $T_C = 50$; $N_X = N_A = 100$ rpm clockwise

Speed of arm = 100 rpm anti-clockwise

First, we have to prepare the table of motions

Table 2.6 Table of motions

Step no.	Conditions of motions	Arm	Gear A	Gear C	Gear B
1	Arm B fixed. Gear A rotates through +1 revolution anti-clockwise	0	+1	$\pm T_A/T_C$	$-T_A/T_C \times T_C/T_B = -T_A/T_B$
2	Arm B fixed Gear A rotates through +x revolutions	0	+x	$\pm x{*}T_A/T_C$	$-x\,{*}T_A/T_B$
3	Add +y revolutions to all motions	+y	+ y	+y	+y
4	Total motion	+y	x + y	$y \pm x\,{*}T_A/T_C$	$y - x\,{*}T_A/T_B$

- The ± sign is given to the wheel *C* since it is in different plane. So we cannot indicate the direction of motion specifically, i.e. either clockwise or anti-clockwise.

Speed of the arm is 100 rpm clockwise, therefore from the fourth row of the table,

$$Y = +100$$

Also the speed of the driving shaft *X* or gear *A* is 100 rpm anti-clockwise, therefore

$$x + y = -100$$

$$x = -y = -100 - 100 = -200$$

Speed of the driven shaft (i.e.) shaft *Y*

$$\begin{aligned} NY &= \text{Speed of gear } B \\ &= y - x\,{*}T_A/T_B \\ &= -100 - (-200 \times 40/30) \\ &= +366.7 \text{ rpm anti-clockwise} \end{aligned}$$

2.2.9 Epicyclic gear train in textile testing (Fig. 2.12)

Epicyclic gear train finds application in testing equipments for determining the linear density of the yarn. In Fig. 2.12, the bevel gears *A*, *B* and *C* are equal in size. Bevel *A* is fixed on stud *S*, bevel *B* is an idler wheel and bevel *C* is free to rotate on stud *S*. The arm *D* rotates around stud *S* and receives its motion from the handle *H*. If the handle *H* is rotated in clockwise direction, bevel *C* also rotates in clockwise direction but it makes two revolutions for each revolution of the handle, one revolution due to the gearing via *B* with *A* and the other due to the rotation of the arm *D*.

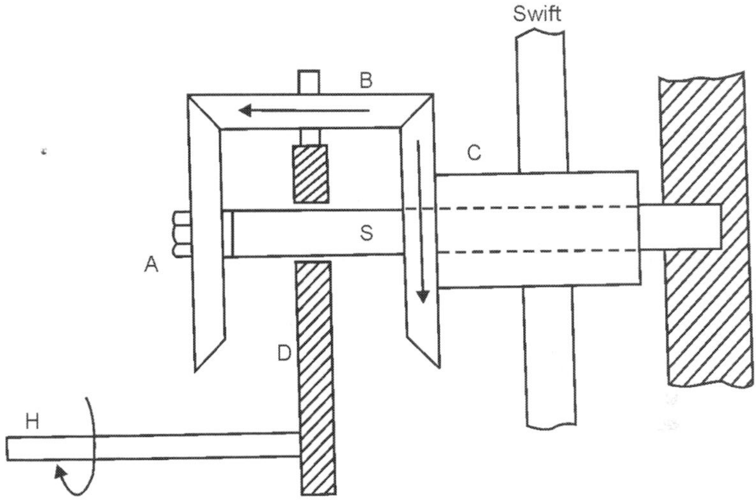

Figure 2.12

2.2.10 Epicyclic gears as transmission gears (Fig. 2.13)

Epicyclic gears can be used as transmission gears which is best illustrated by an example.

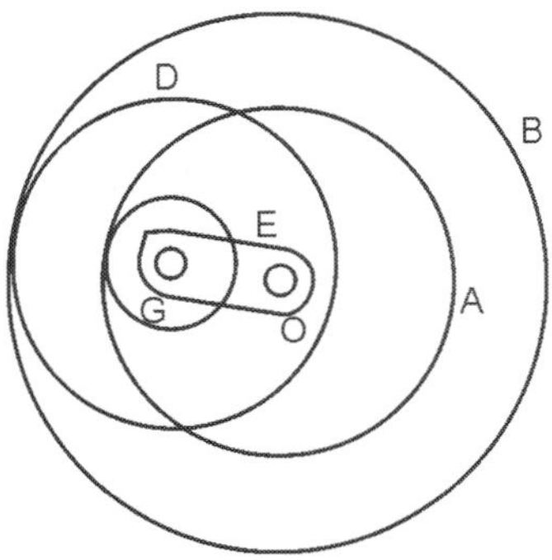

Figure 2.13 Epicyclic gear train as transmission gears

Example 2.6: In Fig. 2.13, wheels A and B are internal gears and B is free to rotate independently about centre O. Wheels C and D form a compound carrier with centre Q, carried by the arm E and meshing with A and B. Wheel A is fixed and has 22 teeth; wheel B is rotated by power source. How many revolutions will the last wheel make?

Solution:

The value of gear train is given by

$$e = +22/19 \times 20/23 = +440/437$$

The value of *e* is positive because clockwise rotation of the first wheel *A* causes clockwise rotation of wheel *B*.

$$e = 1 - a/f - a$$
$$440/437 = 1 - 1/0 - 1$$
$$-440/437 = 1 - 1$$
$$1 - 440/437 = 1 = -3/437$$

The wheel *B* rotates in the opposite direction to that of the arm *E* and the speed reduction ratio is 437/3 = 146:1.

Example 2.7: Figure 2.14 shows a planetary mechanism using four gears. Wheel *A* is fixed and has 40 teeth, wheel *B* has 50 teeth and compounded with it is wheel *C* with 51 teeth. Wheel *C* drives the last wheel *D* which has 50 teeth. The arm *E* rotates about the fixed shaft of wheel *A*. Calculate the reduction ratio.

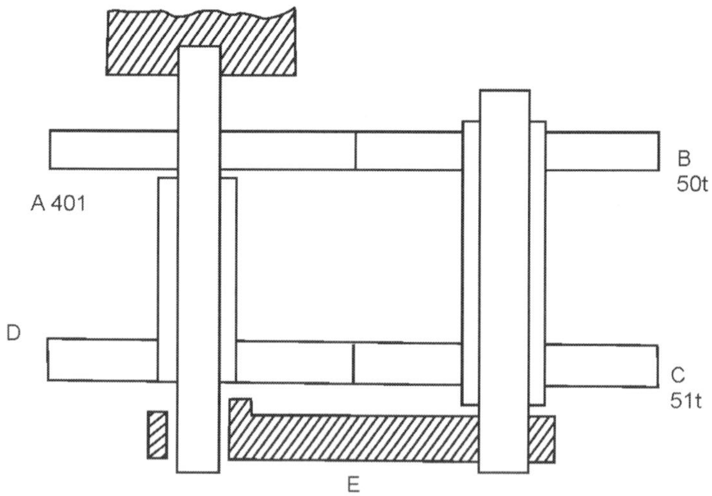

Figure 2.14

Solution:

The value of the gear train is

$$e = +49/50 \times 51/50$$
$$= +2499/2500$$

Let the arm rotates through 1 revolution. Then

$$e = +2499/2500$$
$$= 1 - 1/0 - 1$$
$$-2499 / 2500 = 1 - 1$$

and $$1 - 2499/2500 = 1/2500$$
$$1 = 1/2500$$

Hence with the use of planetary gearing, a high reduction ratio of 2500 : 1 = obtained

2.2.11 Epicyclic gear train in loom let-off motion

Epicyclic gear train is used in Roper's let-off motion in a loom to control the rate at which the warp is released from the loom beam. The epicyclic gear train is shown in Fig. 2.15; there is a ratchet wheel, *R*, rotated intermittently by a pawl, whose movement is dependent on both the warp tension and the loom beam diameter. The ratchet wheel is loose on the shaft S and is compounded with an eccentric *E*. There is a wheel *A* of 35 teeth loosely mounted on this eccentric. Is the first wheel of the train Wheel *A* drives an internal gear *B* of 40 teeth, which is secured to shaft *S*? On the other end of shaft *S* is a small pinion, which drives the beam gear.

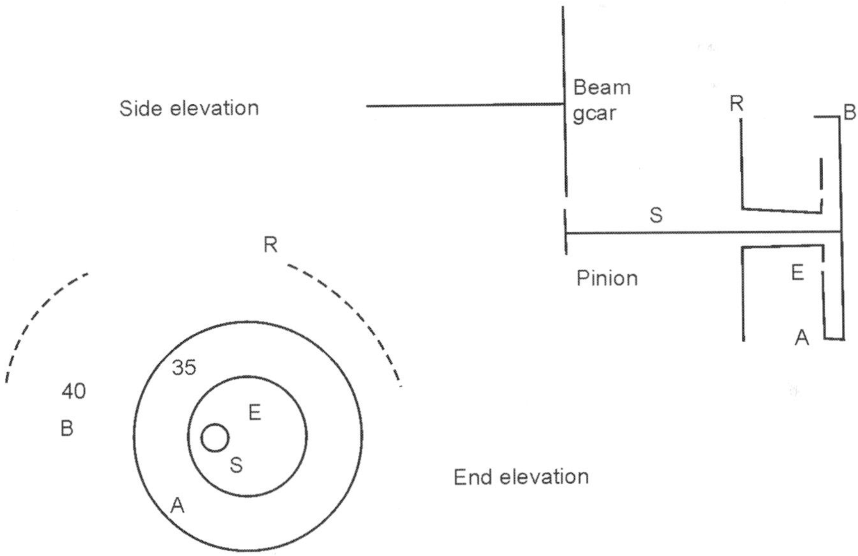

Figure 2.15 Roper let-off motion

The arm of the train is eccentric which is rotating at the speed of the ratchet wheel. The value of the gear train is

$$e = +35/40$$
$$= +7/8$$

If we assume wheel A as first wheel with zero rev/ min. and the arm rotates through one revolution, then:

$$e = 1 - 1/0 - 1$$
$$= +7/8$$

from which $l = +1/8$

Thus a reduction ratio of 8 : 1 is obtained between the ratchet wheel and the pinion driving the beam wheel.

2.2.12 Epicyclic gear train as differentials

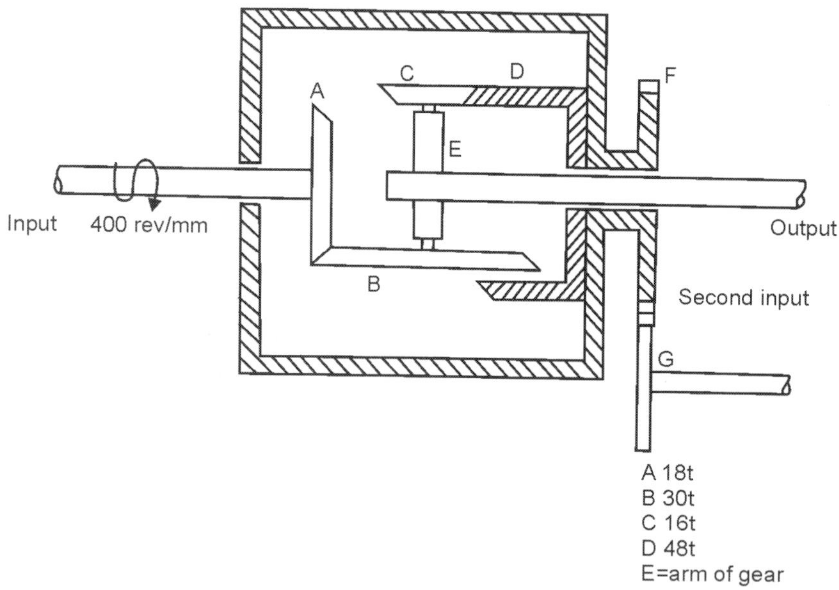

Figure 2.16

In this system, the last wheel D is secured to the housing of the mechanism. The housing also having the outer gear F through which the second input may be fed. In this differential, the first input is through the bevel A, which meshes with bevel B, part of a compound carrier.

The value of the system is

$$e = +18/30 \times 16/48 = +1/5$$

Let the input speed be 400 rev/min and the shell or wheel D remain stationary. Now we have to determine the speed and direction of rotation of output shaft, i.e. rev/min of the arm E. Since

$$e = 1 - a/f\text{-}a$$

$$= 1/5 = 0 - a/400 - a$$

$$a = -400/4 = -100 \text{ rev/min}$$

Therefore, the output shaft rotates at = 100 rpm anti-clockwise.

If through the shell D, the second input is given in clockwise direction at 20 rev/min. How will this affect the output speed? The last wheel now has +20 rev/min, i.e. l = +20. Putting this value in the equation, we have

$$e = 1 - a/f\text{-}a$$

$$1/5 = 20 - a/400 - a$$

$$a = -300/4 = -75 \text{ rev/min}$$

If instead feeding the speed of 20 rev/min. in clockwise direction, we had fed in anti-clockwise direction, the output speed will be given by

$$1/5 = -20 - a/400 - a$$

from which $a = -500/4 = -125$ rev/min

From this, it can be clearly seen that the output speed is controlled basically by the constant input speed of 400 rev/min, but that a second input speed which can be variable also has an effect. This is the great advantage of differential gear which can add and subtract rotary motion.

This can be explained in tabular method as shown in Table 2.7.

Table 2.7 Analysis of differential gear

Step no.	Conditions of motions	Wheel A	Wheel D	Arm E
1	Gears fixed on arm	A	A	a
2	Arm fixed	400 – a	1/5(400 – a)	0
3	Total motions	400	20	A

From Table 2.7

$$20 = 1/5\,(400 - a) + a$$

$$= 80 + 4/5a$$

$$= -60 + 4/5a$$

$$a = -5 \times 60/4 = -75 \text{ rev/ min}$$

Numerical problems (in SI units)

1. A simple gear train consists of two gears only. Each gear is mounted on separate shaft which are parallel. The first gear is driving the second gear and the speed of the first gear is 600 rpm. The number of teeth on first and second gears are 20 and 30 teeth respectively. Determine: (i) speed ratio of the gear train; (ii) Train value of the gear train; (iii) Speed of second gear; (iv) Direction of rotation of the second gear, if first gear is rotating clockwise.
   Ans. (i) 3, (ii) 0.33, (iii) 200 rpm, (iv) Anti-clockwise.
2. A simple gear train consists of three gears mounted on a separate shaft. The shafts are parallel. The gear 1 meshes with gear 2 and gear 2 meshes with gear 3. The gear 1 is the driver rotating clockwise at 1000 rpm and gear 3 is follower. The number of teeth on the gears 1, 2 and 3 are 20, 30 and 50, respectively. Find (i) speed ratio of the gear train, (ii) speed of the follower, (iii) direction of rotation of the follower.
   Ans. (i) 2.5, (ii) 400 rpm, (iii) Clockwise
3. Two parallel shafts are connected with the help of two gears one gear on each shaft. The number of teeth on one gear is 40 and speed of the shaft is 500 rpm. If the speed ratio is 2.5 and circular pitch of the gears is 24 mm, determine: (i) number of teeth and speed of another shaft; (ii) centre distance between two shafts.
   Ans. (i) 100 and 200 rpm; (ii) 534.75 mm
4. In a compound gear train, the power is transmitted from a motor shaft to output shaft. The motor shaft is connected to gear 1 whereas the output shaft is connected to gear 4. The gears 2 and 3 are mounted on the same shaft. The gear 1 meshes with gear 2 whereas gear 3 meshes with gear 4. The motor shaft is rotating at 1200 rpm in the clockwise direction. Find the direction and speed of the output shaft if no. of teeth on gears 1, 2, 3 and 4 are 25, 50, 20 and 40.
   Ans. 300 rpm clockwise

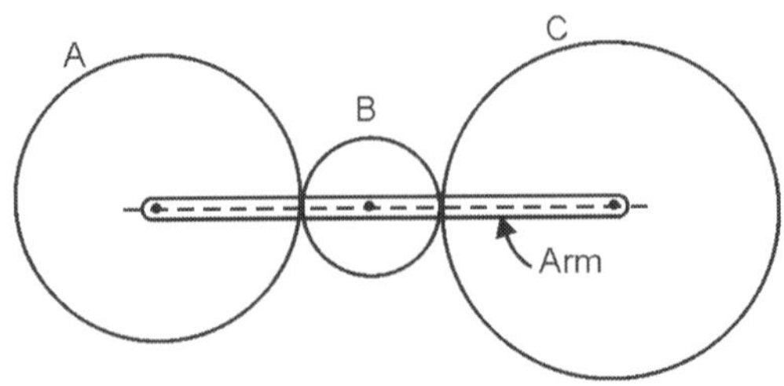

Figure 2.17

5. In an epicyclic gear train shown in Fig. 2.17, the number of teeth on wheels *A*, *B* and *C* are 48, 24 and 50, respectively. If the arm rotates at 400 rpm clockwise, find (i) speed of wheel *C* when *A* is fixed and (ii) speed of wheel *A* when *C* is fixed.

   Ans. (i) 16 rpm clockwise; (ii) 16.67 rpm anti-clockwise

6. In an epicyclic gear train shown in Fig. 2.18, the wheel *C* is keyed to the shaft *B* and wheel *F* is keyed to shaft *A*. The wheels *D* and *E* rotate together on a pin fixed to the arm *G*. The number of teeth on wheels *C*, *D*, *E* and *F* are 35, 65, 32 and 68, respectively. If the shaft A rotates at 60 rpm and the shaft B rotates at 28 rpm in the opposite direction, find the speed and the direction of rotation of arm *G*.

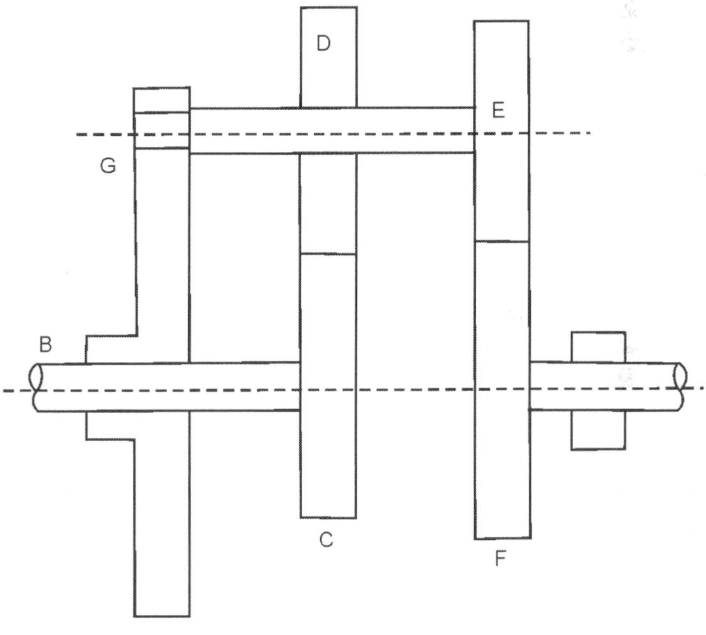

Figure 2.18

   Ans. 90 rpm in the same direction as shaft *A*

7. An epicyclic gear train shown in Fig. 2.19 has a sun wheel *S* of 30 teeth and two planet wheels *P-P* of 50 teeth. The planet wheels mesh with internal gear of a fixed annulus *A*. The driving shaft carrying the sun wheel transmits 4 kW at 300 rpm. The driven shaft is connected to an arm which carries the planet wheels. Determine the speed of the driven shaft and the torque transmitted if the overall efficiency is 95%.

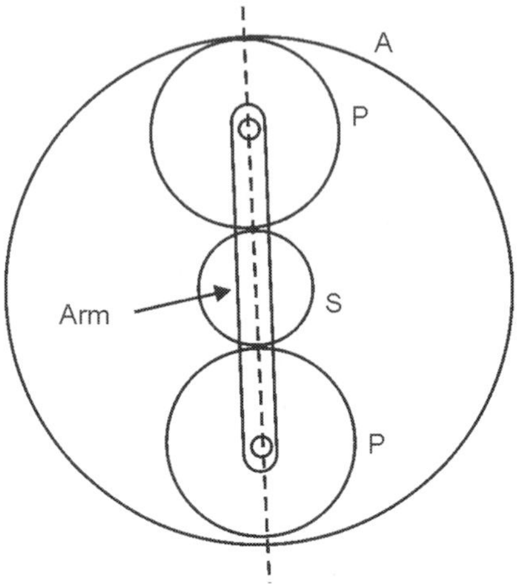

Figure 2.19

Ans. 56.3 rpm; 644.5 N-m

8. An epicyclic gear train shown in Fig. 2.20 is composed of fixed annular wheel *A* having 150 teeth. The wheel *A* is meshing with wheel *B* which drives wheel *D* through an idle wheel *C*. *D* being concentric with *A*. The wheels *B* and *C* are carried on an arm which revolves clockwise at 100 rpm about the axis of *A* and *D*. If the wheels B and *D* have 25 and 40 teeth, respectively, determine the number of teeth on *C* and the sense of rotation of *C*.

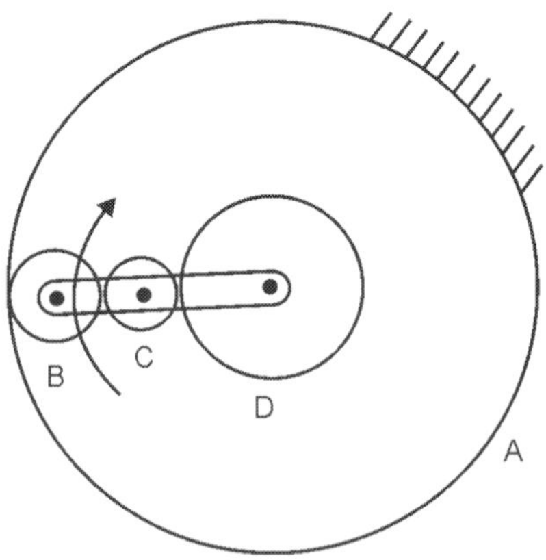

Figure 2.20

Ans. 30; 600 rpm; clockwise

9. An epicyclic gear train consists of a fixed annular wheel of 180 teeth. Meshing with wheel A is a gear B which drives gear D through idle gear C. The gear D is concentric with wheel A. Gears B and C are carried on an arm which revolves at 80 rpm clockwise. About the axis of A or D, if the gears B and D are having 30 and 48 teeth respectively, find the number of teeth of C and speed and sense of rotation of C.
   Ans. 36: 490 rpm clockwise
10. An epicyclic gear train consists of sun and planet gear. The annular gear D has 80 internal teeth whereas the sun and planet gear consists of 40 and 20 external teeth, respectively. The gear B meshes with annular gear D and also with sun gear A. The arm E rotates about the centre of annular gear D. The gear B is carried on the arm E. If the gear D is fixed and the arm rotates at 30 rpm, find the speeds of gears A and B.
    Ans. $N_A = 90$ rpm; $N_B = -90$ rpm

3
Rotary motion

3.1 Introduction

The parts of a machine may have a linear motion or circular motion or motion along a curved path whose radius is changing. For example, in case of a loom, the movement of shuttle is in straight line from one end to the other end, whereas the movement of sley move along with circular arcs with centre as the axis of the crank shaft. This chapter deals with these types of motions and also with the forces which act on these parts.

3.2 Linear displacement

Displacement involves direction as well as magnitude and hence it is a vector quantity. Therefore, it can be represented by a straight line. The magnitude is represented by the length of the line and the direction is represented by the arrowhead placed on the line.

3.3 Angular displacement

Angular displacement is the angle moved by a revolving body. The commonest unit for angular displacement is 1 revolution. Another useful unit is 1 radian which is the angle subtended at the center of the circle by an arc equal to the radius as shown in Fig. 3.1.

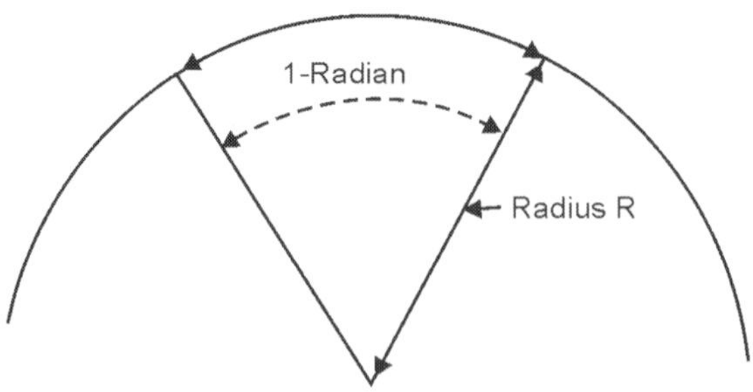

Figure 3.1 Angular displacement

$$1 \text{ revolution} = 2 * \pi \text{ radians} = 360°$$

or

$$1 \text{ radian} = \tfrac{1}{2} * \pi \text{ of a revolution} = 57.3°$$

3.4 Speed

It is the rate of change of position, or displacement in unit time. Linear speed is generally expressed in feet per second, angular speed in revolutions per minute (rpm), revolutions per second or radians per second.

3.5 Transmission of rotary movement in machines

The exact speed of rotation of machine parts are arrived by connecting them together by toothed gears, chains, belt or rope drives. Gears and chains give an exact speed relationship between the parts they connect together. Belts and ropes may slip when transmitting movement.

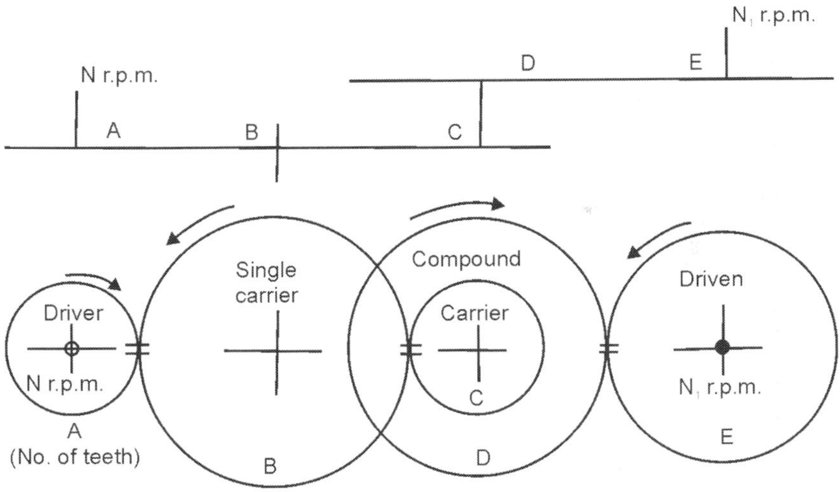

Figure 3.2 Speed calculations for wheel gearing

When two parts are connected by toothed gearing or by chain gearing, the speeds are in inverse proportion to the number of teeth in the two wheels. There is a general rule that the speed of a driven shaft is the speed of the driver multiplied by the product of the number of teeth in the driver wheels, and divided by the product of the numbers of teeth in the driven wheels as shown in Fig. 3.2.

$$N_1 = N * A/B * B/C * D/E = N * A/C * D/E$$

or

$$N = N_1 * E/D * C/A$$

Example 3.1: Speed gear for seamless hosiery machines. Many knitting machines are fitted with a two-speed driving gear, to enable the knitting operations to be carried out at reduced speed.

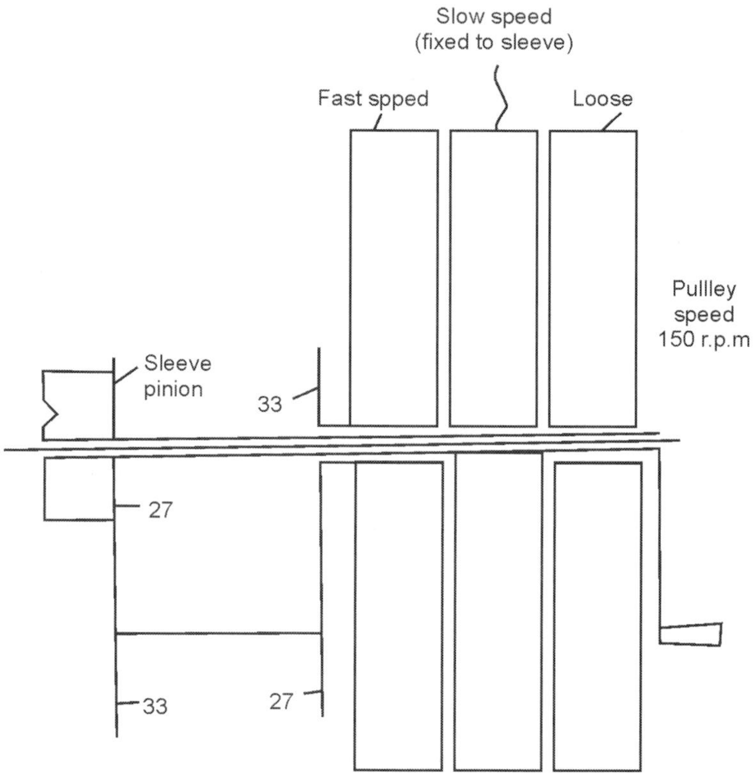

Figure 3.3 Seamless hose knitting machine

From Fig. 3.3,

Slow speed : Sleeve pinion speed = Machine pulley speed = 150 rpm

Fast speed : Sleeve pinion speed = Machine pulley speed * 33/27 * 33/27 = 150 * 33/27 * 33/27 = 224 rpm

Example 3.2: Loom shedding tappet shaft drive

Figure 3.4 shows the usual arrangement of gearing between the crank shaft of the loom and the shafts on which the shedding tappets are placed. Since one pick is inserted in the cloth for each revolution of the crank shaft, the speed of the crank shaft divided by the speed of the tappet shaft must be equal to the number of picks to the repeat of the pattern.

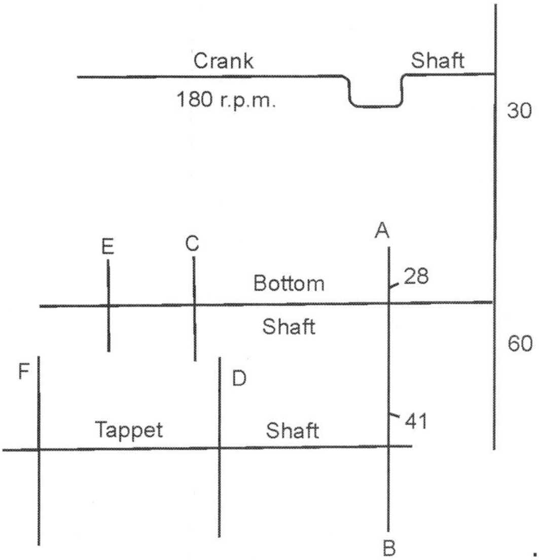

Figure 3.4 Loom shedding drive

Hence, 2 picks per repeat. Tappets are placed on the bottom shaft.

Speed of the bottom shaft = 180 * 30/60 = 90 rpm

3 picks per repeat, tappets on tappet shaft, wheels *A* and *B* in gear.

Speed of tappet shaft = 180 * 30/60 * 28/42 = 60 rpm

Speed of crank shaft/Speed of the tappet shaft = 180/60 = 3

4 or 5 picks per repeat.

With *C* and *D* in gear, for 4 picks per repeat, the speed of the tappet must be ¼ of the crank shaft speed, i.e. speed of the tappet shaft = 180/4 = 45 rpm; since the speed of the bottom shaft is 90 rpm, the wheels *C* and *D* must be in the proportion 1:2.

Example 3.3: A loom is driven by a V-rope drive from an individual motor. The motor speed is 960 rpm and the effective diameters are 3 and 16 in. for motor and loom, respectively.

Answer: Loom speed = Motor speed * 3/16

= 960 * 3/16 = 180 rpm (Assuming no slip)

If the wear on the ropes cause the rope to reduce the pulley diameters by 0.1 in., the effective speed will be

960 * 2.9/15.9 = 175 rpm

Such an adjustment has to be provided for the distance between the pulley canters to compensate for belt wear.

3.6 Classification of gears

The motion from one shaft to another shaft may be transmitted by belts, ropes and chain drives. These methods are used when the two shafts are having long centre distance. If the distance between the shafts is small, then gears are used to transmit motion from one shaft to another. In case of belts and ropes, the drive is not positive and slip in the motion always takes place. This slip will reduce the velocity ratio. But, gear drives are positive and transmit exact velocity ratio.

The gear is defined as a toothed element which is used for transmitting motion from one shaft to another.

The following are the important classifications of gears.

(i) Classification based on the position of the axes on the shafts. The axes of the two shafts between which motion is transmitted may be.

(a) Parallel, (b) intersecting, (c) non-parallel and non-intersecting

(a) Parallel shafts

The following are the main types of gears to join parallel shafts.

(i) Spur gears, (ii) Helical gears, (iii) Double-helical gears

(i) *Spur gears* – The gears used to connect two parallel shafts and having straight teeth parallel to the axes of the wheel as shown in Fig. 3.5(a) are known as spur gears. In Fig. 3.5(a) the gears have external teeth on the outer surfaces and the two shafts rotate in opposite direction.

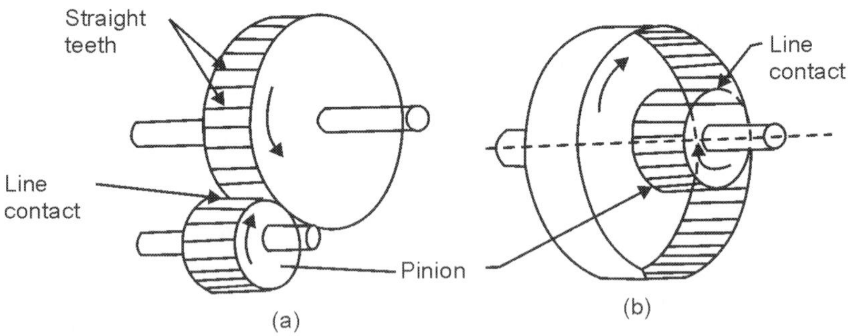

Figure 3.5 Spur gears

In Fig. 3.5(b), the internal teeth are formed over the outer wheel and external teeth are formed over the inner wheel. The inner wheel having external gears (smaller gear) is known as pinion. The two shafts will rotate in the same direction. In spur gears, the contact occurs across a line. Hence, spur gears are having line contact.

(ii) Helical gears

The gears used to connect two parallel shafts and having their teeth inclined or curved to the axes of the shafts as shown in Fig. 3.6 are known as helical gears. In helical gears, each tooth is helical in shape. The two mating gears will have the same helix angle but have teeth opposite.

In helical gears, the contact occurs at a point of curved teeth at the beginning of engagement and afterwards extends along a diagonal line across the teeth.

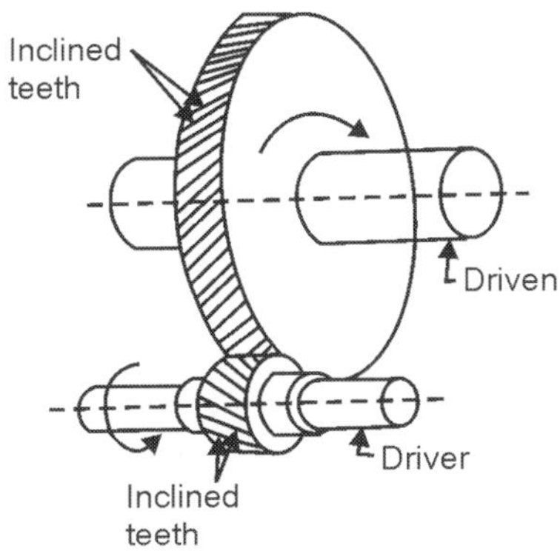

Figure 3.6 Helical gears

(iii) Double helical gears

A pair of helical gears secured together, one having a right-hand helix and the other having a left-hand helix, are known as Double-helical gears (Fig. 3.7). This pair is mounted on one shaft and the similar other pair, which is having opposite teeth, is mounted on other parallel shaft which is to be connected.

The double helical gear is also known as herringbone gear if the right and left inclinations of the double helical gears meet at a common apex and there is no groove in between them as in Fig. 3.7.

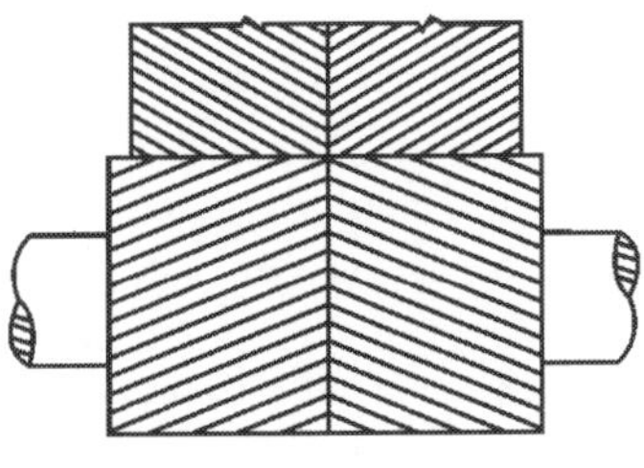

Figure 3.7 Double helical gears

(b) Intersecting shaft

When two intersecting shafts are to be connected, bevel gears are used. If the teeth on the gears are straight radial to the point of intersection of shaft axes, then gears are known as straight bevel. But if teeth are inclined, then gears are known as helical bevel or spiral bevel as shown in Fig. 3.8(a) and (b).

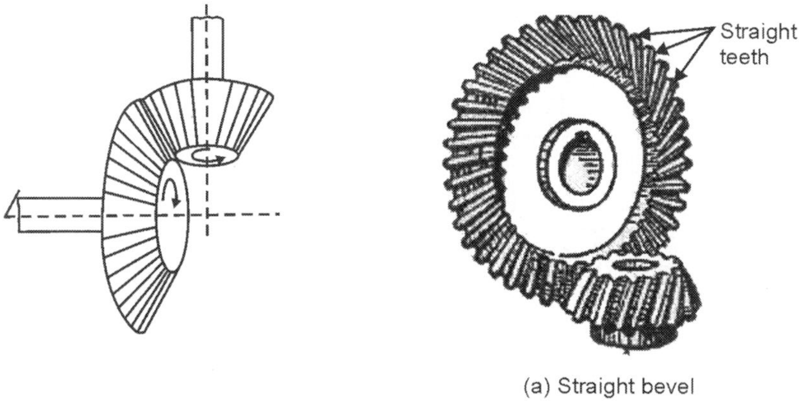

Figure 3.8 (a) Straight bevel; (b) Spiral bevel

(c) Non-parallel and non-intersecting shafts

When two non-parallel and non-intersecting shafts are to be connected, skew bevel gears or skew spiral gears are used.

3.7 Gears in textile machines

Gears like spur, helical, worm and worm gearing, rack and pinion, ratchet wheels are mostly used in spinning and weaving machines. They are discussed in the following sections.

3.7.1 Worm and worm gearing

This type of gearing is used where the speed of the driven shaft must be less than the speed of the driver shaft.

From Fig. 3.9, it is seen that the worm must drive the wheel: the wheel will not drive the worm. The worm is really a screw or thread suitably shaped to fit the teeth of the wheel. The worm in its simplest form has a single thread cut helically from one end to the other. The worm is prevented from moving endways by suitable collars on the driving shaft and as the worm rotates it moves the wheel slowly round.

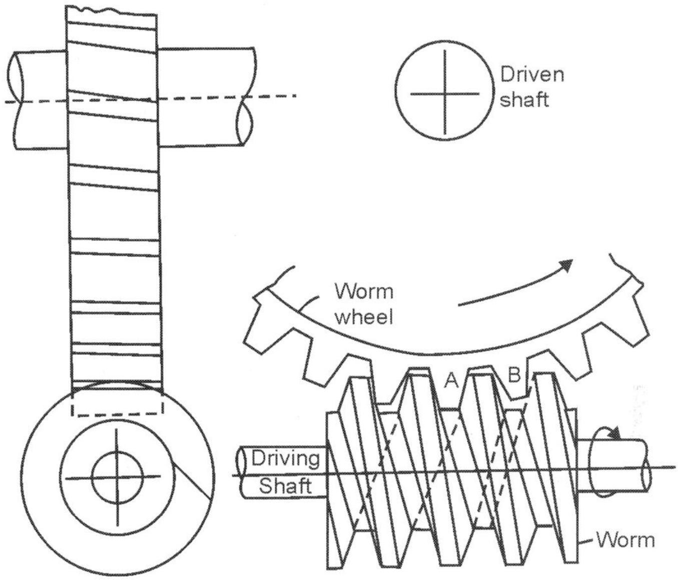

Figure 3.9 Single worm and wheel

In one revolution of the worm as the direction indicated in Fig. 3.9, tooth A will be forced to remain in the groove from A to B, and at the end of the revolution it will occupy position B, i.e. one revolution of the worm moves the wheel one tooth. If the wheel has, say, 100 teeth, it will require 100 revolutions of the worm to give the wheel one revolution. The single worm is considered as a driving wheel having one tooth.

Worms are sometimes made "double" or "triple", in which case instead of there being a single screw thread on the worm there are two or three, as shown in Fig. 3.10.

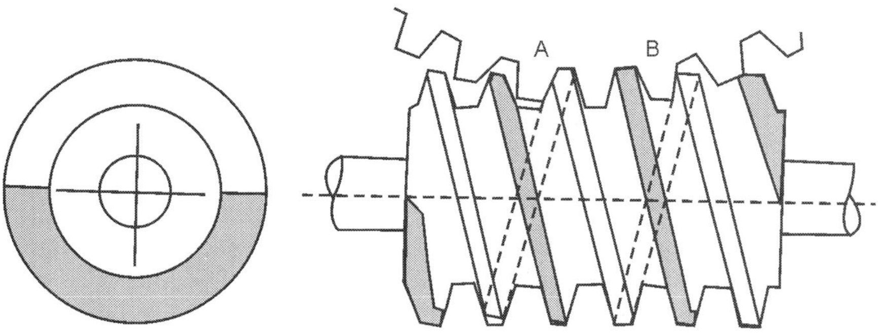

Figure 3.10 Double worm

Figure 3.10 shows a double worm, the threads are distinguished from the other by dark lines. It will be seen that now a tooth at A will move in one revolution of the worm to position B, hence the double worm moves in wheel 2 teeth per revolution and it may be considered the speed calculations as a driving wheel of 2 teeth. Likewise, a treble worm may be considered as a driving wheel of 3 teeth.

The best way to distinguish between single, double and treble worm is to look at the end of the worm, when the beginning of each thread can be seen.

3.7.2 Ratchet wheels

Ratchet wheels are generally used to convert a reciprocating or to-and-fro movement in to an intermittent rotary movement.

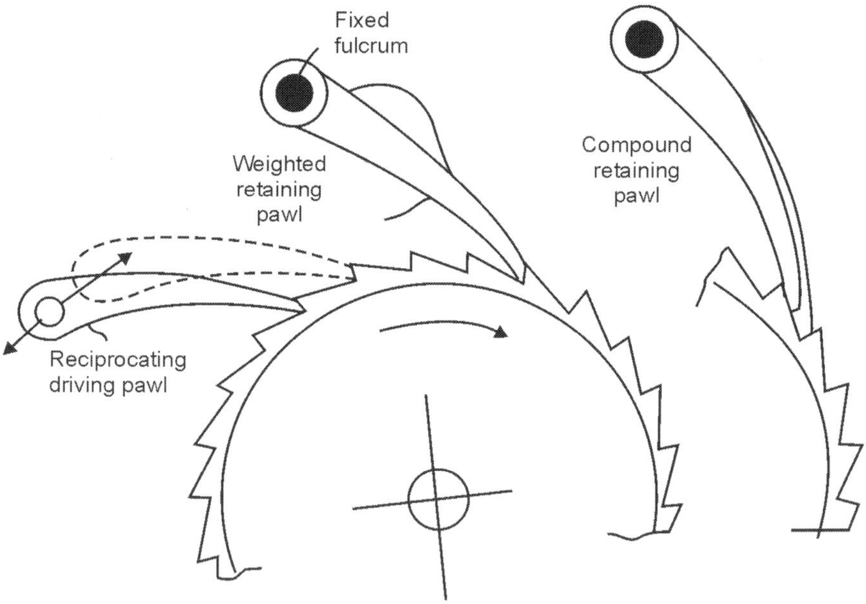

Figure 3.11 Ratchet wheel

The wheel as shown in Fig. 3.11 receives its movement from the driving pawl which has a to-and-fro movement. On its driving movement, the pawl pushes the wheel in the direction of the arrow, usually through a displacement of one tooth, sometimes the displacement for 2 or more teeth. On its backward movement, the driving pawl slips back over one or more teeth ready for its next forward movement. To ensure that the wheel will not turn backwards during the backward movement of the driving pawl, a second pawl called the retaining or holding pawl is used. This retaining pawl can move freely on a fixed fulcrum pin so that the wheel can turn easily in the forward direction, and at the same time, the retaining pawl prevents any backward movement of the pawl.

In certain applications, the amount of movement given by the driving pawl is variable and then it is necessary to have a composite retaining pawl to ensure that the movement given to the wheel is not lost. This is illustrated in Fig. 3.11 in detached form where three pawls of different lengths are mounted side by side.

It can be seen that any movement of the wheel over one-third or a tooth displacement, will be retained by one or other of the pawls.

Ratchet wheels are commonly used in spinning machineries, and in weaving and knitting machines.

A common example in weaving is the use of ratchet wheel is in the taking up motion of the loom, as shown in Fig. 3.12.

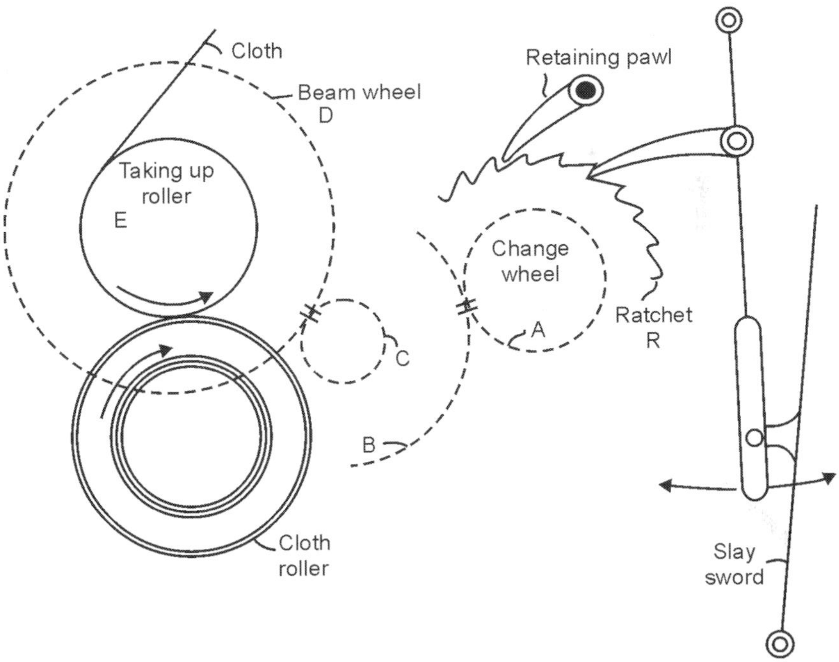

Figure 3.12 Ratchet wheel

A ratchet wheel R tooth receives usually 1 tooth movement for every reciprocation of the loom, i.e. for every pick of weft inserted in the cloth. R is compounded with a gear wheel A, driving a compound carrier BC, the latter driving the beam wheel D fixed to the taking up roller E, which draws the cloth forward and also drives the cloth roller on which the cloth is wound. The wheel A is the change wheel, by changing which the picks per inch in the cloth is altered.

Picks per inch in the cloth = Picks inserted divided by inches of cloth drawn forward in the same time.

When the ratchet wheel has been given one complete turn,

Picks inserted = R (considering R to represent the number of teeth in wheel R.)

Inches of cloth drawn forward = 1 * A/B * C/D * circumference of roller E in inches.

Picks per inch = $R/$ 1 * A/B * C/D * circumference of E

= R/A * B/C * $D/$ circumference of E

Assuming wheel sizes are: R = 50 teeth, A = change wheel, B = 75 teeth = 15, D = 120T, E = 15 inches circum.

Picks per inch = 50/A * 75/15 * 120/15 = 2000/A.

If for example, change wheel of 35 teeth is used, picks per inch will be

2000/35 = 57.14. This is approximately the picks per inch in the cloth when the cloth is under tension in the loom. When the cloth is taken from the loom and the tension is relaxed, the picks may come closer together to the extent of 1 to 3 percent or even more. To allow for this, it is customary to add 1.5 percent to the picks as calculated above.

Therefore, picks per inch

= 2000/A + 1.5 percent = 2000/A * 101.5/100 = 2030/A

or Change wheel A = 2030/picks per inch required

The number 2030 is called "dividend" of the taking up motion of the loom.

That is, the number which when divided by the number of teeth in the change wheel gives the number of picks per inch in the cloth.

It will be obvious that these calculations give approximate results, owing to the difference in cloth tension during weaving and also the chances of cloth slippage over the taking up roller surface.

In knitting machines, a similar arrangement for drawing-off and winding-up of knitted fabric is made operated from a ratchet wheel. The number of courses per inch is regulated in knitting process and not the rate of drawing the cloth forward as in weaving.

3.7.3 Hunting cog motions

Hunting cog motion principle forms the basis of widely used stop motions for measuring lengths as in Scutcher in blow room and in warping machines.

The principle of hunter cog motion is explained as follows:

From Fig. 3.13, two wheels, gear together, one of the wheels having different number of teeth from the other. Fixed to the side of each wheel is a projection, the outer point of one in line with a tooth *P*, whereas the other P_1, is opposite to the gap between two adjacent teeth. The tooth at *P* can gear in the usual way with any tooth of the other wheel except that at P_1. When *P* and come together the projections meet and the wheels are forced apart. Provision has been made in such a way that operates the stop motion every time *P* and P_1 come together.

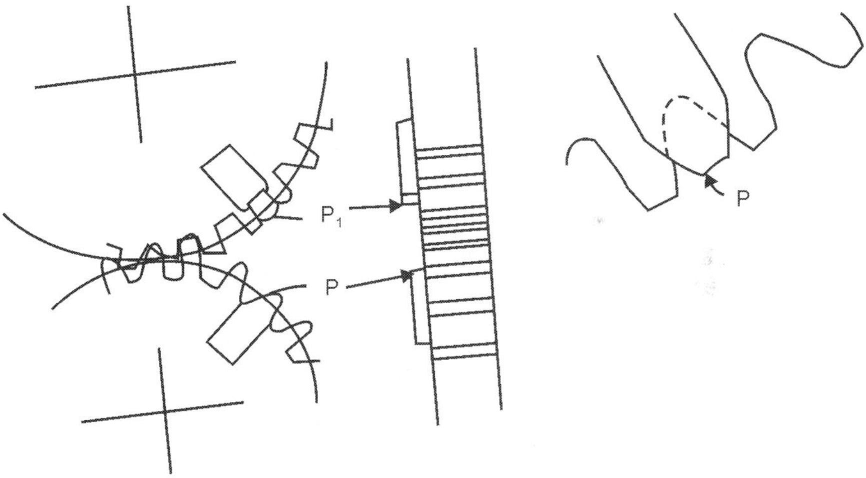

Figure 3.13 Hunter and cog motion

The time that elapses after *P* and P_1 come together until again depends upon the number of teeth in the two wheels. Both wheels must make an exact number of revolutions. If, for example, one wheel has three times as many teeth as the other, the hunting cogs will come together every revolution of the larger wheel, the smaller one making three revolutions.

In general, if the diameters of the wheels are in the proportion *A/B*, then * in any given time being in the proportion *B/A*: the hunting cogs will come together every *B* revolutions of wheel *A* or every A revolutions of *B*.

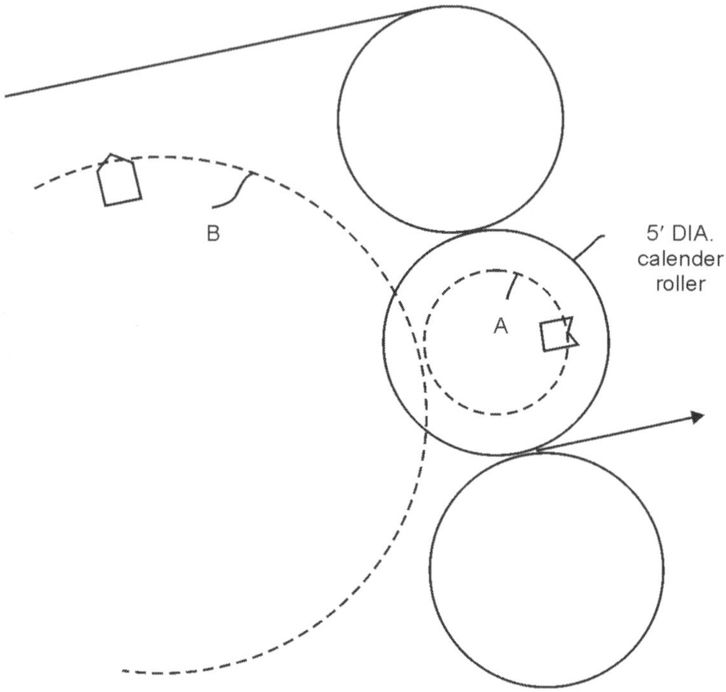

Figure 3.14 Hunting cog motion

Example 3.4: A hunting cog motion on a cotton Scutcher has one of the cog wheels *A* on a 5 in. diameter calendar roller, round which the cotton passes. The other, *B* is on the stop motion lever which stops the roller every time the hunting cogs come close together.

If Wheel *A* has 21 teeth and Wheel *B* has 71 teeth, what length of cotton will pass round the roller between stoppages?

Solution: The lowest whole number proportion between the teeth is 21/71; the cogs will come together every 71 revs. of wheel A.

The length of cotton passing round the calendar roller will be

B/A * diameter of calendar roller * $\pi/12$

$= 71/21 * 5 * \pi/12 = 93$ ft

3.7.4 Bevel gears

Bevel wheels are used to connect two shafts which are at right angles to each other. Figure 3.15 illustrates the general method of driving.

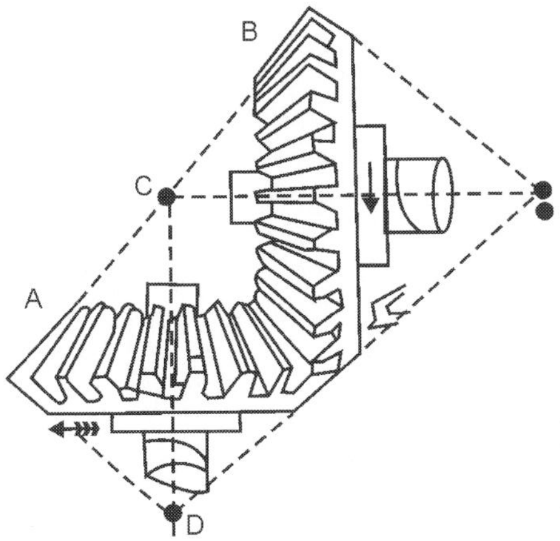

Figure 3.15 Bevel gears

The direction of rotation and the calculations connected with bevel wheels are similar to other wheels. For instance, if A with 20 teeth is driving B with 30 teeth, then

$$A/B = \text{revolution of B}$$
$$20/30 = 0.66 \text{ revs. of B}$$

3.7.5 Spiral gears

Spiral gears are used to drive shafts at right angles to each other. As the teeth are set at 45° to the wheel sides, the gearing is quarter than ordinary gears.

3.7.6 Rack and pinion

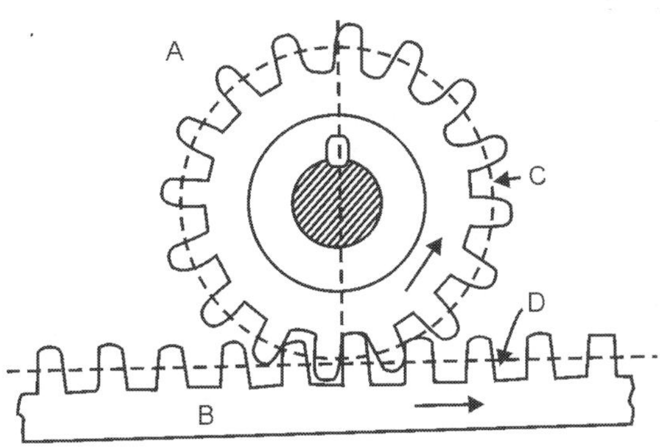

Figure 3.16 Rack and pinion

The rack and pinion is an important element in the gearing of textile machinery. A straight-toothed rack is used in to which gears a wheel: the rack slides in, guides and is attached to some part of the mechanism that requires to be moved in a straight line at a definite speed for a certain distance.

An elementary type of this is shown in Fig. 3.16. Wheel *A* is the pinion and *B* is the rack. The wheel A revolves on a fixed centre in the direction of the arrow, and by virtue of gearing with B it moves the rack forward as shown in Fig. 3.16; if A has 16 teeth, one revolution of the wheel will move the rack a distance equal to 16 teeth on the rack so that to know the exact distance moved by the rack it is necessary to know the pitch of the teeth or the space occupied by the whole of the teeth that are moved.

If the pitch of the teeth in Fig. 3.16 is half-an inch, then three revolutions of A will move the rack 3 x 16 = 48 teeth and the distance moved by the rack will be 48 × ½ in. = 24 in.

3.7.7 Star wheel

This arrangement sometimes forms part of the gearing of a machine especially to produce an intermittent feed to the rollers whilst other parts are running continuously.

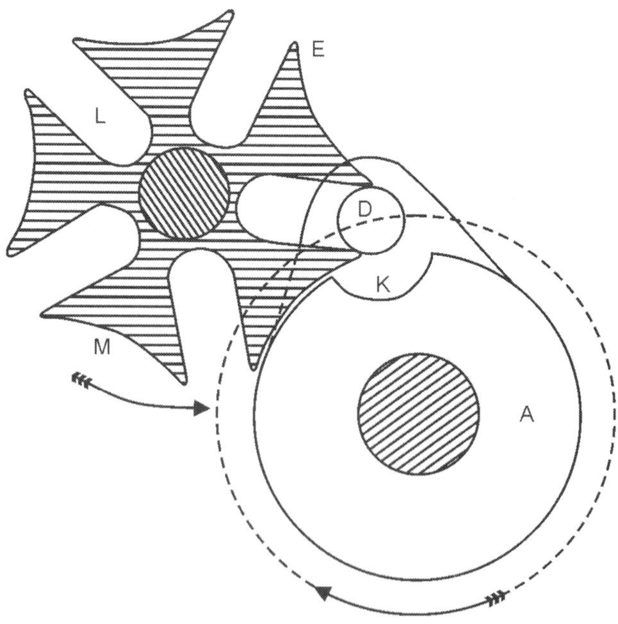

Figure 3.17 Star wheel

Figure 3.17 gives its essential features. An arm carrying a pin or stud *D* is fixed to the shaft. This arm is also provided with turned boss *A* having a recess cutout at *K*. On the shaft to be driven intermittently is keyed a specially formed wheel *E*, having slots *L* and the surface between the slots cutout to the same radius as the boss *A*. As the pin *D* revolves, it passes in to one of the slots l and carries the star wheel *E* along with it until it emerges from the slot. The recess at *K* permits *E* to revolve by allowing room for the points of the star to describe a portion of a circle. On coming out of the slot of *E*, the pin *D* will continue its motion for the greater part of a revolution without any effect on the star wheel, which thus remains stationary, and *E* is only moved again when *D* enters the next slot. If *E* has five slots, it will require five revolutions of *A* to produce one revolution of *E*. For purpose of calculation, the pin *D* may be considered as a wheel of one tooth and the star-wheel *E* as a wheel of five teeth. The arrangement may be made for more than one pin *D* and carrying slots to suit the characters of the intermittent motion required.

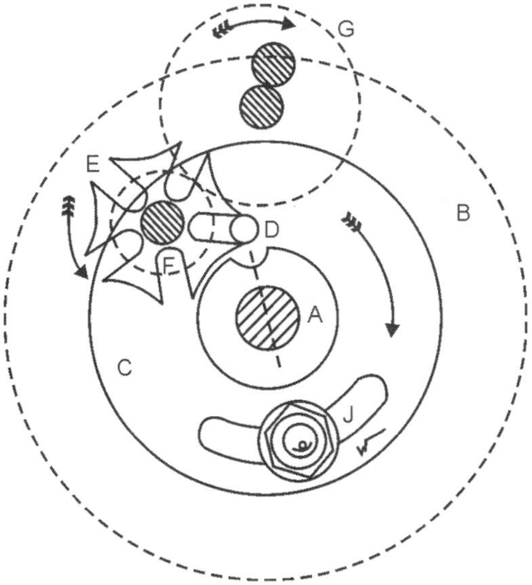

Figure 3.18 Star wheel in comber machine

Figure 3.18 shows the application of star wheel in comber machine in spinning preparatory. The object is to revolve the roller *H* intermittently so that a certain length of cotton will pass between the two rollers at definitely fixed times in the machine's cycle of actions. This effect is obtained by gearing *H* through *G* to the wheel *F*; on the same shaft as *F* is fixed, the star-wheel

E, and in this wheel is geared a pin *D* attached to a disc. The disc *C* revolves continuously, but the star-wheel E will only move round a fifth of a revolution at that part of the revolution of *C* when the pin *D* gears with the slots in *E*. The rest of the movement of *C* will have no effect on *E*, so that the star wheel remains idle during this period.

Example 3.5: A machine shaft is revolving at the rate of 180 revolutions per minute. What is the speed in revolutions per second, radians per second and radians per minute?

Solution:

1 revolution = 2 * π radians = 360°

or 1 radian = ½ * π of a revolution = 57.5°

(a) Revolution per minute of the machine shaft = 180

Revolution per second = 180/60 = 3

(b) Radians per minute

1 revolution = 2 * π radians

For 180 revolutions = 180 * 2 * π = 1130.4 radians per minute

(c) Radians per second

For 180 revolutions, the shaft has 1130.4 radians per minute

For 1 second, radians per minute/60

= 1130.4/60 = 18.84 radians per second

Example 3.6: Calculate the spindle speed in an upright spindle winding frame, if the spindle wharves are 1.25 in. diameter driven by bands from a 5 in. diameter tin roller, revolving at 220 rpm. Allow for 4 percent loss by slip.

Solution:

Speed of the tin roller = 220 rpm

Diameter of the tin roller = 5 in.

Diameter of the spindle wharves = 1.25 in.

Speed of the upright spindle =

Speed of the tin roller * Tin roller diameter/Spindle wharve diameter

220 * 5/1.25 = 880 rpm

If the slip is 4%, then the actual speed of the spindle is

96/100 = 845 rpm

Example 3.7: At what speed must the tin roller of a ring spinning machine be driven to give a spindle speed of 900 rpm? The tin roller diameter is 10

in., the wharve on the spindle is 7/8 in. diameter and the slip is estimated as 5 percent.

Solution:

In order to drive the spindle speed at 9000 rpm,

Spindle speed = Tin roller speed/spindle wharve diameter

Tin roller speed = Spindle speed * Spindle wharve diameter

= 9000 * 7/8 in.

= 7875 rpm

Tin roller diameter = 10 in.

Therefore, tin roller speed = Tin roller speed/Tin roller diameter

= 7875/10 = 787.5 rpm

If the slip is estimated as 5 percent, the tin roller must rotate

787.5 * 105/100 = 0.828 rpm

The tin roller must rotate 828 rpm to give a spindle speed of 9000 rpm.

Example 3.8: A 7-wheel take-up motion has the following gearing: ratchet wheel of 24 teeth is compounded with a standard wheel of 36 teeth. This gears with the change wheel, compounded with a 24 tooth wheel. The latter drives a wheel of 89 teeth, compounded with one of 15 teeth, driving a beam wheel of 90 teeth. The roller circumference is 15.05 in.

Find the dividend or constant number for the motion, slowing 1.5 percent for contraction. The standard wheel may be changed to 27 or to 18 teeth. With these wheels, what will be the relation between picks per inch and change wheel?

Solution:

Ratchet wheel (*A*) = 24T; Standard wheel (*B*) = 36T; *CW*(*C*) =?; *D* = 24T; *E* = 89T; *F* = 15 T; *G* = 90T; *H* = 15.05 in.

Constant = *A*/*B* * *C*/*D* * *E*/*F* * *G*/*H* * contraction percent

= 24/36 * *C*/24 * 89/15 * 90/15.05 * 101.5/100 = 0.98 OR 1

With a standard wheel of 27 teeth, picks per inch inserted is

= 24/27 * 1/24 * 89/15 * 90/15.05 = 1.31

Picks per inch inserted = 1/1.31 = 0.76 picks per in.

With a standard wheel of 18 teeth, picks per inch inserted is

= 24/18 * 1/24 * 89/15 * 90/15.05 = 1.97

Picks per inch inserted = 1/1.97 = 0.50 picks per in.

Example 3.9: A taking-up motion on a jute loom has the following gearing: a 44 tooth ratchet wheel is compounded with the change wheel. This gears with a 70 tooth wheel, compounded with one of 27 teeth, which drives the beam wheel of 80 teeth. The beam is 15.05 in.

Calculate the constant number or dividend and the number of picks or shots per inch with a 39 tooth change wheel. No allowance is made for contraction in jute weaving.

Solution:

Ratchet wheel (*A*) = 44T; *CW* (*B*) = ?; *C* = 70t; *D* = 27T; *E* = 80T; *F* = 15.05 in.

Dividend or Constant number = A/B * C/D * E/F

= 44/B * 70/27 * 80/15.05

= Dividend is 606/B

With 39 tooth change wheel, shots per inch is

Dividend/change wheel = 606/39 = 15.54 shots/in.

Example 3.10: A taking-up motion on a woollen loom is arranged as follows:

An 18 tooth chain sprocket wheel on the dobby, revolving once per pick, drives a sprocket of 18 teeth, compounded with a gear wheel of 20 teeth. The latter drives the change wheel, which is compounded with a pinion of 18 teeth driving one of 84 teeth. This wheel is compounded with a single worm driving a worm wheel of 85 teeth on the end of taking-up roller or beam, which is 20 inches in circumference. It is intended that the number of teeth in the change wheel should be equal to the picks per inch in the cloth. What allowance has been made for contraction?

Solution:

Percentage contraction allowed is =

A = 18t; *B* = 20T; *C* = ?; *D* = 18t; *E* = 84T; *F* = 1T; *G* = 85T; *H* = 20 in. circumference

*A/B * C/D * E/F * G/H*

= 18/20 * *C*/18 * 84/1 * 85/20

Change wheel (*C*) = 17.85 or 18

Actual change in wheel is 17.85

Contraction 5 = 18-17.85/18 * 100 = 0.83

Example 3.11: The following taking-up motion on an automatic loom for cotton weaving has been arranged so that the number of teeth in the change wheel should be equal to the number of picks per inch in the cloth. A ratchet of 42 teeth is compounded with a wheel of 28 teeth. The latter drives the change wheel, compounded with a wheel of 12 teeth. This wheel drives through a single carrier, the beam wheel of 110 teeth, the beam circumference being 14.14 in. roller shaft. What cloth contraction has been allowed in working out the gearing?

Solution:

A = 42T, B = 28T, Change wheel 9(C) = ?; D = 12T; E = 110T; F = 14.14 in.

$A/B * C/D * E/F = 42/28 * C/12 * 110/14.14$

$C = 0.97$

Contraction% = $100 - X/100 = 0.97$

Therefore $X = 0.97 * 100 = 100 - X$

X = 3 percent

Numerical questions

1. The traverse on an upright spindle winding frame winding flanged warper's bobbins operated from a mangle wheel motion. The mangle wheel has 126 teeth or pins and the pinion has 8 teeth. The pinion is compounded with a wheel of 100 teeth, which drives one of 90 teeth on the roller shaft. The tin roller of 5 in. diameter drives the spindle wharves of 1.25 in. diameter; the bobbin flanges resting on friction driving discs on the spindles. Find the number of coils wound between the bobbin flanges during one traverse, allowing for 3 percent slippage at the spindle drive.

   Ans. 68 coils

2. A taking-up motion of a rayon loom is arranged as follows:

   The ratchet wheel is the change wheel and is compounded with a single worm which drives a worm wheel of 60 teeth on the take-up roller. If the number of teeth in the ratchet wheel is to be equal to the number of picks per inch in the cloth, calculate the circumference of the cloth roller allowing 1.5% contraction.

   Ans. 30.45 in.

3. In a semi-positive let-off motion, the warp beam is connected to the ratchet wheel which turns the beam by the following gearing. A 96 tooth wheel on the beam gears with one of 24 teeth, which is compounded with a 20 tooth worm wheel. A double worm driving

the worm wheel is compounded with the ratchet wheel, which has 80 teeth. Calculate the average number of teeth taken per pick by the ratchet pawl when the beam is 20 and 50 in. circumference, the cloth having 40 picks per inch.

Ans. 4 teeth at 20 in.; 11 teeth in 8 picks at 50 in.

4. A measuring and stop motion on a beam warping machine consists of a measuring roller, connected by gearing to a cam which stops the machine on completion of each revolution of the cam. The measuring roller is turned by the yarn being in contact with the surface of the roller, which has a circumference of 0.5 yard. The gearing from the measuring roller to the cam is a single worm driving a worm wheel of 70 teeth, compounded with another single worm driving a wheel of 100 teeth on the cam shaft. Calculate the length of "warp" or length warped between the stops by the cam.

   Ans. 3500 yards

5. The yarn passing through a sizing machine is measured and marked by a cam mechanism, the cam operating a marking hammer which puts a "cut" mark on the warp after every revolution of the cam. The cam is driven by gearing from a measuring roller of 2/5 yard circumference, which is turned by the warp. The gearing to the cam is as follows: A change wheel (roller wheel) on the measuring roller drives a second change wheel (stud wheel) which is compounded with a single worm. The worm drives a worm wheel of 45 teeth on the cam shaft. Suggest suitable size of the wheels for a cut length of 90 yards.

   Ans. Stud wheel; Roller wheel = 5:1, 100 teeth and 20 teeth.

6. A cotton conveyor lattice consists of a flexible belt of 1/8 in. thick, on which are fixed lags of 3/8 of an in. thick. The belt passes over a pulley of 4.5 in. diameter, driven as follows: a wheel of 96 teeth on the lattice pulley shaft gears with a pinion of 12 teeth, compounded with a worm wheel of 88 teeth, which is driven by a single worm revolving at 450 rpm. At what rate in inches per minute, does the lattice convey the cotton?

   Ans. 9.3 yards per minute

7. In a cotton opener, the feed and the pedal rollers which feed the cotton to the beater get their movement from the top cone drum through the following gearing: A single worm on the top cone shaft drive a worm wheel of 62 teeth, compounded with a gear wheel of 34 teeth. The latter drives a wheel of 40 teeth on the pedal roller, and another wheel of 40 teeth on the pedal roller drives one of 33 teeth on the feed roller.

The pedal roller is 2 * 3/4 in. and the feed roller is 3 in. in diameter. What is the draft between the pedal and feed roller? If the top cone drum speed is 640 rpm, at what rate in yards per minute does the pedal roller deliver cotton to the beater?

Ans. Draft = 1.12; Speed of feed = 2.1 yards per minute.

8. Allowing for belt thickness of ¼ in. and for 2 percent loss of speed by slip, find, to the nearest half-inch, suitable diameters of line shaft pulleys, revolving at 100 rpm.
   (a) A revolving cylinder plain web knitting frame which is required to have a cylinder speed of 30 rpm. The cylinder wheel has 288 teeth and is driven by a wheel of 48 teeth on the driving shaft, which has a 10 in. diameter driving pulley.
   (b) A French circular machine to have a needle ring speed of 24 rpm. The wheel on the needle ring has 32 teeth and is driven by one of 16 teeth. This is compounded with a meter wheel, gearing with another on the machine shaft. The machine driving pulleys are 8 in. diameter and are driven by a 6 in. diameter pulley on a countershaft. The pulley on the counter shaft, which is driven from the line shaft pulley is 18 in. diameter.

Ans.

(a) Belt thickness = ¼ in.; Diameter of the line shaft pulley = 18.57 in.

(b) Belt thickness = ¼ in.; Diameter of the line shaft pulley = 11.63 in.

9. In a drum winding machine, the warper's bobbin are driven by frictional contact between the yarn and the bobbin and the driving drums are 7 in. diameter. The traverse cam which operates the thread guide is driven from the drum shaft by the following gearing. A wheel of 12 teeth on the drum shaft drives one of 24 teeth which is compounded with a single worm. The worm wheel of 32 teeth is fixed to the cam.
   (a) Calculate the winding rate assuming 2% slip between drum and bobbin, when the drum speed is 180 rpm.
   (b) Find the constant number, if there is no slip which is equal to the product of bobbin circumference and number of coils in a double traverse.
   (c) If when the bobbin circumference is 4 in., the number of coils in a double traverse is found to be 334. What is the percentage slip between the bobbin and the drum?

Ans. (a) 128 yards/min.; (b) 1408; (c) 5.1 percent

10. A quick traverse cheese winding machine has a cheese driving roller of 3 in. in diameter. The thread guides are reciprocated by a cam, driven by gearing at half the speed of the machine shaft. The latter drives the rollers by a change wheel A of range 20 to 58, driving through a 34 tooth carrier another change wheel of range 20 to 30. On the roller shaft, find the greatest and the least number of coils that can be wound on double traverse when the cheese diameters are (a) 3 in. and (b) 6 in.

    Ans. (a) 5.8 coils and 1.33 coils; (b) 2.9 and 0.66 coils

4

Equations of movement

4.1 Acceleration

Acceleration is the rate at which the velocity of a body changes, or the change in velocity in unit time. The unit for linear acceleration is 1 foot per square second (ft/s^2), and for angular acceleration the unit is radian per square second (radian/s^2).

4.2 Retardation

When the velocity of the body is decreasing, the acceleration is considered to be negative and is often called as retardation. When the body is moving with a constant acceleration a, the relationship between the velocities of the body at the beginning and at the end of *a* time period t and the space or distance *S* passed over by the body is given by the following equations:

(a) $V = u \pm at$

(b) $S = u\,t \pm ½\,\mathrm{at}^2$

(c) $V^2 = u^2 \pm 2$

(d) S = average velocity * time

Where V is final velocity, u is initial velocity, a is acceleration or retardation due to gravity and S is the distance or space moved by the body.

The plus sign is used to denote acceleration and the minus sign denotes retardation.

The acceleration must be uniform, which very often is not the case in moving machine parts. The familiar example of a body moving with uniform acceleration is a falling body. The velocity of the body increases approximately 32.2 ft/s^2; it is denoted by the letter "*a*" or "*g*". If again, the body is thrown upwards in to the air vertically, it is subject to retardation due to gravity and its velocity is reduced by 32.2 ft/s.

Familiar examples in machines where the acceleration is approximately uniform, as such in the movement of the parts of most belt-driven machines when starting up or stopping, the movement of the loom shuttle while it is passing through the warp shed. The acceleration of the shuttle by the picker, however, is not generally uniform and the equations of motions cannot be applied directly in that case.

Examples:

1. A stone is dropped from a bridge and is observed to take 3 s to reach the bottom. What is the height of the bridge and the maximum velocity of the stone? Assume $a = 32.2$ ft/s.

   Solution:

$$V = u + at$$
$$= 0 + 32.2 * 3 = 96.6 \text{ ft/s}$$
$$S = u\,t + \tfrac{1}{2}\,at^2$$
$$= 0 + \tfrac{1}{2} * 32.2 * 3 * 3$$
$$= 144.9 \text{ ft}$$

2. A shuttle is moving at 40 ft/s when it enters the warp shed and moves 5 ft before it is through the shed. During its passage, it is subjected to a retardation, which may be considered uniform of 30 ft/s^2. Find the time it takes to traverse the shed and its velocity when leaving.

   Solution:

   Initial velocity u =40 ft/s

   Distance moved $S = 5$ ft

   Leaving velocity $V = ?$

   Time = ?

   We know that, $V^2 = u^2 - 2\,aS$

$$= 40^2 - 2 * 30 * 5 = 36.05 \text{ ft/s}$$
$$\text{Time} = ?$$
$$V = u - at$$
$$36.05 = 40 - 30 * \text{t}$$
$$\text{Time } (t) = 0.133 \text{ s}$$

3. A machine is started from rest to full speed of 180 rpm with a uniform acceleration of 36 radian/s^2. Find the time taken to reach full speed and the angle moved through by the driving shaft during acceleration.

Solution:

The final angular velocity is 180/60 = 3 rev/s

$$V = u + at$$

$$3 * 2 * \pi = 18.84 \text{ radian/s}$$

$$18.84 = 0 + 36\, t$$

$$t = 0.523 \text{ s}$$

$$S = ut + ½\, at^2$$

$$= 0 + ½ * 36 * 0.523^2 = 4.92 \text{ radians}$$

$$= 4.92 * 57.3° = 283°$$

Angle moved by the shaft during acceleration is 283°

4. A cricket ball is thrown vertically upwards with a velocity of 64 ft/s. How high it will rise, what time it will elapse before it strikes the ground and how fast will it be moving? Assume $a = 32$ ft/s^2.

Solution:

$$V = u + at$$

$$64 = 0 + 32.* t$$

$$t = 64/32 = 2\, s$$

$$S = u\, t + ½\, at^2$$

$$= 0 + ½ * 32 * 4$$

$$= 64 \text{ ft.}$$

Again, $S = u * 2 - ½ * 32 * 4$

$$64 = u * 2 - ½ * 32 * 4$$

$$64 = 2\, u - 64$$

$$u = 128/2 = 64 \text{ ft/s}$$

5. A loom shuttle is required to pass through the warp shed at a distance of 4 ft in 1/12th of second. What is its average speed during the passage? It is subjected to a retardation of 32 ft/s^2 during the passage. What are its speeds at the beginning and at the end of the passage?

Solution:

We know that S = average velocity * time

4 = average velocity * 1/12th of second

Average velocity =4/0.083 s = 48 ft/s

Average velocity $= u \pm ½\, at$

$$48.01 = u - \tfrac{1}{2}\, 32 * 0.083 \text{ s}$$
$$= 48.01 + 1.33 = 49.03 \text{ ft/s}$$
$$\text{Initial velocity} = 49.3 \text{ ft/s}$$
$$\text{Leaving speed} = V^2 = u^2 - 2\, a\, S$$
$$= 49.3 * 49.3 - 2 * 32 * 4$$
$$V = 46.63 \text{ ft/s}$$

6. The shuttle of a heavy loom enters the warp shed at a speed of 36 ft/s and it is retarded at the rate of 12 ft/s^2. What time will it take to pass through the shed, a distance of 5 ft, and with what speed will it leave the shed?

Solution:

Leaving velocity (V) – $V^2 = u^2 - 2\, a\, S$

$$u = 36 \text{ ft/s};\ S = 5 \text{ ft}$$
$$V^2 = 36^2 - 2 * 12 * 5$$
$$= 34.29 \text{ ft/s}$$

Time (t):
$$V = -\,u - at$$
$$34.29 = 36 - 12 * t$$
$$t = 0.142 \text{ s}$$

7. A spinning mule carriage is accelerated from rest to a speed of 30 in./s in 1.5 s. It then moves at a uniform speed for 0.5 s and finally it is retarded and stopped in a further 1.5 s. Assuming constant acceleration and retardation find their values. Also find the distances moved by the carriage: (a) during acceleration; (b) during retardation; (c) when the speed is constant.

Solution:

$$V = 30 \text{ in./s};\ t = 1.5 \text{ s}$$

Therefore,
$$V = u + at$$
$$30 = 0 + a * 1.5$$
$$a = 20 \text{ ft/s}$$
$$V^2 = u^2 - 2aS$$
$$30^2 = 0 - 2 * 20 * S$$
$$S = 22.5 \text{ in.}$$

Movement during constant speed =

$$S = u\,t + \tfrac{1}{2}\, at^2$$

$$S = 30 * 2 + \tfrac{1}{2} * 22.5 * 2^2$$
$$S = 15 \text{ in.}$$

8. A carding engine was observed to take 24 s to take a full speed of 160 rpm. What is the acceleration and how many revolutions does the card make during the acceleration?

Solution:

$$V = u + at$$
$$160/60 = 0 + a * 24$$
$$a = 160/60 = 2.66 \text{ rev/s} \div 24 \text{ s} = 0.111 \text{ rev/s}^2$$

or 0.111* 2 * π radians = 0.697 radians/s

No. of revolutions at the time of acceleration =

$$S = ut + \tfrac{1}{2} at^2$$
$$= 0 + \tfrac{1}{2} * 0.111 * 24 * 24$$
$$= 31.96 \text{ revolutions}$$

Numerical questions

1. A carding engine was observed to make 120 revolutions in coming to rest when the belt is put on loose pulley. If the running speed is 160 rpm and the retardation is uniform, what is the retardation and how long does it take the card to stop?

   Ans. Retardation = 0.186 radians/s^2; Time to stop = 1.5 min

2. The rim shaft of a spinning mule has to be brought to rest in 1.5 s from a speed of 960 rpm. What is the retardation and how many revolutions does the shaft make before coming to rest? After the stop, the shaft has to be accelerated in reverse direction to a speed of 48 rpm. If the value of acceleration is the same as the retardation during the stop, how long does it take and how many revolutions are made during acceleration?

   Ans. Retardation = 67 radian/s^2; Retardation during stop = 12; Time to accelerate = 0.075 s; Revolution during acceleration = 0.03.

3. In a loom running at 220 picks/min, the shuttle begins to move at 100° past beat-up and comes to rest at 225°. How long is the shuttle moving?

   Ans. 49.39 km/h

4. The yarn width in the reed is 100 cm and the shuttle length with the curved ends neglected is 30 cm. The loom speed is 200 picks/min and 100° of crank shaft rotation is available for shuttle movement through the shed. If the retardation is 950 cm/s^2, determine (a) mean velocity, (b) maximum velocity, and (c) minimum velocity.

   Ans. Mean velocity = 15.66 m/s; Maximum velocity = 16.05 m/s; Minimum velocity = 15.21 m/s

5. A beam warping machine with a normal warping speed of 480 yards/min is stopped with uniform retardation in 2 s. What length of yarn will run on to the beam during stoppage? If the beam diameter is 1 foot, what is the angular retardation of the beam in radian/s^2?

   Ans. 8 yards; Angular retardation = 24 radians/s^2

6. A loom shuttle is moving at 30 ft/s when it leaves the warp shed. It has to be brought to rest in a further 9 in. of movement and ought to be stopped before the crankshaft moves a further 60 degrees. Loom speed is 180 picks per minute.

   (a) If the retardation is uniform, what is its value? Will the shuttle be stopped in time?

   (b) Assuming that the shuttle is just stopped as the crank completes the 60 degrees movement, what is the average retardation?

   Ans. (a) Retardation = 600 radians/s^2; shuttle stops in 54 degrees of crank movement. (b) Retardation = 594 ft/s^2

7. The slur cock's on a cotton frame move at 54 in./s during ½ revolution of the cam shaft whose speed is 60 rpm. How far do they move during that time?

   They are given a speed of 54 in./s in 1/8th s. If the acceleration is uniform, what is the value and the distance moved by the slur cock's during the acceleration?

   Ans. 27 in.; Acceleration = 432 in./s^2. Distance moved during acceleration = 3.375 in.

5

Parallelogram of velocities

5.1 Introduction

A velocity having magnitude and direction can be represented by a straight line, the length of the line representing the magnitude of the velocity and the direction in which the line is drawn representing the direction of the velocity.

This is shown in Fig. 5.1(a), if *AB* is 6 units long, represents the velocity of the train running at 60 miles per hour in a direction 30 degrees to the east of North, then *AC* is 4 units long, will represent a velocity of 40 miles per hour towards the South.

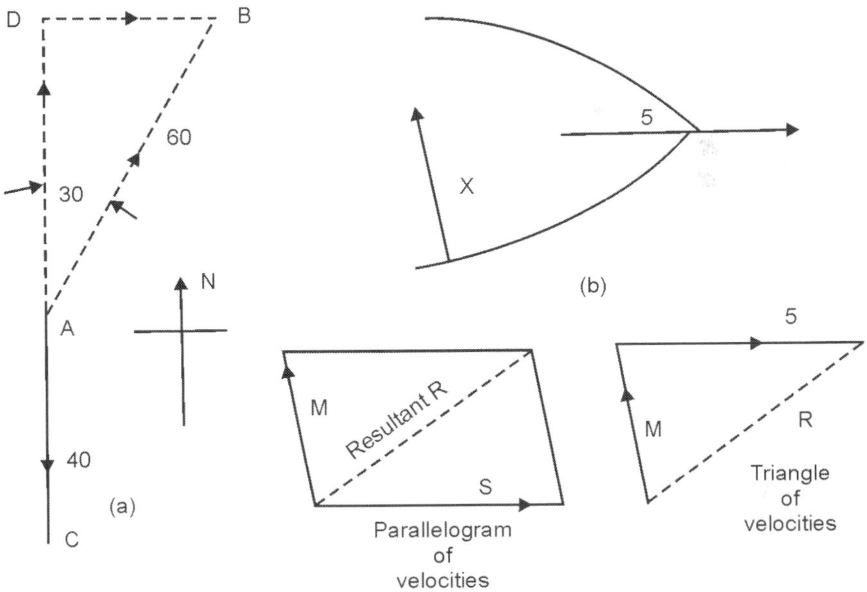

Figure 5.1(a) and (b)

If a body has a movement which is the result of two separate velocities, for example the velocity of a man who walks across the deck of a moving ship, the resultant velocity is found by the parallelogram of velocities. This is shown in Fig. 5.2(b), the resultant *R* of the man's velocity *M*, relative to the ship, and the ship's velocity *S*, showing the actual direction and the rate of movement of the man relative to the water. The resultant can be found also by drawing a triangle, as illustrated.

M and *S* are the components of man's resultant velocity *R*. Conversely, any velocities can be split up in to two or more components in any given directions. Generally, but not always, the components required are those at right angles to each other, and an example is given in Fig. 5.1(a), where the velocity is 60 miles per hour represented by *AB* has been resolved into two components *AD* and *DB*; *AD* represents the northerly components of the train's velocity and by calculation or measurement it will be found to be 52 miles per hour. *DB*, 3 units long, shows that the train is moving towards the East at 30 miles per hour.

5.2 Velocity

The rate of change of displacement with respect to time is known as the velocity of a particle or body. If the direction of displacement is known, then the direction of velocity can be easily obtained. If the displacement is constant in direction, the velocity will be in the same direction. But if the displacement is not constant in direction, then the velocity at a given instant will be in the same direction as the displacement at that instant. As displacement at that instant is in tangential direction, hence the direction of velocity be tangential to the path of the particle or body.

Exercises:

1. A shuttle 3 ft above the floor level and moving at 30 ft/s flies out of the loom at an angle of 10° to the horizontal. How high will it rise, how fast will it move horizontally before it strikes the floor, how fast and in what direction is it then moving as shown in Fig. 5.2?

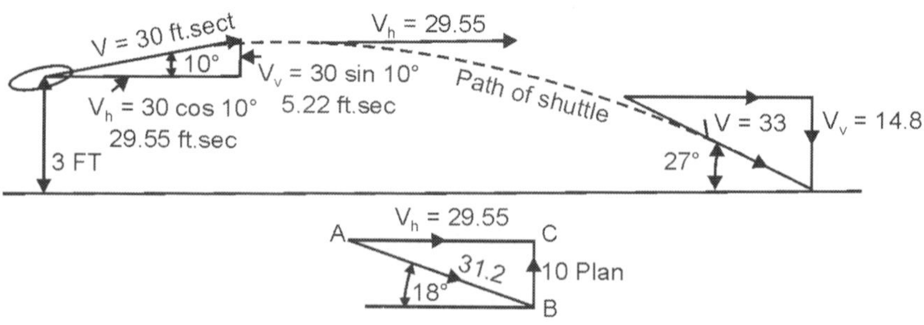

Figure 5.2

Solution:

Consider the vertical movement.

Vertical velocity V_v is given by

$$30 * \sin 10° = 5.20 \text{ ft/s}$$

Horizontal velocity V_h is given by

$$30 * \cos 10° = 29.54 \text{ ft/s}$$

The shuttle has initial velocity of 5.29 ft/s, retardation is −32.2 ft/s; space passed over by shuttle is 3 ft, since upward movement has been considered positive by taking the initial velocity positive.

$$V^2 = u^2 + 2as$$
$$= 5.20^{2*} + 2 * 32.2 * 3$$
$$V = \pm 14.8 \text{ ft/s}$$

The vertical velocity when the shuttle strikes the floor is downwards, hence the shuttle vertical velocity is given by $V = \pm 14.8$ ft/s

Total time of flight can be given by

$$V = u = at$$
$$-14.8 = 5.20 + 32.2 * t$$
$$t = 0.620 \text{ s}$$

In other words, upward and downward movement can be considered separately.

Upward movement,

$$V = u + at$$
$$0 = 5.20 + 32.2 * t$$
$$t = 5.20/32.2 = 0.161 \text{ s}$$
$$V^2 = u^2 - 2as$$
$$0 = 5.20^2 - 2 * 32.2 * s$$
$$s = 0.42 \text{ ft or 5 in.}$$

The shuttle will rise approx to 5 in. or 0.42 ft

Downward movement,

$$S = 3 \text{ ft} + 0.42 \text{ ft} = 3.42 \text{ ft.}$$
$$V^2 = u^2 + 2\,as$$
$$= 0 + 2 * 32.2 * 3.42$$
$$= 14.84 \text{ ft/s}$$
$$V = u + at$$
$$14.8 = 0 + 32.2 * t$$
$$t = 14.8/32.2 = 0.460 \text{ s}$$

Total time taken for the shuttle to reach the floor = 0.161 + 0.460 = 0.622 s

Final shuttle velocity:

The final velocity of the shuttle when it strikes the floor is the resultant of horizontal velocity and final vertical velocity downwards. Therefore, the final shuttle velocity is

$$V^2 = \sqrt{14.8^2 + 29.55^2}$$

$$V = 33 \text{ ft/s}$$

Angle of shuttle flight is

$$\tan\theta = 14.8/29.55 = 0.5 \text{ or } 27°$$

2. A bale comes down from a shoot inclined at 60° from the horizontal, and is projected from it at a velocity of 24 ft/s. How fast is it moving (a) horizontally, (b) vertically, as it leaves the floor? During a drop to the lorry 16 ft below the bottom of the shoot, the vertical velocity is increased by the acceleration of gravity 32 ft/s, but the horizontal velocity remains unchanged. What is the vertical velocity when it reaches the lorry, the magnitude and direction of the actual velocity, the time taken from the bottom of the shoot to the lorry and the horizontal distance from bottom of the shoot to where the bale strikes the lorry?

Solution:

Vertical velocity $V_v = 24 \sin 60° = 20.78$ ft/s

Horizontal velocity $V_h = 24 \cos 60° = 12$ ft/s

Final vertical velocity:

$$V^2 = u^2 + 2\,as$$

$$V^2 = 0 + 2 * 32 * 16$$

$$V = 38.15 \text{ ft/s}$$

Time taken (t) is

$$V = u + at$$

$$38.15 = 20.78 + 32 * t$$

$$t = 0.542 \text{ s}$$

Actual velocity is

$$V^2 = \text{Horizontal velocity}^2 + \text{Final vertical velocity}^2$$

$$= \sqrt{12^2 + 38.15^2}$$

$$V = 39.99 \text{ ft/s}$$

Horizontal movement of the bale to the lorry is

Horizontal velocity * time

$$12 * 0.542 = 6.504 \text{ ft}$$

3. In a bobbin-changing automatic loom, the shuttle at the time of transfer is moving horizontally forward at 1.5 ft/s. If the transfer hammer moves the new bobbin downwards at 9 ft/s, then in what direction should the bobbin approach the shuttle so that it will enter parallel to the vertical sides of the shuttle?

Solution:

$$\tan\theta = 9/1.5 = 6$$

$$\tan^{-1} = 80.53 \text{ or } 90 - 80.53$$

$$= 10° \text{ from the vertical.}$$

4. The picking stick of an under pick loom, swinging about a fixed fulcrum from a rocking shaft, has passed the vertical position by 6° when the shuttle is about to leave the picker. If the upper end of the stick, where it passes through the picker is moving at 40 ft/s, then how fast is it moving the picker horizontally?

Solution:

Horizontal velocity V_h = 40 ft/s * cos 6° = 39.9 ft/s

5. During the run in of a spinning mule carriage, the counter faller moves horizontally with the carriage. The yarn passes from the rollers to the counter faller makes an angle of 4° and of 45° with the horizontal at the beginning and at the end of the run, respectively. Find, for each position, the rate at which the counter faller is moving towards the roller as a fraction of the speed of the carriage.

Solution:

Since the counter faller is moving horizontally, the rate of movement at the beginning is

$$\cos 4° = 0.997$$

At the end of the run,

cos 45° = 0.707

6. A knitting needle in a cotton's frame moves downwards at an angle of 15° with the vertical. What are the relative rates of movement of the needle bar due to (a) the needle in and out motion, (b) the needle up and down movement.

Solution:

Knitting needle in and out motion:

(a) sin 15 °/cos 15° = tan 15° = 0.267

(b) Knitting needle up and down movement = sin 90° = 1.00

Numerical questions

1. A man walks at 4 miles per hour (a) towards the front, (b) towards the rear of a bus travelling at 10 miles per hour. What is his velocity in each case, relatively to the earth?

   Ans. (a) 14 miles; (b) 6 miles per hour

2. A man jumps from a bus at 10 miles per hour at right angles to the direction in which the bus is travelling. If the speed of the bus is 17.3 miles per hour, in what direction and with what velocity does he strike the ground?

   Ans. 30° with the direction of the bus; 20 miles per hour

3. In what direction must a stone be thrown at 60 miles per hour from a train moving at 42.5 miles per hour, so that it will hit a tree directly opposite the window from which it is thrown?

   Ans. Backwards at 45° with the direction of the train

6
Force, mass and momentum

6.1 Introduction

The fundamental importance of a force is that it is the cause of movement or change of movement in bodies. Thus, an aero plane in flight, a loom shuttle passing through the warp shed, action of knitting needle in knitting machine, the movement of the ring rail in spinning machines all have their movements as the result of action of their forces or the stopping of shuttle at the end of the traverse, or the change in the direction of a motor car turning a corner requires the application of force.

Again, if a body is at rest it does not follow that no forces are acting on it: it simply means that the forces acting on it in any direction are exactly balanced by equal forces acting on it in opposite direction.

6.2 Definitions and units

Mass is the quantity of matter in a body. Quantities of matter are compared by the effect of gravity on them. The usual unit of mass is 1 pound (lb) or kilograms (kg). Another common unit of mass is the Engineers Unit of mass, which is equal to "*g*" *lb*, where "*g*" is 32.2, the value of acceleration due to gravity. Hence the number of Engineer's unit of mass in a body = no. of lb/g.

Force produces or tends to produce changes in motions of a body. Its nature can be explained by the Newton's laws of motion:

1. Everybody continues in its state of rest or of uniform motion in a straight line, except in so far as it is compelled by impressed forces to change that state.
2. Motion takes place in the direction of the applied force, and the change of motion is proportional to the applied force.
3. For every force in action, there is an equal and opposite reaction.

A familiar example is the shuttle in a loom. The shuttle remains at rest in the shuttle box, until the picker applies force to move the shuttle in to the warp shed. The picker applies force to the shuttle to give sufficient velocity to carry it through the warp. The picker must apply the force in the direction the shuttle is moving – horizontally and parallel to the reed. As the shuttle passes through the warp shed, it would travel at a constant speed but for the frictional resisting forces at the race board and reed. These forces act in the direction opposite to that in which the shuttle is moving and reduces the speed of the shuttle. Finally, the shuttle enters the shuttle box, striking the swell and picker and applying force to these parts. At the same time, the picker and swell apply equal and opposite forces, acting on the shuttle in the direction opposite to that in which it is moving, bringing it again to rest.

6.3 Units of force

(a) *The Poundal* – The force that acts for 1 second on a mass of 1 *lb* and changes its velocity by one foot per second.

(b) *The pound (lb)* – The force that is required to support a mass of 1 *lb* against the attractive force of gravity.

The Poundal is known as the absolute unit of force, and the pound is the gravitational unit. It is generally easier to measure the force in pounds than in poundals; and except for scientific work, the pound is the common unit.

Since a force of 1 pound, acting for 1 second on a mass of 1 *lb*, changes its velocity by "*g*" foot per second, as when 1 *lb* of mass is allowed to fall freely under the influence of gravity, it follows that

$$0 \text{ pound} = g \text{ poundals}$$

Pounds of force are frequently written as pounds (weight) or *lb wt* to distinguish them from pounds of mass.

6.4 Momentum in quantity of motion

If a force acts on a mass for a certain time and changes its rate of movement, it changes its quantity of motion or momentum. When a given force is applied for a given time to a body, the change of momentum is the same whatever the mass of the body, but the change in the rate of movement of the body, if of large mass, is less than that of a smaller body. Momentum thus depends on both rate of movement and mass and is equal to the product

Mass * Velocity, the unit of momentum is 1 unit of mass * 1 unit of velocity: generally 1 Engineers unit of mass (g, lb) * 1 foot per second

Relation between force, time and momentum

The change in velocity of a body caused by a force acting on it is directly proportional to the magnitude of the force, directly proportional to the time during which the force acts, and inversely proportional to the mass of the body.

Hence, change of velocity V = Force F * Time T/ Mass M

or $V = F * T/M$

Correct units are V is foot per sec, F in poundals, Time in seconds, M is *lb*, for 1 poundal acting on 1 *lb* of mass for 1 second changes its velocity by 1 foot per second.

Fundamental formula: Force * Time = Mass * Velocity

The product, force * time is termed as Impulse

Hence Impulse = Momentum

Force (*lb*) = Change in momentum per second

Or, Force (*lb*) = Mass (Engineer's units) * Acceleration

Since change of velocity/time = acceleration

Example 1:

The shuttle of a loom weighing 0.75 lb is given a velocity of 50 ft/s in the time the loom crankshaft revolving at 200 rpm turns 50°. What is the average force acting on the shuttle during that time to change its velocity?

Solution:

The time during which the force acts is the time for 50/360 revolution at 200 rpm.

i.e., 50/360 * 60/200 s = 0.041 s

e = Mass * change in velocity/time

= 0.75/32 * 50/0.041 = 28.58 lb (wt)

This is the average force acting during the time. The force will be constant only if the acceleration is uniform.

Example 2: In the above problem, the shuttle leaves the picker and moving at 50 ft/s and passes through the warp shed in 0.1 s. During the passage, a retarding force acts on the shuttle. What is the retarding force and the shuttle velocity if the average retarding force is equal to the weight of the shuttle?

Solution:

Force = Mass*Retardation

Retardation = Force/Mass

= 0.75/32/0.75 = 32 ft/s

Shuttle velocity (u) = $V - at$

u = 50 − 32 * 0.1 = 46.8 ft/s

Example 3: A mule carriage weighing 6400 lb is given a velocity of 30 in./s in 1.5 s. What average force has acted on it to give this velocity?

Solution:

$$\begin{aligned} \text{Force} &= \text{Mass} * \text{Velocity/time} \\ &= 6400/32 * 30/12/1.5 \text{ s} \\ &= 333.33 \text{ lb} \end{aligned}$$

Example 4: A shuttle of a heavy loom weighs 3.2 lb and is moving at 40 ft/s when it has passed through the warp. It must be stopped before the crank shaft, running at 150 rpm, has turned 60°. What is the least average force required to stop the shuttle?

Solution:

Time for the shuttle to pass through the warp shed is

$$\begin{aligned} 60/360 * 60/150 &= 0.066 \text{ s} \\ \text{Force} &= 3.2/32 * 40/0.066 = 60 \text{ lb} \end{aligned}$$

Example 5: The griffe of a single lift jacquard loom weighs 128 lb and is moved up and down vertically with a simple harmonic motion, with a maximum displacement of 4 in. from its bottom position. It is operated through a long connecting rod from a crank rotating at 120 rpm. In a 4 in. diameter circle, a resistance of 30 lb due to friction at the guides is constantly operating in the opposite direction to that of the griffe movement. What force and in what direction, up or down movement, must be applied to the griffe?

(a) At the beginning, centre and end of the up movement. (b) At the beginning, centre and end of the down movement? At the start or either up or down movement an acceleration of v^2/r is required: at the end of down movement, retardation of v^2/r is required, where v = crank pin speed in ft/min and r = radius of the crank circle in feet. In the centre position, acceleration is zero.

Solution:

Speed of the crank pin in ft/s = 120/60 * 4/12 * π = 2.09 ft/s

Acceleration = v^2/r = 2.09*2.09*12/2 = 26.20

Maximum accelerating force = Mass * acceleration

= 128/32 * 26.20 = 104.8 lb

(a) At the beginning of the up movement, the force required is

Mass + resistance + inertia force

= 128 + 30 + 104.8 = 263 lb

At the middle acceleration is zero, therefore

Mass + Resistance

= 128 + 30 = 158 lb

At the end of the up movement,

Mass + Resistance − Inertia resistance

= 128 + 30 − 104.8 = 53 lb.

(b) At the start of the down movement,

Mass − Resistance − inertia resistance

128 − 30 − 104.83 = 6.83 lb

At the middle,

Mass − Resistance

= 128 − 30 = 98 lb

At the end of the down movement,

Mass − Resistance + inertia resistance

= 128 − 30 + 104.83 = 203 lb

Numerical questions

1. The thread guide bar of a high speed winding machine reciprocates with a speed of 5 ft/s. At the ends of its stroke, it has to get up this speed in 1/100 s. Find the average force required to reverse the movement per lb of the mass of the bar.

   Ans. Force = 15.62 lb

2. A loom sley is equivalent to a mass of 96 lb concentrated on the loom sword pin at the beat-up position, it is stationary. After one quarter of a revolution of the crank shaft, the sley sword pin has speed equal to that of the crank pin, which is revolving in a 5 in. diameter circle at 140 rpm. What average force has acted on the sley during that time to change its velocity?

   Ans. Force = 141.3 lb

3. A mule carriage weighing 6400 lb and moving at 30 in./s has to be brought to rest in 0.1 s. What average force is required to stop it?

   Ans. Force = 500 lb

7
Work and power

7.1 Introduction

Whenever a force acts on a body and the body undergoes some displacement, then work is said to be done. For e.g., if a force (P) acting on a body causes it to move through a distance(s) as shown in, then work done by the force P is

Force × Displacement

$= F \times x$

If the force varies uniformly from zero to a maximum value of F, then

Work done $= 0 + F/2 * x = \frac{1}{2} \times F * x$

When a couple or torque (T) acting on a body causes angular displacement (θ) about an axis perpendicular to the plane of the couple, then

Work done = Torque × Angular displacement $= T\,\theta$

The unit of work depends upon the unit of force and displacement. In the S.I. system of units, the practical unit of work is N-m. It is the work done by a force of 1 Newton, when it displaces a body through 1 m. The work of 1 N-m is known as joule, normally written as J such that 1 N-m = 1 J.

7.2 Units of work (in SI units)

We have already discussed that the work is the product of the force and distance through which the body moves due to the action of the force. Then the units of work depend on the units of work and distance. The units of work are

1 N-m – It is the work done by a force of 1 N, when it displaces the body through 1 m, it is called joule (briefly written as J). Mathematically,

$$1 \text{ Joule} = 1 \text{ N-m}$$

1 kN-m – It is the work done by a force of 1 k N, when it displaces a body through 1 m. It is also called kilo joule (briefly written as kJ). Mathematically,

$$1 \text{kilo- joule} = 1 \text{ k N-m}$$

Note: Sometimes, the force stretches or compresses a spring or penetrates into a body. In such cases, the average force is taken as half of the force for the purpose of calculating the work done.

7.3 Graphical representation of work

The work done during any operation can be represented by a graph, by plotting distance along *X-X* axis and the force along *Y-Y* axis.

Since the work done is equal to the product of the force and distance, therefore the area of the figure enclosed represents the work done to some suitable scale. Such diagrams are called force-distance diagrams. If the force applied is not constant and varies uniformly with the distance, the force distance diagram is not a rectangle but a trapezium.

Example 1: (In SI units) A spring is stretched by 50 mm by the application of force. Find the work done if the force required to stretch 1 mm of the spring is 10 N.

Solution:

Spring is stretched by the application of force (s) = 50 mm

Force applied = 10 N

Work done to stretch the spring by 50 mm

= Force × Distance

= 10 × 50 = 500 N

Average force = 500 / 2 = 250 N

Work done = Average force × Distance

= 250 × 50 = 12500 N-m

= 12.5 J

7.4 Useful definitions

Work: Work is done when a resistance is overcome through a displacement. The amount of work done is the product of the resistance, or the force overcoming the resistance and the displacement.

The usual unit of the work is *I* foot pound (lb wt), the work done when a resistance of 1 lb is overcome through a displacement of 1 foot (in FPS units).

Power: Power is defined as the rate of doing the work. It is thus the measure of performance of engines. For e.g., an engine doing a certain amount of work in 1 s will twice as powerful as an engine doing the same amount of work in 2 s. The usual unit is 1 HP (Horse Power). In S.I. units, the unit of power is watt, which is written as W. This is equal to 1 N-m/s or 1 J/s. Sometimes it is also mentioned in kW or MW.

The rate of doing work is equal to 33,000 ft lb/min.

Example 2: A loom beam weighing 100 lb is moved down from beam store to loom at a distance of 100 ft by means of a trolley, which when loaded offers a resistance of 20 lb. If the centre of gravity of the beam is 9 in. from the floor level in the beam store and 1 ft. 9 in. from the floor when in loom, how much work is done by the over looker in moving the beam?

Solution:

Work done in lifting beam against gravity = 100lb *1 ft (rise of centre of gravity)

= 100 ft lb

Work done in moving the truck = Resistance * Displacement

= 20 lb * 100 ft lb

= 2100 ft lb

Total work done = 2100 ft lb

If the work is done in 3 min., Horse power expended is

HP = Work done per minute/time = 2100/33000*3 = 0.021 HP

Example 3: A carding engine requires one horse power to drive at 180 rpm. If the belt pulley is 18 in. in diameter, what resistance does the driving belt has to overcome?

Solution:

Belt speed = 180 * π * 18/12 = 847.8 ft

Work done by the belt per minute = resistance at the pulley rim * belt movement per minute

= 33000 ft lb

Resistance at the pulley rim = 33000/847.8 = 38.92 lb

Resistance of machine parts may be due to one or more of several reasons:

1. *Resistance due to gravity:* In lifting of a hoist, raising the lifter rail in speed frame, lifting of griffe in a jacquard loom.
2. *Resistance due to friction:* Bearings of the machines, the loom shuttle and the reed and the race board, the resistance offered by the movement of the flats in a carding machine.
3. *Strain resistance:* The resistance offered by the springs to extension or compression, as in springs of heald shaft reversing motions. In this case, the resistance varies in direct proportion to the stretch or compression of the springs.

4. *Inertia resistance:* The resistance offered by a body to having its velocity changed, i.e. being accelerated. If the acceleration is uniform, the resistance has a constant value.

Work may be positive or negative. A familiar example is the lifting of griffe in Jacquard loom, as shown in Fig. 7.1.

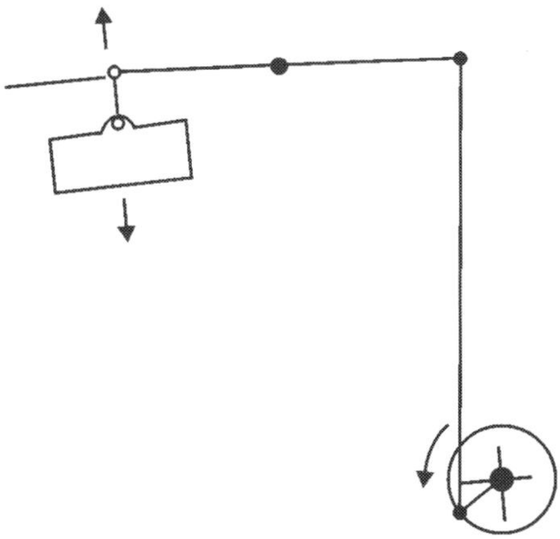

Figure 7.1

As shown in Fig. 7.1., positive work has to be done in overcoming the resistance of gravity acting downwards on the griffe. This positive work is done by the driving crank pulling down at the lower end of the connecting rod. During the downward movement of the griffe, the force of gravity is now available to lower the griffe; and in doing so, it applies an upward force from the connecting rod to the crank, helping to turn the crank shaft. If the friction is not there, the net work done by the crankshaft, per revolution, in lifting and lowering the griffe would be zero, since the positive work done during the upward movement is being exactly balanced by the negative work done during the downward movement.

The same thing applies during stretching or compression of the springs; positive work is done during stretching or compression of the springs and negative work is done when the springs remain in its original length. Also, during acceleration, positive work is done in overcoming the inertia resistance and negative work is done during retardation.

7.5 Work done by a varying force

When work is done in overcoming a varying resistance, the amount of work done is the product of the average resistance and the displacement.

Work can always be represented by the area of a diagram, one ordinate of which represents the resistance or the force overcoming it and the other is the displacement. For a uniform resistance, the work diagram is a rectangle.

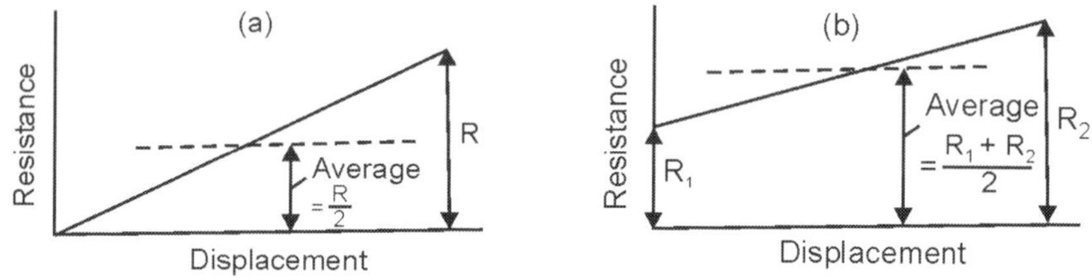

Figure 7.2 (a) and (b) Representation of the work done by graphical method

When resistance varies uniformly from zero to a maximum, the work diagram is a triangle. In Fig. 7.2(a), when the resistance varies uniformly from R_1 to R_2, the work diagram is a trapezium. In Fig. 7.2(b), when the resistance varies uniformly from zero to R, the average resistance is R/2 and the work done is $R/2$ * displacement.

When the resistance varies uniformly fro to , the average resistance is +/2 and the work done is $R_1 + R_2/2$ * displacement.

Example 4: The two springs for reversing the motion of a heald shaft, as shown in Fig. 7.3., have their ends to be extended 5 in. to get them in position with the heald shaft down. If the stiffness of the spring is 1 lb wt per in., find the work done (a) in putting the springs in position (b) in stretching the springs to additional 4 in. when the shaft is lifted.

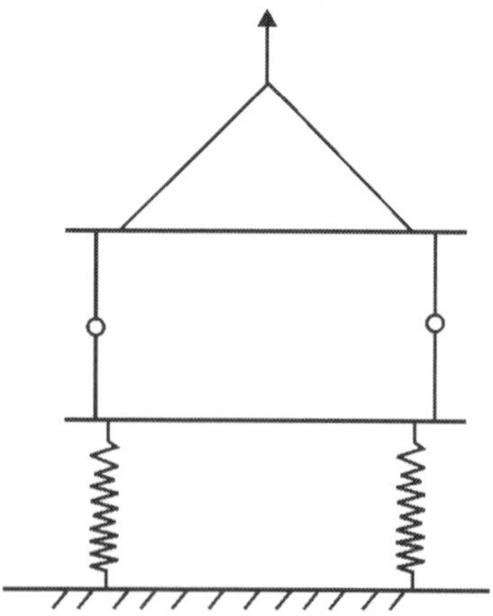

Figure 7.3

Solution:

The spring offers resistance which varies uniformly with the stretch. The term “stiffness” refers to the force required to stretch the spring 1 in., in this case 1 lb.

(a) The work done on each spring in overcoming the resistance which varies uniformly from 0 to 5 lb, the displacement being 5 in.

Work done per spring = 5/2 * displacement

= 5/2 * 5 = 12.5 in lb

Work done on two springs=12.5 * 2 = 25 in lb

(b) Each time when the shaft is lifted, the springs have to be stretched to additional 4 in. to 5 in. The resistance offered by the springs varies from 5 lb at the beginning and 9 in. at the end.

Work done per spring = 5 + 9/2 * displacement

= 7 * 4 = 28 in. lb

Work done in stretching two springs during lift of the shaft

= 28 * 2 = 56 in. lb

Numerical questions

1. A hoist, with its loads, weighs 2 tons. How much work is done in lifting it 80 ft? If it goes up in 2 min., what average horse power is required?

   Solution:

   Work done = 160 tons

   HP required = 5.34

2. At what horse power does a man weighing 140 lb work:

   (a) If he climbs a mountain 3300 ft high in 2 h?

   (b) If he climbs 15 ft rope in 12 s?

   (c) If he pushes a lawnmower against a resistance of 25 lb at 3 miles per hour?

   Solution: (a) 0.117, (b) 0.318, (c) 0.2 HP

3. A ring frame traveler moving in a circle of 1.8 in. diameter at 9000 rpm offers a resistance to movement of 250 grains. What HP is expended, if the frame has 400 spindles, in moving the travelers?

   Ans. HP = 1.83

4. Cloth is drawn forward by the taking up motion of a loom against a cloth tension of 300 lb. If there are 60 picks per inch in the cloth and the loom speed is 180 picks per minute, what work is done per minute in taking up the cloth?

   Ans. 75 ft lb

5. A loom shuttle offers an average resistance, during picking of 30 lb over a distance of 6 in. How much work is done per minute in picking the shuttle, if the loom speed is 180 picks per minute?

   Ans. 3600 ft lb

6. An endless chain of flats in a carding engine is driven by a 16 tooth chain wheel (one revolution of the wheel moving the chain 16 flats). The chain wheel is compounded with a 56 tooth worm wheel, driven by a single worm, which is compounded with a 32 tooth worm wheel. This in turn is driven by a single worm, compounded with a rope driving pulley of 13 in. diameter.

   (a) How many revolutions does the driving pulley make per revolution of the chain wheel and how far is the driving rope displaced?

   (b) If a pull of 8 lb wt is required at the rim of the driving pulley to move the flats, how much work is done per revolution of the chain wheel?

   (c) If the pitch of the flats is 1.5 in. and the speed of the chain is 3.5 in. per minute, what HP is required in driving it?

   Ans. (a) 1792 revs, (b) 6100 ft, (c) 48800 ft lb, (d) HP = 0.316.

7. The resistance offered by a rayon thread in stretching was found to increase uniformly with the stretch up to a load of 100 g. The stretch was then 5 mm. How much work is done in cm g units?

   Ans. 25 cm g

8. The inertia resistance of the shuttle and the picker of an under pick loom increases uniformly with displacement from 0 to 54 lb during a shuttle displacement of 8 in. How much work is done in overcoming the resistance? In addition to inertia resistance, the shuttle is constantly acted by a frictional resistance of 3 lb during the 8 in. displacement and the loom speed is 165 ppm. What horse power is expended in the shuttle?

   Ans. 18 ft lb, HP = 0.1

9. A spiral spring used for loom let-off motion has a stiffness of 30 lb/in. and has to be compressed to 3 in. by the over looker in adjusting the warp tension. How much work is done in compressing the spring? When working, the spring compression is increased from 3 to 3.1 in. How much work is done during the increase in compression?

   Ans. 135 in. lb, 9.15 in. lb

10. The driving belt from a motor to a line shaft is removed from the motor pulley. With the machine belts on the loose pulleys, it was found with a spring balance that a force of 20 lb is required on the driving belt to turn the shaft and machine belts. Assuming that the resistance remains constant, what HP is required to run the shaft at 200 rpm, if the driving pulley on the line shaft is 3 ft diameter?

    Ans. HP = 1.14

8
Resolution of forces

8.1 Introduction

Any exertion such as a push or a pull is a force, and this force will have magnitude and will be exerted in some direction. The magnitude or amount of the force may be represented by a straight line drawn to a convenient scale say 10 lb = 1 in. Such lines are shown in Fig. 8.1. In Fig. 8.1, the lines are represented by their lengths of forces mark; they are also shown by their position of the directions of the force. But none of the lines tell us the sense of the direction of the force; it means we do not know whether the force is from *A* to *B* or from *B* to *A*.

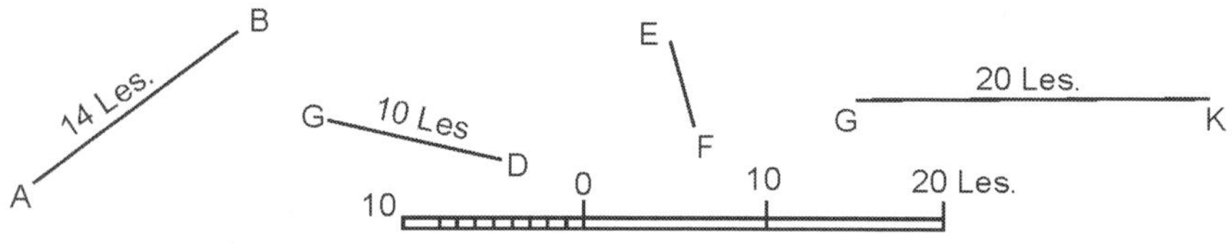

Figure 8.1

To represent forces by lines, we must represent amount, direction and the sense of direction. The sense of direction is represented by an arrow-head somewhere on the line itself as shown in Fig. 8.1(a).

Figure 8.1(a)

Such lines are sufficient to represent forces so that we can put down our thoughts for graphical solution to a problem.

8.2 Lines of representation of forces

Forces can be represented by lines just as we represent the distances. Suppose if number of forces act on a body and tend to move it in various directions, it will be equally simple to show the direction in which it will tend to move and the magnitude of the resultant force which is actually exerted in this direction. This can be best illustrated by an example shown below.

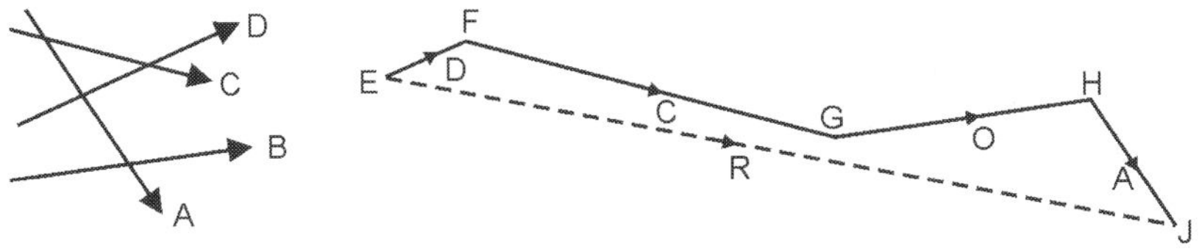

Figure 8.2

Example 1: A body is acted upon by several forces whose directions are shown in Fig. 8.3. If $A = 12$ lb, $B = 20$ lb, $C = 30$ lb, and $D = 7$ lb, then find the resultant force.

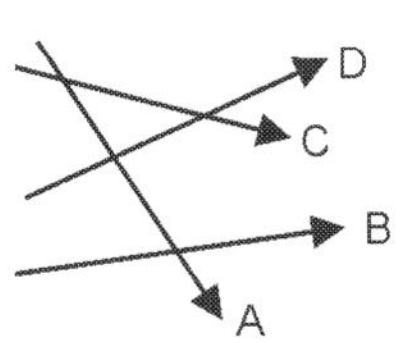

Figure 8.3

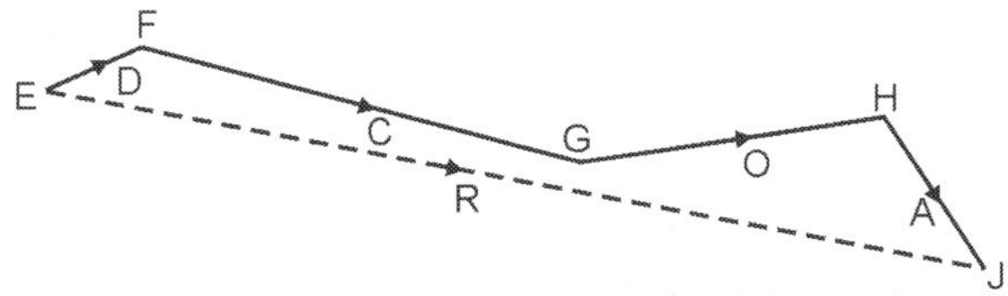

Figure 8.4

Solution:

Take a scale, say 10 lb = 1 in. and commence at a point *E* and draw a line parallel to any of the lines in Fig. 8.3. In Fig. 8.4, we have started by drawing *EF* parallel to *D* and in the same sense as shown by the arrow-head and the length *EF* equal to 7 lb on the scale. From *F* draw a line parallel to any of the lines in Fig. 81, except *D*, and again in the same sense we have chosen *C*, and so made *FG* parallel to *C* and equal to 30 lb on the scale. From *G* draw a line parallel to *B* or *A*. We have chosen *B* and made *GH* parallel to *E*, again noting the direction of arrow-head and equal to 20 lb on the scale. Now draw *HJ* parallel to the remaining line A and make its length equal to 12 lb on the scale. By joining *EJ*, we obtain a line which represents the resultant direction of the four forces acting on the body. When measured by the scale, we get its amount; and of course, the sense of direction must always be away from the starting point, as shown by the arrow on *EJ*. If we measure the length *EJ* on the scale, it will found to be 63.25 lb.

There is no hard and fast rule to start with the point *D*; the force *D* is taken as an example. Any of the forces may be chosen as the starting point. The next example takes the same diagram of the position of the forces as in Fig. 8.3.

Example 2: In this example with the same diagram, *A* is chosen as the starting force. It is to be noted that (i) the arrow heads of the components *ABC* must all run in the same direction; (ii) the arrow head on the resultant runs in the opposite direction.

Proceed by drawing *EF* parallel to *A*. Now draw *FG* parallel to *C*, then *GH* parallel to *D*, then *HJ* parallel to *B*. Join *EJ* and we obtain the resultant and it will be found when measured on the scale is 63.25 lb, which is exactly same as obtained in Example 1.

The forces shown in the above examples show the direction in which they are acting, however large or small. If the body is fixed or if there is great frictional resistance, the body will not move; but the resultant is the force which tends to move the body.

8.3 Law of Polygon of forces

If any number of forces acting at a point can be represented in direction and magnitude by the sides of the polygon taken in order, then the forces are in equilibrium.

Let the forces P_1, P_2, P_3, P_4 and P_5 acting at *O* be represented in direction and magnitude by the sides *AB, BC, CD, DE, EA* of the polygon *ABCDE*. Join *AC, AD.*

Then

$$\begin{aligned} AB + BC &= AC \\ AC + CD &= AD \\ AD + DE &= AE \\ AE + EA &= 0 \end{aligned}$$

Hence we may say that the system is in equilibrium.

From the diagram, it is clear that the resultant of the forces represented by *AB, BC, CD, DE* is represented by *AE.*

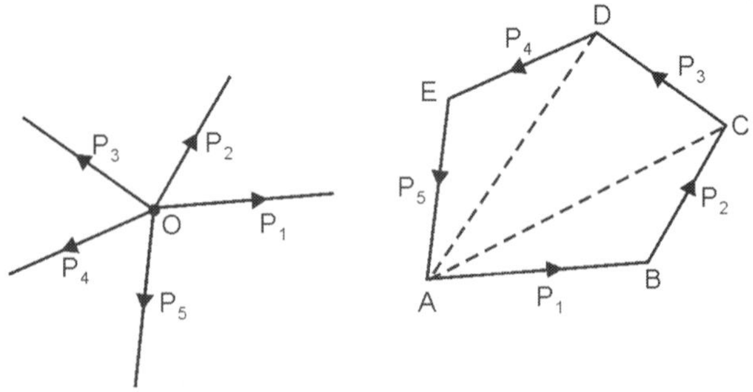

Figure 8.5

8.4 Law of Triangle of Forces

If three forces acting upon a body be represented by magnitude and direction by the sides of the triangle taken in order, they will be in equilibrium.

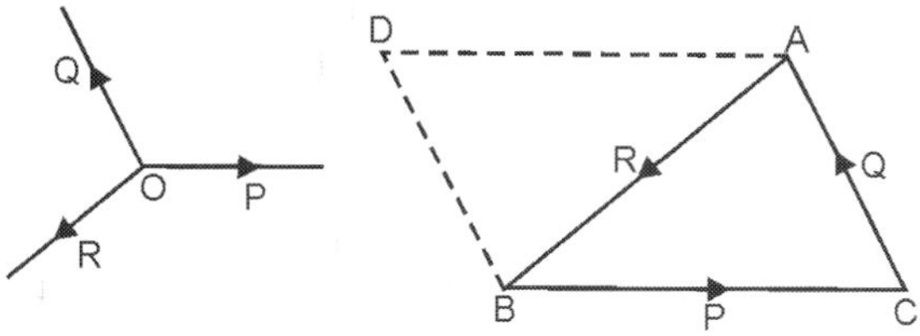

Figure 8.6

Let forces *P, Q, R* acting at *O* be represented in magnitude and direction by the sides of the triangle taken in order.

Complete the parallelogram *BCAD*. Then *BD*, being equal and parallel to *CA* represents *Q*. By the law of parallelogram of forces, the resultant of *BC* and $BD = BA = -AB = -R$, i.e. the resultant of *P* and $Q = -R$. Hence the resultant of *P*, *Q* and $R = -R + R = 0$.

Hence the system is in equilibrium.

8.5 Tension and compression

When a string neglects its weight and is pulled with a force *P*, then at every point of the string equal and opposite forces *P* act. Each of these forces is called the tension of the string. It must be remembered that tensions at every point of a string are equal and opposite. This is shown by arrow showing the direction of tension at any part of the consideration (Fig. 8.7).

Figure 8.7

Let *AB* be a string whose end *A* is fixed to some point and let a weight *W* be attached to *B*. It is evident that the string is pulling at *A*; hence if the equilibrium of *A* is considered, the arrow is showing the tension *T* in the part adjacent to *A* points downwards. Again the string is clearly preventing the weight at *B* from falling down, i.e. the string is pulling *B* upwards. Hence the arrow near *B* points upwards.

Rod under compression

Figure 8.8

We will assume that when a light string passes round a smooth peg or pulley, its tension remains unchanged. When a bar is pulled at its ends by equal and opposite forces, it is in tension and the above remarks apply to it. A bar under tension is called a tie. When a bar is pushed at both ends by equal and opposite forces, it is said to be under compression.

Let *AB* be a bar subjected to compressive forces *P* at *A* and *B*. At every point of *AB* equal and opposite forces *P*' (where *P*' = *P*) act. P; acting at *A* must balance the external force *P*; hence, when considering the equilibrium of *A*, the arrow showing *P*' must point towards *A*. So when considering the equilibrium of *B*, the arrow points towards *B*.

A bar under compression is called strut.

8.6 Graphical methods

The triangle of forces and polygon of forces can be used to obtain graphically the resultant forces acting at a point.

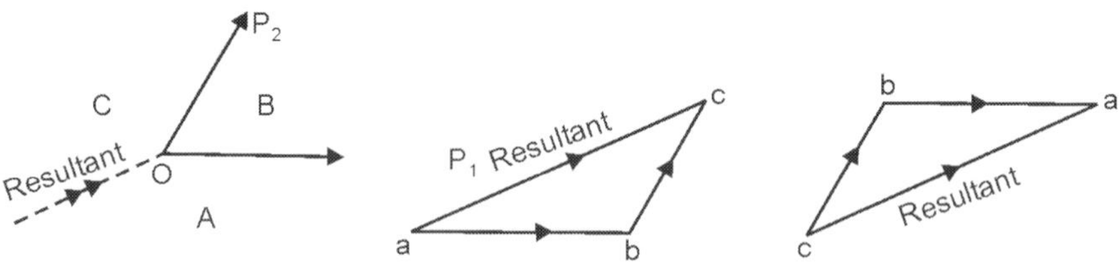

Space diagram · Vector diagram · Vector diagram

Figure 8.9(a) **Figure 8.9(b)** **Figure 8.9(c)**

(a) Let the forces P_1 and P_2 act at a point. To find their resultant, draw a line *AB* parallel and proportional to P_1 on a suitable scale. Then *AB* represents P_1 in direction and magnitude. Similarly draw *BC* to represent P_2. Join AC. Then AC represents the resultant of the forces in direction and magnitude. AC can be measured using the scale.

(b) In Fig. 8.9(a) are shown two forces , acting at O. The spaces about these forces are named as A,B,C. The force P_1, which lies between spaces A and B, is then named as AB. The force P_2 is then named as BC. This diagram showing the position of the forces is called space diagram.

(c) In Fig. 8.9(b), draw a vector to represent P_1 (A vector is a straight line representing a force in magnitude and direction). Since in the space diagram, P_1 has been named as AB, the vector representing by corresponding small letters *ab*. Similarly, the force P_2 is represented by the vector bc just as BC in space diagram. Join ac. Then ac represents the resultant. This diagram which shows the direction and magnitude of the forces is called a *vector or force diagram* (Fig. 8.9c).

This method of naming the forces in the space and vector diagram is called Bow's notation.

Thus, we can see from the above diagrams that we can move either clockwise or anti-clockwise but the movement must be uniform throughout. If P_1 is named as AB, then must be named as BC and not CB.

8.6.1 Graphical methods for more than two forces

In the previous article, we have seen only two forces. Now, let it be required to find the resultant of three forces. Let the three forces be P_1, P_2, P_3 acting at O. Draw the space diagram as shown in Fig. 8.10(a). Draw vectors *ab, bc, cd* to represent AB, BC, CD. Join *ad*. Then *ad* represents the resultant; ad is the line joining the initial point of the first force and terminal point of the last force. The position and direction of the resultant have been shown in the space diagram by a dotted line with two arrow heads which has been drawn parallel to ad. In the space diagram, the forces are AB, BC, CD and the resultant is AD.

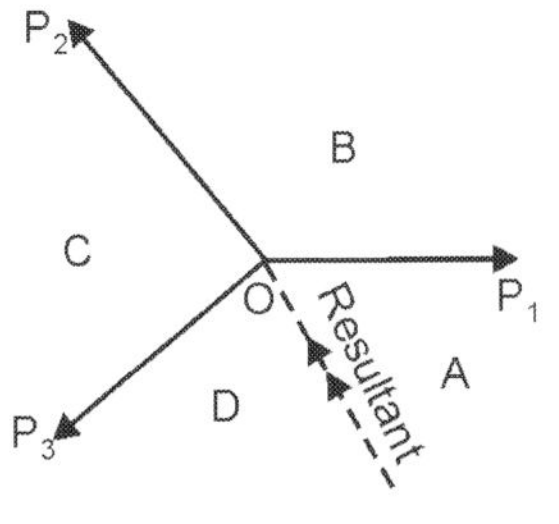

(a) Space diagram

Figure 8.10(a)

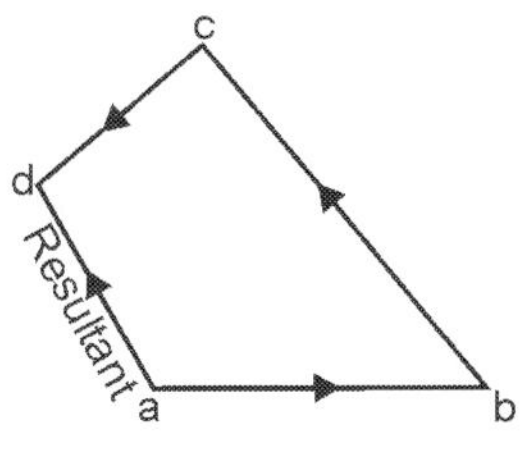

(b) Vector diagram

Figure 8.10(b)

The force that is applied at *O* will produce equilibrium, which is called the equilibrant of the system. The equilibrant is equal and opposite to the resultant and in the diagram it is represented by *DA* in the space diagram and da in the vector diagram (Fig. 8.10b).

The method explained above is applicable to any number of forces.

8.6.2 Jib crane

The essentials of a jib crane are shown in Fig. 8.11(a) and (b).

AD is a vertical post. *AC* is a rod called *jib* hinged at *A* so that it can turn around *A* in a vertical plane. It is supported by a chain or rod *DC* called tie which is attached to a point *D* in *AD*. At *C* there is a pulley over which passes a chain *ECF*; the load *W* to be lifted is attached to the end *F*, while the end *E* is wound round a drum. The effort is applied at *E* or to lift the load. *ED* is horizontal and *EC* coincides with it.

Supporting the pulley at *C* to be smooth, the tension in *EC* = tension in *CF* = *W*.

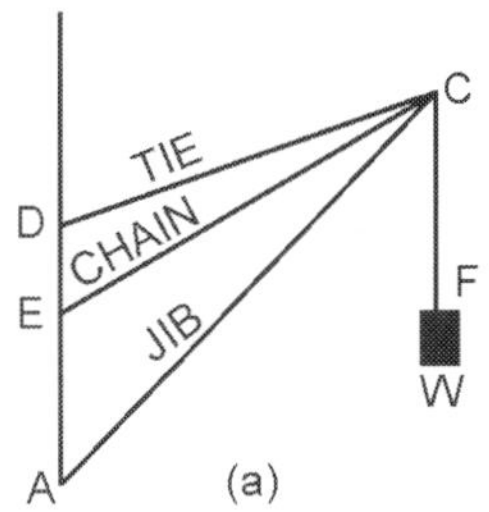

Figure 8.11(a)

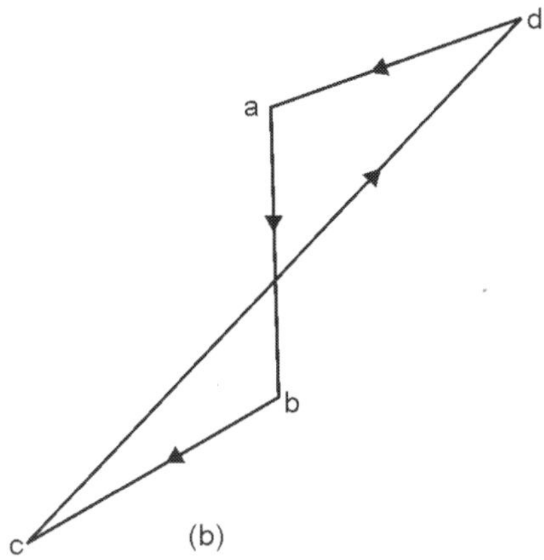

Figure 8.11(b)

Figure 8.11(b) shows the vector diagram of the forces acting at *C*; *ab* represents *W*, *bc* represents the tension in *EC* which is also *W*; *cd* has been drawn parallel to *AC* and ad parallel to *DC*.

The diagram is drawn with the following data:

The *jib AC* = 5 m, the *tie CD* = 3.6 m, *AD* = 2.4 m, *AE* = 1.6 m, *W* = 20 kN

By measurement with suitable scale value, it is found that

Force in *AC* = 50 KN thrust

Force in *CD* = 18 KN tension

Example 3: In a jib crane, the height of the vertical post, as measured with its joints with the jib and the tie is 2.1 m. The *jib* is 4.5 m and the tie is 3 m long. A load of 50 KN is suspended from a chain passing over a smooth pulley supported on the crane head and the chain is fixed to the post at a point 0.9 m above its junction with the jib. Find graphically the stresses on the jib and the tie. Find also: (i) if the supporting chain is taken parallel to the tie, and (ii) it is parallel to the jib

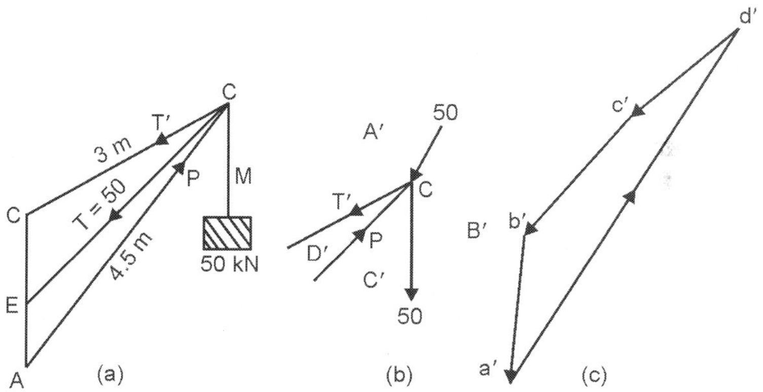

Figure 8.12

Solution:

The crane and the forces acting at the crane head *C* are shown in Fig. 8.12(a). Let us consider the thrust in the jib *AC* be *P KN*, and let tension *DC* be *T*. The tension *T* in the chain *EC* is 50 KN, being equal to the weight. The tension in the portion *CM* of the chain is also 50 KN. By applying bow's notation provided the two known forces in *CM* and *CE* are together, there being no unknown forces between them.

The force $T = 50$ KN in *CE* tends to move the point *C* towards *E*. Make *EC* towards the right hand side. Then *T* can be replaced by an equal force acting along the line and pushing at *C*. The free body diagram of the crane head *C* can be drawn as shown in Fig. 8.12(b), which may be taken as space diagram. The spaces have been named as *A' B' C'* and *D'*. Figure 8.12(c) shows the vector forces of a polygon.

By measuring *c'*, *d'*, and *a' d'* we get *P* and *T*.

$$P = 141.5 \text{ KN thrust}$$

$$T = 55.5 \text{ KN pull}$$

(i) Let *BE* coincide with *CD*. Then the total force in the direction *CD* is $T + T = 50 + T$. At *C*, there are only three forces, viz. 50 KN along

CM, *P* along *AC* and 50 + *T* along *CD*. Now the sides of the triangle *ACD* are parallel to these forces.

$$P/AC = 50 + T/CD = 50/DA$$

$$P/4.5 = 50 + T/3 = 50/2.1$$

Then $P = 107.14$ KN (compression)

$T = 21.43$ (tension)

(ii) Let *CE* coincides with *AC*. Then the forces acting at *C* are 50 KN along *CM*, *P* − 50 along *AC*, and *T* along *CD*.

$$P - 50/4.5 = T/3 = 50/2.1$$

Then $P = 157.14$ KN (compression)

$T = 71.43$ KN (tension).

8.6.3 Force and its characteristics

Force produces or tends to produce a change in the state of rest or of uniform motion in a straight line of a body.

Consider a horizontal force *P* (Fig. 8.13) applied to a body placed on a rough horizontal plane. When the applied force *P* is small, the body does not move. When *P* is increased, the body will start moving in a straight line if the line of action of *P* passes through the centre of gravity(cg) of the body. Hence there will be a motion of translation as well as rotation if the line of action of P does not pass through the centre of gravity of the body.

Thus we know that the effect of force depends on three characteristics.

(i) Magnitude

(ii) Direction

(iii) Position or line of action

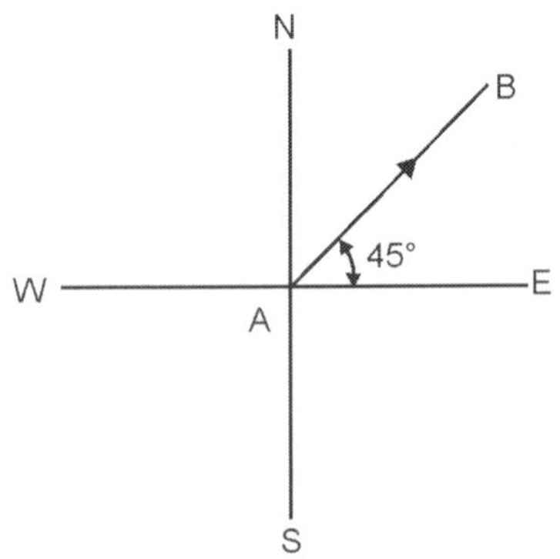

Figure 8.13

If we draw a straight line parallel to the line of action of the force, whose length is proportional to the magnitude of the force, the line is said to represent the force in *magnitude and direction.*

Thus let a force P of 15 lb acting in the north-east direction and 1 in. length represents 5 lb then a straight line AB of length 3 in. drawn in the north-east direction will represent the force P in direction and magnitude. An arrow is placed on the line with the arrow-head pointing north-east represents the sense of the force, i.e. the force is acting from A towards B. The force represented by the line AB is written as AB.

Vectors are also represented by a single letter P or a etc. The magnitude of the vector a is represented by (a) or a.

8.7 Resolution and composition of a force

Resolution of a force means 'finding components of a given force in two given directions'.

Let a given force be R which makes an angle θ with X-axis as shown in Fig. 8.14. It is required to find the components of the force R along X-axis and Y-axis.

Components of force along X-axis = $R \cos \theta$

Components of force along Y-axis = $R \sin \theta$

Hence the resolution of forces is the process of finding out components of forces in specified directions.

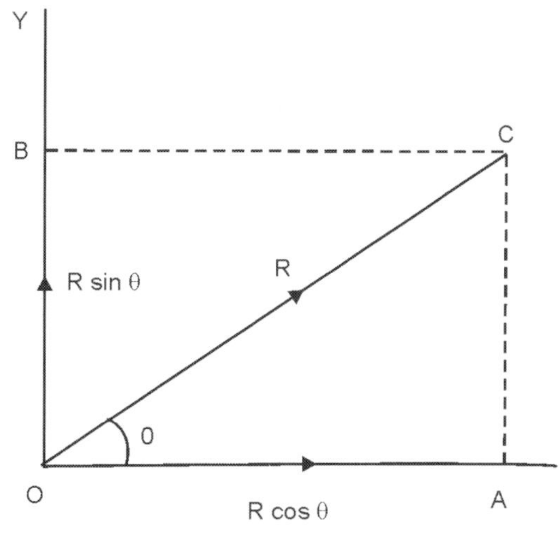

Figure 8.14

8.8 Resolution of a number of coplanar forces

Let a number of coplanar forces R_1, R_2, R_3... are acting at a point as shown in Fig. 8.15 (forces acting in one plane are called coplanar forces).

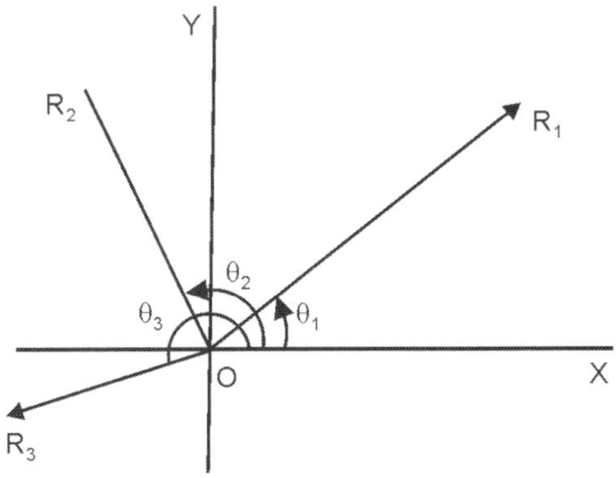

Figure 8.15

Let θ_1 = Angle made by R_1 with X-axis

θ_2 = Angle made by R_2 with X-axis

θ_3 = Angle made by R_3 with X-axis

Let H = Resultant components of all forces along X-axis

V = Resultant components of all forces along Y-axis

R = Resultant of all forces

θ = Angle made by resultant with X-axis

Each force can be resolved in to two components, one along X-axis and other along Y-axis.

Component of along X-axis = $R_1 \cos \theta_1$

Component of along Y-axis = $R_1 \sin \theta_1$

Similarly, the components of R_2 and R_3 along X-axis and Y-axis are ($R_2 \cos \theta_2$, $R_2 \sin \theta_2$) and ($R_3 \cos \theta_3$, $R_3 \sin \theta_3$), respectively.

Resultant components along X-axis = sum of components of all forces along Y-axis

$$V = R_1 \sin \theta_1 + R_2 \sin \theta_2 + R_3 \sin \theta_3 + \ldots$$

Then resultant forces, $R = \sqrt{H^2 + V^2}$

The angle made by R with X-axis is given by $\tan \theta = V/H$.

Example 4: Two forces are acting at a point O as shown in Fig. 8.16. Determine the resultant in magnitude and direction.

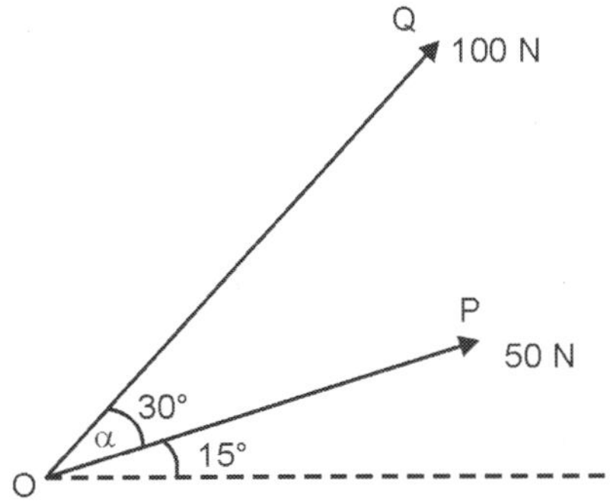

Figure 8.16

Solution:

The problem can be solved by resolution of forces.

Force = 50 N and force Q = 100 N.

Let us first, find the angles made by each force with X-axis.

Angle made by P with X-axis = 15°

Angle made by Q with X-axis = 15 + 30 = 45°

Let H = Sum of components of all forces along X-axis

V = Sum of components of all forces along Y-axis

The sum of components all forces along X-axis is given by

$$H = P\cos 15° + Q\cos 45°$$
$$= 50 \times \cos 15° + 100\cos 45°$$
$$= 119 \text{ N}$$

Sum of components of all forces along Y-axis

$$V = P\sin 15° + Q\sin 45°$$
$$= 50\sin 15° + 100\sin 45°$$
$$= 83.64 \text{ N}$$

The magnitude of the resultant force is given by

$$R = \sqrt{H^2 + V^2}$$
$$= \sqrt{1132 + 83.64\ 2} = 145.46 \text{ N}$$

The direction of the resultant force is given by

$$\text{Tan}\ \theta = V/H$$
$$\theta = \tan^{-1} 83.64/119 = 35.10°$$

Example 5: *A* heald shaft is connected as shown in Fig. 8.17 to a dobby jack at *B*. If the jack has to exert an upward force of 20 lb to support the shaft, find the forces acting on the cords *AD* and *AC*.

Solution:

In the above figure, three forces act at the point *A*.

(i) The upward force of 20 lb

(ii) Force in the cord *AD*

(iii) Force in the cord *AC*

The problem can be solved graphically.

Set of *xy* = 20 units long to represent the force *AB*. From the ends of this line, setoff lines parallel to *AD* and *AC*. To meet in *z*, then *yz* represents the force in *AC* and *zx* represents the force in *AC*. By calculation or measurement, these forces are found to be 14.1 lb.

In order to represent the force in the triangle *xyz*, arrow heads show the directions in which the forces act; *xy* represents an upward force. The arrow heads must follow round the triangle so that the forces *yz* and *zx* are downward, showing that the forces in the two cords *AD* and *AC* are away from *A*.

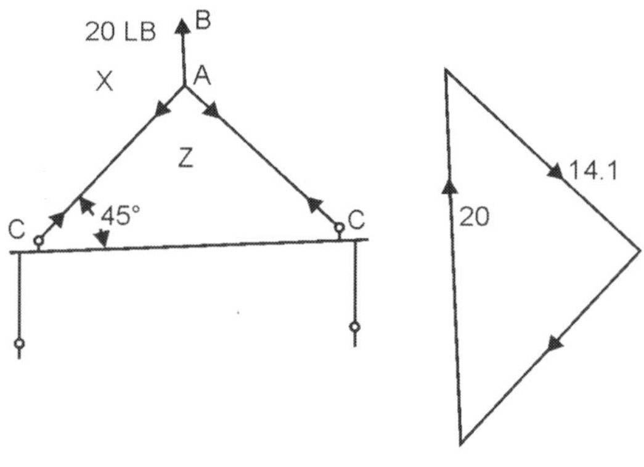

Figure 8.17

We have seen from the diagram that the cords *AD* and *AC* are said to be in *tension, since the forces acting on them tend to stretch the cords.* This means the upper end of *AD* is applying a downward force of 14.1 lb to help to balance the downward force of 20 lb. Hence, the lower end of *AD* is applying the same force of 14.1 lb upwards at *D* to help to support the shaft.

It is to be noted that the total tension or force required to break the cord is 14.1 lb, and not 28.2 lb. It means one force is the action and the other is the reaction.

This can be clearly explained by another example.

Consider a vertical cord attached to a hook supporting a weight of 10 lb. The lower end of the cord is applying the upward force of 10 lb needed to support the weight. The upper end is applying a downward force of 10 lb to the hook. Hence the total force to break the cord is 10 lb and not 20 lb.

Example 6: The picker of an over pick loom is offering a horizontal resistance of 100 lb (Fig. 8.18). Find the pull required to apply to move the picker, if the strap is in same vertical plane as the picker and makes an angle of 30 degrees with the picker spindle.

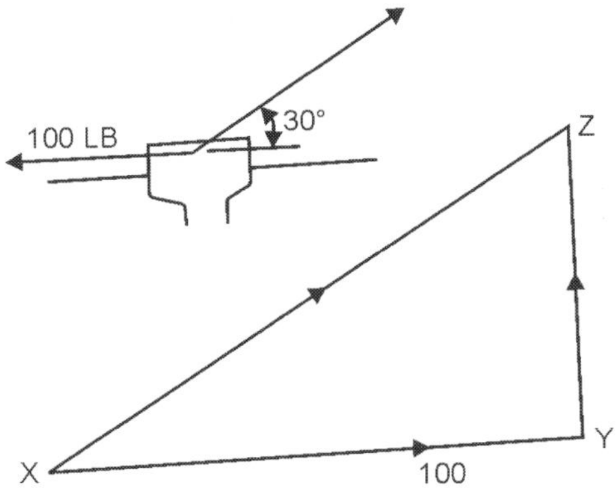

Figure 8.18

Solution:

The problem can be solved graphically.

Set off *xy* = 100 lb horizontally, which represents the horizontal resistance of the pull of the strap. This component must be 100 lb to overcome the picker resistance.

Set off *xz* to represent the picker parallel to the shaft.

Complete the right-angled triangle *xyz*.

Measure the distance of *xz* which is 115 lb and gives the necessary pull of the strap.

(or) horizontal resistance = 100 lb

Angle made by the picker on the spindle = 30°

Pull exerted by the picker = 100/sin 30°

= 100/0.866 = 115 lb

Example 7: A horizontal weighting lever *AB* shown in Fig. 8.19 is 9 in. long, pivoted at *A* and carrying a weight of 2 lb at *B*. At *C* is a link *CP* attached to the lever, the angle $ACP = 60$ degrees and $AC = 1.5$ in. Find out the force CP that must apply to the lever to keep it in equilibrium. And what is the direction and magnitude of the force at the pivot A? Ignore the weight of the lever itself.

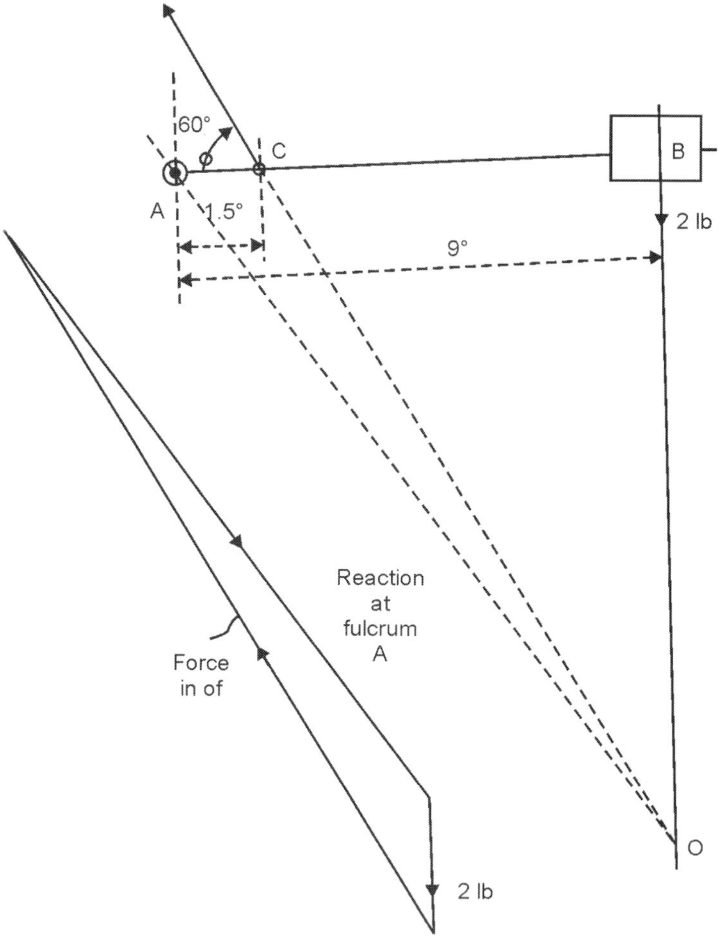

Figure 8.19

Solution:

In this problem, three non-parallel forces act on the lever

(a) The force of gravity 2 lb acting vertically downwards

(b) The force in the supporting link CP

(c) Reaction at the pivot or fulcrum A

It may be noted that all these three forces meet at point obviously at D. The direction of the reaction must then be in the direction of the line DA. The magnitude of the forces can be found by the triangle of forces by taking suitable scale and solving it graphically. From which the force in CP is approximately 13.83 lb and the reaction at the pivot or fulcrum A is 12. 2 lb. Another method is to solve by Principle of Moments.

Taking moments about the pivot or fulcrum A, we have

$$W \times 9 = AC \sin 60° \times CP$$

$$CP = 18/1.5 \times \sin 60°$$

$$= 13.83 \text{ lb}$$

$$CP = 13.83 \text{ lb}$$

Reaction at pivot A =

$$\sqrt{CP^3 - AB^3} = \sqrt{13.83\ 2 - 282} = 187.28, \text{ but } AC = 1.5 \text{ in.}$$

$$13.48 - 1.5 = 12.19 \text{ lb}$$

Example 8: A ring frame traveler rotates and the forces acting on it are shown in Fig. 8.20. Find the resultant force acting on the traveler.

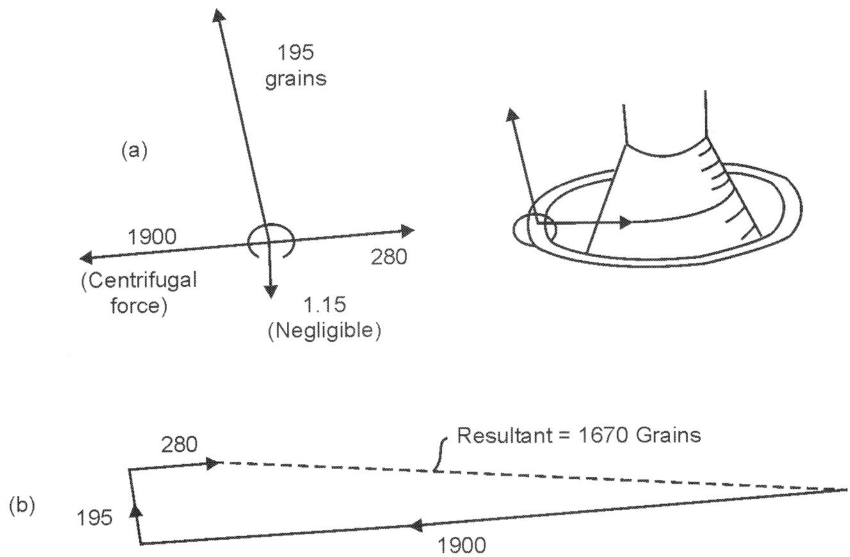

Figure 8.20

Solution:

From Fig. 8.20, the forces acting on the traveler are

(i) Horizontal inward pull of 280 grains

(ii) Horizontal outward pull or centrifugal force of 100 grains

(iii) Vertical yarn tension of 195 grains

From the polygon law of forces, the problem can be solved graphically.

Consider 140 grains = 1 cm

Therefore, 195 grains = 1.39 cm

280 grains = 2 cm

1900 grains = 13.57 cm

When the measurements are drawn to scale and the resultant force is measured, it is found to be 11.9 cm.

Resultant force = 11.9 × 140 = 1666 grains

Example 9: Figure 8.21(a) shows the load applied to a scutcher lap during lap formation. Find the pressure between the roller and the lap at the start and at the end of the lap.

Solution:

The pressure between the roller and the lap acts along the line joining the centers' of roller and lap, and the resultant of the pressures applied by the two rollers must be equal to the downward load acting on the lap. In other words, the sum of the vertical components of the pressures must be equal to the downward load.

(i) Start of the lap

The triangle of forces acting on the lap is shown in Fig. 8.21(b)

Hence, the distance between the rollers and the lap is

+ 1.5 = 6.25 in

Distance $xy = \sqrt{6.25 * 6.25 - 5.125 * 5.125} = 3.56$ in.

From similar triangles,

$$P/650 = 6.25/3.56$$

$$P = 1140 \text{ lb}$$

(ii) End of the lap

The triangle of forces is shown in Fig. 8.21(c)

The distance between the centers of lap and roller is

= 4.75 + 6 = 10.75 in.

Distance $xy = \sqrt{10.75810.75 - 5.12585.125} = 9.45$ in.

From similar triangles,

$$P/650 = 10.75/9.45$$

$$P = 753 \text{ lb}$$

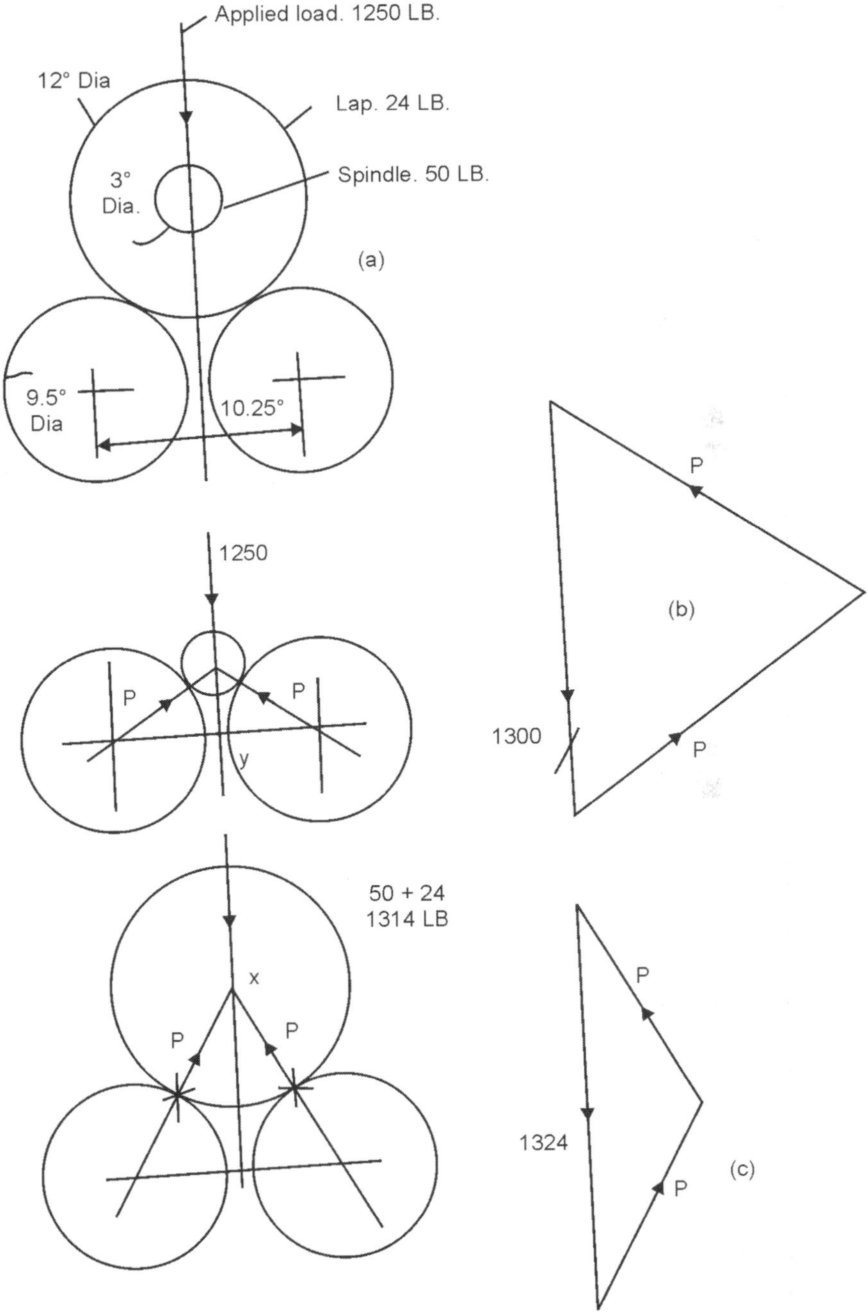
Applied load. 1250 LB.
12° Dia
Lap. 24 LB.
3°
Dia.
Spindle. 50 LB.
(a)
9.5°
Dia
10.25°
1250
P
P
y
P
P
1300
(b)
50 + 24
1314 LB
x
P
P
P
1324
P
(c)

Figure 8.21

Example 10: Figure 8.22 shows a bracket carrying a rope drive for a ring spinning frame subjected to loads, horizontal and vertical loads, due to ropes, each of 220 lb and the weight of the pulleys is 20 lb. What is the magnitude and direction of the resultant load applied to the bracket?

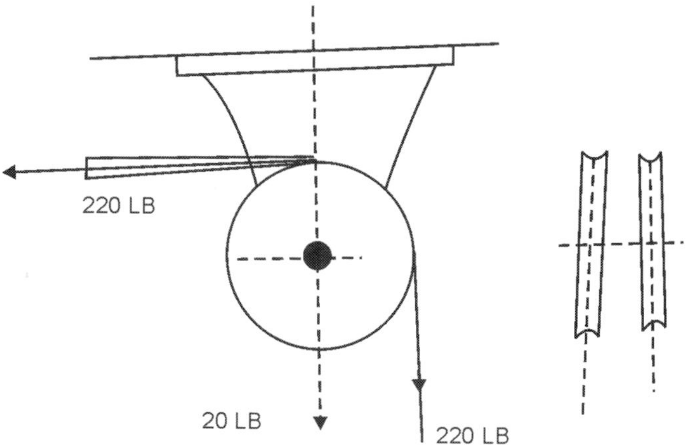

Figure 8.22

Solution:

Horizontal pull on the pulley = 220 lb

Vertical pull on the pulley = 220 lb

Load on the pulley = 20 lb

Therefore, the vertical load acting downwards = vertical pull = load on the pulleys

$$= 220 \text{ lb} + 20 \text{ lb} = 240 \text{ lb}$$

Construct a right angled triangle to show the forces.

$$\text{Resultant } AC = \sqrt{AB^2 + BC^2}$$
$$= \sqrt{220^2 + 240^2}$$
$$= 325.57 \text{ lb}$$

Angle (α) to the vertical =

$$\text{Sin } \alpha = AB/AC$$
$$= 220/325.57 = 0.67$$

$\alpha = \sin^{-1} 0.67 = 42°$ to the left of vertical acting through the centre of the pulleys.

Example 11: A hinged horizontal loading platform projects 6 feet from a wall and is supported by two chains, attached to its other corners, and to the wall above, each chain making an angle of 45° with the horizontal. If the platform weighs 100 lb. and a load of 500 lb. is placed centrally on it, what is the tension in each chain and the reaction at the hinge?

Solution:

Weight of the platform = 100 lb

Load applied to the plat form = 500 lb centrally

Let the tensions on the chains on both sides be .

Total load on the platform =

Weight of the platform + Load applied on the platform

= 100 lb + 500 lb = 600 lb

Since, the two chains are making an angle of 45° to the vertical respectively,

Tension on the chains =

$$2\ T_l \times 2 \sin 45° = \text{total load on the platform}$$

$$T_1 = 600/1.414 \times 2$$

$$= 212.16 \text{ lb.}$$

Reaction at each hinge =

Total load on the platform = 600 lb

Since the load is acting at the centre of the platform, the reaction R at the each hinge is

$R/\sin 45° = 300$ lb (Since the load is equally shared by the two chains)

$$R = 300 \times \sin 45°$$

$$R = 300 \times 0.707$$

$$= 212\ 7 \text{ lb}$$

Numerical questions

1. A war per's beam weighing 500 lb is rolled up an inclined plane of 30° with the horizontal on to the rail board of a truck. The beam is pulled up by ropes passed round the gudgeon pins and the rope makes an angle of 60° with each other as shown in Fig. 8.23. What pull P is required, parallel to the incline, to roll the beam up and what is the tension in the ropes? Ignore the friction between the gudgeon pins and the beam flanges.

   The rail board is supported by chains at each side of the truck making an angle of 45 degrees with the horizontal. If the weight of the rail board is 40 lb, find the tension in each chain when the beam is at the centre of the rail board.

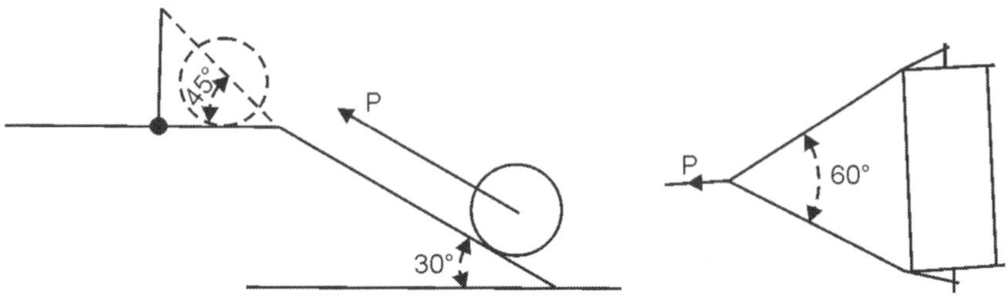

Figure 8.23

Ans. P = 250 lb; rope tension = 144 lb; chain tension on the truck = 191 lb

2. A cross-wound cylindrical ball of yarn is driven by resting it on two 9 in. diameter calendar rollers; distance of 10 in. horizontally between centers. Find the pressure applied by each calendar to the ball.

(a) When the ball is 18 in. diameter, its weight, including the centre shaft is 112 lb and an additional load of 30 lb acts vertically on the ball centre shaft.

(b) If the ball centre shaft is 4 in. diameter and weighs 70 lb, find the downward load applied to the centre shaft when starting the ball to give the same pressure from the calendar rollers, as when the ball is 18 in. diameter.

Ans. (a) 76.5 lb; (b) 78 lb

3. A flat clipping machine weighing 2240 lb is lifted by tackle to the first floor of the workshop and is pulled through a loading door by ropes. The ropes apply a horizontal pull. What pull is required when the tackle chain makes an angle of 20° with the vertical and what is the load on the tackle chain?

Ans. Horizontal pull = 817.6 lb

Load on tackle = 2385.6 lb

4. A warp thread is deflected downward from the horizontal by a vertical heald shaft, until it makes an angle of 10° with the horizontal. If the tension on the warp thread is 1 oz, what force is required at the heald shaft to pull 1000 such threads?

Ans. 21.6 lb

5. Figure 8.24 shows the method of tensioning yarn in weaving machine. Assuming the guide rollers are frictionless, what warp tension T will keep the weighting lever in equilibrium?

Ans. T = 175 lb; reaction at the pivot = 304 lb

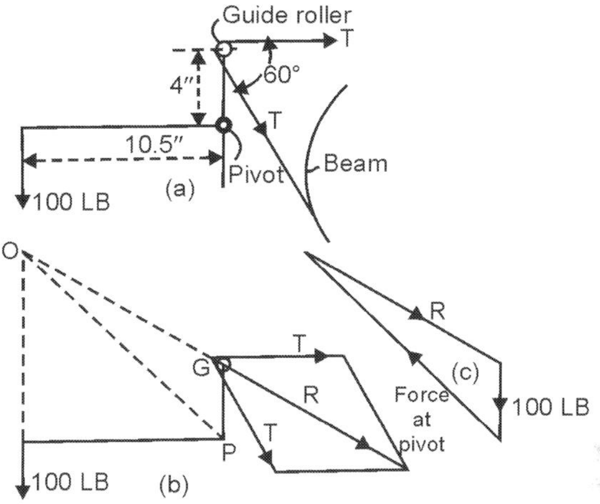

Figure 8.24

6. The under picking stick of a loom is shown in Fig. 8.25. The picking stick is under the action of three forces, viz. the shuttle resistance of 50 lb, the vertical force *P* applied by the side lever and the supporting force or reaction *R* at the fulcrum. Find the magnitude of *P* and magnitude and direction of *R*.

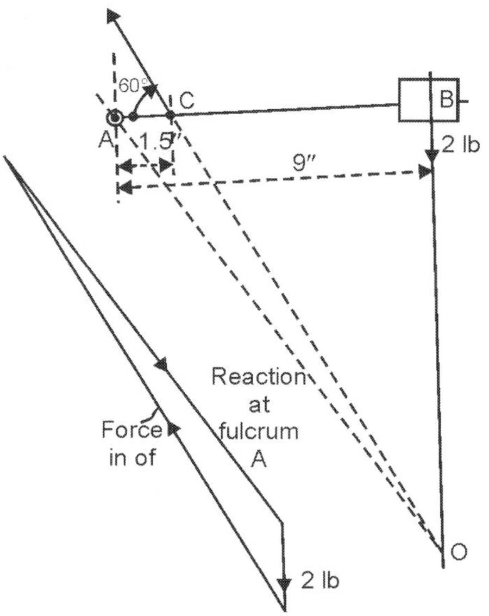

Figure 8.25

Ans. $P = 300$ lb; $R = 304$ lb

7. A line shaft bearing has to carry a vertical load of 200 lb including the weight of shafts and pulleys. In addition, load due to the pull of the belts, each of 100 lb are applied as shown in Fig. 8.26 making an angles 30° to the right and left of the vertical. All the forces are considered to be acting in the same plane. What is the total load carried by the bearings?

   Ans. 373 lb

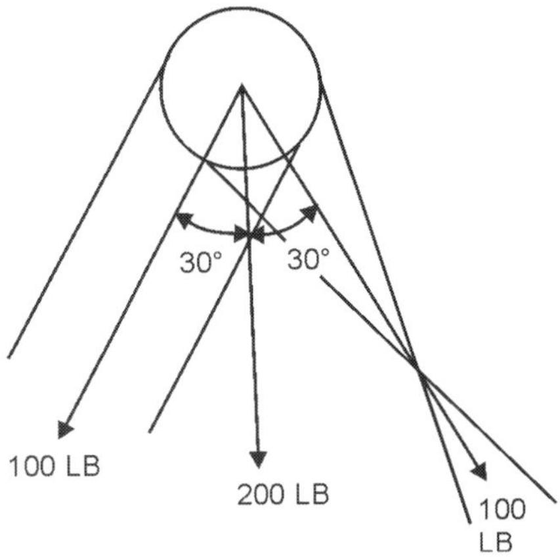

Figure 8.26

7. Give illustrations of triangle and polygon law of forces.
8. A string *OA* is attached to a fixed point *O* and carries a weight of 100 *N* at *A*. A horizontal force *X* is applied to the weight. If in the position of equilibrium *OA* is inclined to the vertical at 30°, find graphically the magnitude of *X* and the tension in the string.

   Ans. X = 57.8 N; tension = 115.5 N
9. Find the resultant of following system of forces acting at a point.

   40 N towards east

   100 N towards 30 degrees north of east

   150 N towards north

   200 N towards south-west

   Ans. 61 N inclined 75 degrees to north of west.

9
Centre of gravity

9.1 Introduction

Every particle of a body is acted upon by the force of gravity. The force of attraction which is proportional to the mass of the particle acts vertically downwards and is known as weight of the body. As the distance between the different particles of a body and the centre of the earth is the same, therefore these forces may be taken to act along parallel lines. This point through which the whole weight of the body acts irrespective of its position is known as centre of gravity, written as CG. It may be noted that everybody has one and only one centre of gravity.

The centre of gravity is therefore that point in a body where the resultant of the force of gravity acts. In other words, when a body is acted upon by gravity alone, the centre of gravity is the point upon which the body will balance; if supported at that point, the body will be in equilibrium.

Let us take some practical examples of centre of gravity.

A thin symmetrical plate of uniform material will have its centre in the centre; for instance, a square plate, a circular or a uniform rod of any section will have its centre of gravity at its middle point.

9.2 Centre of gravity of irregular bodies

In case of irregular bodies or bodies that are not homogenous in structure, the centre of gravity may be found by balancing the body on a knife edge support or other test, and in many cases by calculation or graphics.

Pulleys and other revolving bodies, no matter how carefully they are made, may have their centre of gravity out of the centre of the pulley. They are used to be out of centre and out of balance and liable to cause serious troubles while running. Such bodies have to be carefully balanced by adding to or removing some portions of the material until they will be at rest in any position when supported on their geometrical centre.

9.3 Centre of gravity of levers

If a uniform lever is supported at its centre, its weight will act directly over the support and will not affect the balance, but if such a lever is fulcrumed on one side of the centre of gravity, then the weight of the lever must be taken into account.

Example 1: A uniform lever 18 in. long is pivoted at a point 3 in. from one end. From the short arm hangs a weight of 36 lb. What weight in pounds will be required at the long arm so as to obtain equilibrium?

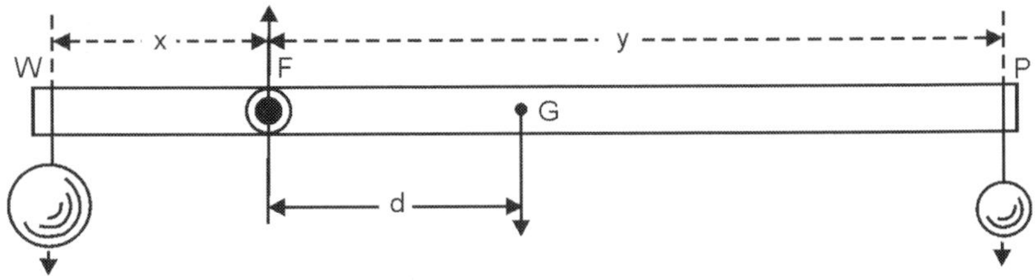

Figure 9.1

Solution:

The centre of gravity is in the centre of the lever, so the weight of the lever will act 6 in. from the fulcrum

The moments of W round F will be = W_X

The moments of P round F will be = P_Y

The moments of G round F will be = G_d

where F is the fulcrum of the lever.

Taking moments about the fulcrum of the lever,

$$\begin{aligned} P_y + G_d &= W_x \\ P \times 15 + 6 \times 6 &= 36 \times 3 \\ P \times 15 + 36 &= 108 \\ P &= 108 - 36 = 72 \\ P &= 72/15 = 4.8 \text{ lb} \end{aligned}$$

P must carry 4.8 lb to maintain equilibrium conditions of the lever.

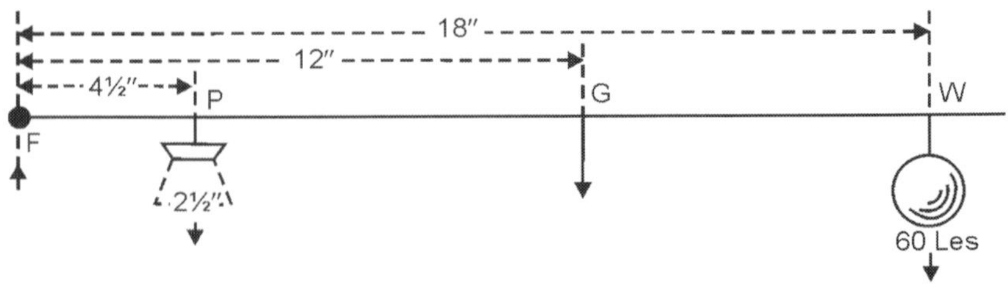

Figure 9.2

Example 2: A safety valve is of 2.5 in. diameter. A lever 22 in. long is pivoted 4.5 in. from the centre of the valve and a weight of 60 lb is hung 18 in. from the fulcrum. What is the total pressure on the valve, and the pressure per sq. in.? The weight of the lever is 6 lb and the centre of gravity acts at 12 in. from the fulcrum.

Solution:

The pressure P is the unknown quantity.

Taking moments about the fulcrum of the lever, we have

$$P \times FP = W \times WF + G \times GF$$

$$P \times 4.5 = 60 \times 18 + 6 \times 12$$

$$P \times 4.5 = 1080 + 72 = 1152 \text{ lb}$$

$$P = 1152/4.5$$

$$P = 256 \text{ lb, the total pressure on the valve}$$

$$\text{Area of the valve} = \pi \times r^2/4$$

$$= \pi \times 2.5^2/4$$

$$= 4.9 \text{ sq in.}$$

$$\text{Pressure per sq in.} = \text{Total pressure/ Area}$$

$$= 256/4.9$$

$$= 52.24 \text{ lb per sq in.}$$

Numerical questions

1. A long lever of 20 in. long is pivoted at 3 in. from one end. A weight of 6 lb is hung from the short arm and balances the lever. At what point in the lever does the centre of gravity act and what is the weight of the lever?

   Ans. Centre of gravity of the lever = 7 in; Weight of the lever = 2.57 lb

2. A uniform beam 11 ft long weighs 392 lb. Two tons are placed on the beam at a point 3 ft from one support. What are the reactions in the supports?

   Ans. Reaction at A = 1.44 tons; Reaction at B = 0.505 ton

10

Principle of moments

10.1 Moment of force

When a force is applied to a body, it has the tendency to turn the body about some point. The turning tendency of a force about a point is called the moment of the force about the point; it is measured by the product of the force and perpendicular distance of its line of action from the point. The length of the perpendicular from the point on the line of action of the force is called arm of the moment.

If the tendency of the rotation is clockwise, the moment is called clockwise; if the tendency of the moment is anti-clockwise, the moment is called anti-clockwise. In general, counter-clockwise moment is called positive.

10.2 Geometrical representation of moments

Let O be the point and let AB represent the force completely, i.e. in direction, magnitude and line of action. Draw OM perpendicular to AB and join OA and OB. Then moment of force AB about O

$= AB \times OM$

= twice the area of the triangle OAB numerically.

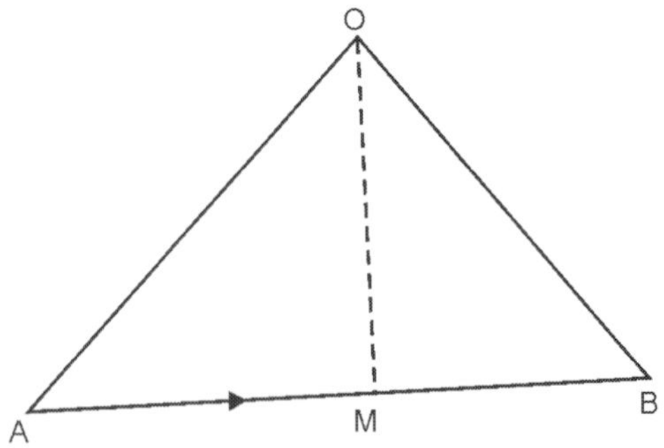

Fig. 10.1

10.3 Types of moments

Broadly speaking, the moments are of the following two types.

1. Clockwise moment
2. Anti-clockwise moment

10.4 Clockwise moments

It is the moment of a force whose effect is to turn or rotate the body, about the point in the same direction in which hands of a clock move.

10.5 Anti-clockwise moments

It is the moment of a force whose effect is to turn or rotate the body, about the point in the opposite direction in which the hands of a clock move.

Note: The general convention is to take the clockwise moments as positive and anti-clockwise moments negative.

10.6 Varignon's Principle of Moments or Law of Moments

It states, "If a number of coplanar forces are acting simultaneously on a particle, the algebraic sum of the moments of all the forces about any point is equal to the moment of their resultant force about the same point."

10.6.1 Parallel forces

Two parallel forces are said to be like when they act in the same sense; they are said to be alike when they act in the opposite sense.

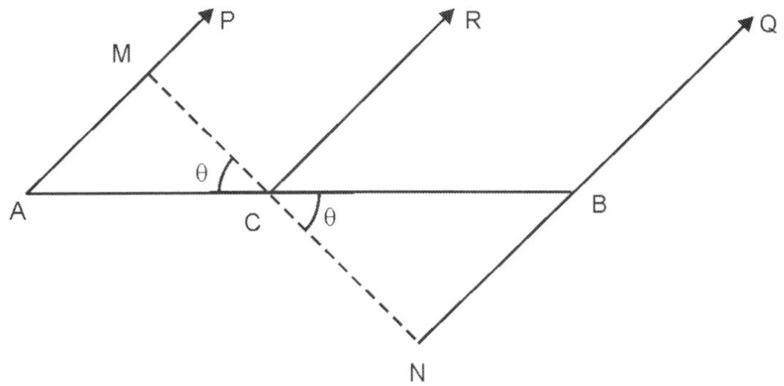

Fig. 10.2

(a) Resultant of two like parallel forces.

Let the like parallel forces P and Q act at points A and B, and their resultant R cut AB at C. Let R be inclined to P and Q at angle α.

By resolving parallel and perpendicular to P, we get

$$R \cos \alpha = P = Q; R \sin \alpha = 0.$$

$$\alpha = 0 \text{ and } R = P + Q$$

Therefore, R is parallel to P and Q.

The moments of R about C is zero, so that the algebraic sum of the moments about P and Q about C must be zero.

Through C draw MN perpendicular to P and Q.

Then
$$CM = Ac \cos \theta$$
$$CN = BC \cos \theta$$

Taking moments about C

$$P \times CM = Q \times CN$$

or
$$P \times AC \cos \theta = Q \times BC \cos \theta$$
$$P \times AC = Q \times BC$$

or
$$P/BC = Q/AC$$

Therefore C divides AB internally in the inverse ratio of the forces.

(b) Resultant of two unlike parallel forces.

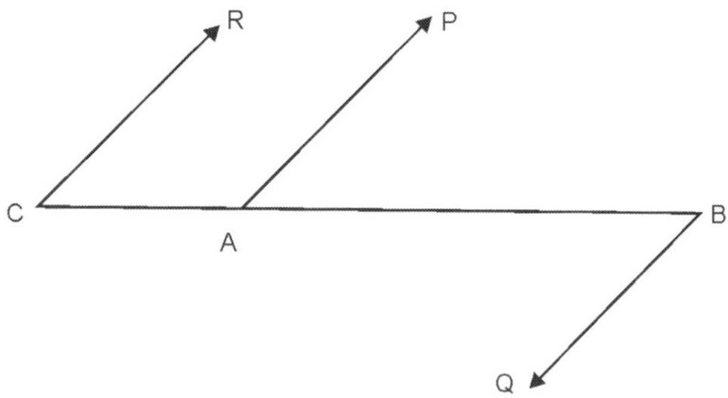

Fig. 10.3

Let the two unlike parallel forces P and Q act at A and B, and their resultant R meet AB at C.

Let P be greater than Q.

As shown in part (a), R is parallel to I or Q acting in the same sense as P.

Hence, the algebraic sum of the moments of P and Q about C must be zero so that these moments must be equal and opposite. Hence C must be outside AB, and it must be nearer to A than to B.

By taking moments about C, we get

$$P \times AC = Q \times BC$$
$$P/BC = Q/AC$$

Here C divides AB externally in the inverse ratio of the forces.

The point C is called the centre of parallel forces. Also it is clear that the position of C is independent of the inclination of the forces to AB.

Example 1: A uniform rod AB of weight 5 kg and length 100 cm is simply supported at A and B in a horizontal position. It carries a uniformly distributed load of 50 g/cm over a distance of 10 cm from A and a concentrated load of 200 g at a distance of 30 cm from A. Find the reactions at A and B.

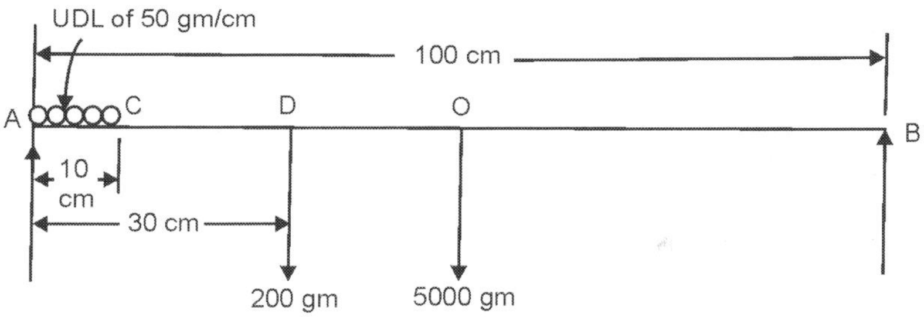

Fig. 10.4

Solution:

Let O be the middle point of AB and $AC = 10$ cm, $AD = 30$ cm.

The weight of the rod is 5 kg or 5000 g acting at O, where $AO = 50$ cm.

The load of 200 g is placed at D.

The uniformly distributed load extends from A to C.

The total value of the load is $50 \times 10 = 500$ g and its centre of gravity is at the midpoint of AC, i.e. at a distance of 5 cm from A, so that its moment about $A = 500 \times 5$ g-cm.

Let the reactions at A and B be RA and RB.

The forces on the rod are equilibrium. Hence the algebraic sum of the moments about any point, say A, in their plane must be zero. In other words, the clockwise moments about A must balance the anti-clockwise moments.

$$RB \times 100 = 500 \times 5 + 200 \times 30 + 5000 \times 50$$

$$RB = 2585 \text{ g-wt or } 2.585 \text{ kgf}$$

Similarly, the algebraic sum of the moments of the forces about B must vanish.

Distance of O from $B = 50$ cm

Distance of D from $B = 100 - 30 = 70$ cm

Distance of middle point of AC from $B = 100 - 5 = 95$ cm.

Taking moments about B,

$$RA \times 100 = 5000 \times 50 + 200 \times 70 + 500 \times 95$$

$$RA = 3115 \text{ g-wt or } 3.115 \text{ kgf}$$

Example 2: A light rod AB 200 mm long rests on two pegs whose distance apart is 100 mm. How must it be placed so that the reactions of the pegs may be equal when weights of $2W$ and $3W$ are suspended from A and B?

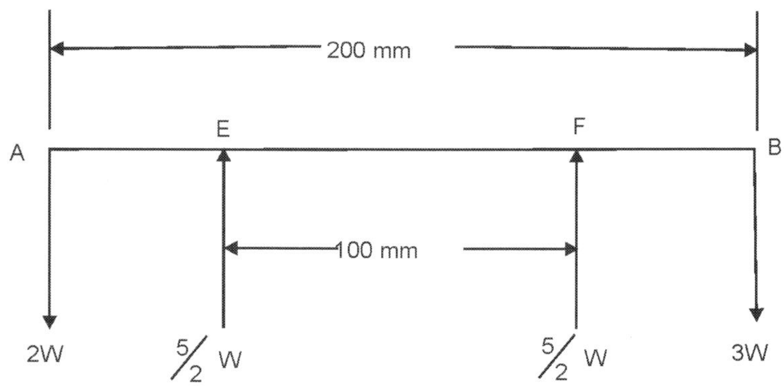

Fig. 10.5

Solution:

Let the pegs be E and F

Since the reactions at E and F are equal, each reaction

$$= \tfrac{1}{2}(2W + 3W) = 5W/2$$

Let $AE = x$

Then $BE = 200 - x$

The rod being in equilibrium, the algebraic sum of the moments of the forces about E must be zero.

The moments about E of $2W$ at A is $2Wx$ anti-clockwise.

The moment of the reaction at F about E is $5W/2 \times 100 = 250\ W$ anti-clockwise

Total anti-clockwise moments = $2Wx + 250W$

The moment of $3W$ at B about E is $3W(200 - x)$ clockwise.

The moment of reaction at E about E is zero.

Equating the clockwise moments and anti-clockwise moments

$$3W(200 - x) = 2Wx + 250W$$

Solving $\quad x = 70$ mm

$$AE = 70 \text{ mm}, BF$$
$$= 100 - 70 = 30 \text{ mm}$$

Example 3: A beam of 12 ft long is supported at both ends. If a weight of 5 tons is placed 4 ft from one end, what is the pressure on each support? Neglect the weight of the beam.

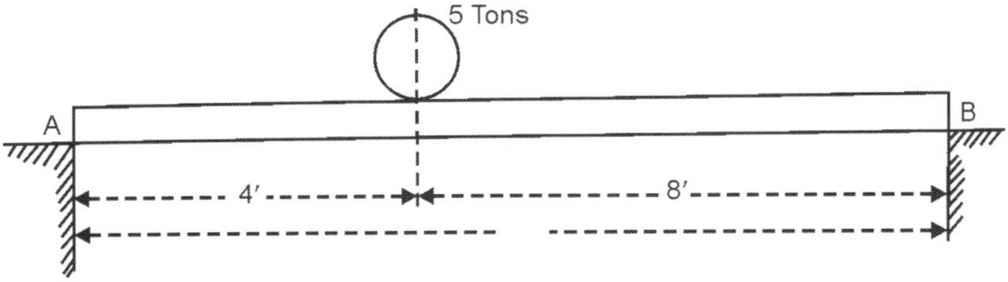

Fig. 10.6

Solution:

(a) Let us find the pressure on support A and x = the pressure.

Then $5 \times 8 - x * 12 = 0$

$X = 5 \times 8/12 = 3.33$ tons

Pressure on A = 3.36 tons

Let us find the pressure on B, X = the pressure

Then $5 \times 4 - x * 12 = 0$

$X = 5 \times 4/12 = 1.66$ tons

The second calculation is unnecessary since the total load on the beam

$$P = 5 \text{ tons}$$
$$PA + PB = 5 \text{ tons}$$
$$PA = 3.34 \text{ tons}$$
$$PB = 5 - \text{PA}$$
$$= 5 - 3.34 = 1.66 \text{ tons}$$

Example 4: A beam is 16 ft long (neglect beam weight) between its supports. Weights of 5, 3 and 7 tons are placed so that 5 tons is 4 ft from the left-hand support, the 3 tons is 6 ft from the left-hand support, and the 7 tons is 7 ft from the right-hand support. What is the pressure on each support?

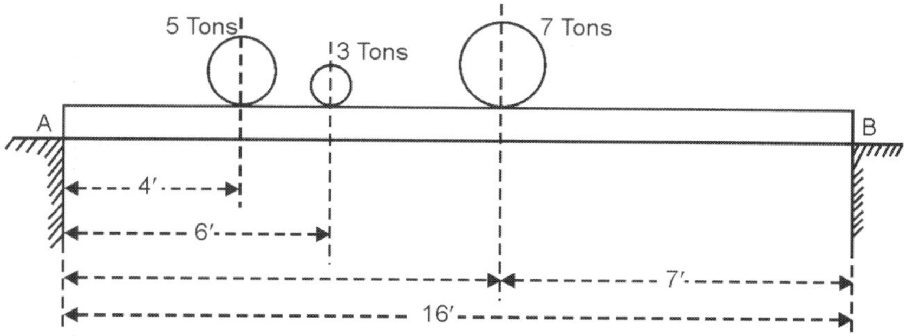

Fig. 10.7

Solution:

Let us first find the pressure on weight A. Let x = the pressure. This means that we must find the moments of the forces about B. As the pressure on A acts upwards, it will tend to turn the beam clockwise around B, so it will be positive (+). The three weights will tend to turn the beam anti-clockwise round B, so that the moments will be negative (−).

$$\begin{aligned}
\text{Moments of } A \text{ will be} &= x * 16 \text{ ft} = 16x \text{ ft tons} \\
\text{Moment of 5 tons} &= 5 * 12 = 69 \text{ ft tons} \\
\text{Moment of 3 tons} &= 3 * 10 = 30 \text{ ft tons} \\
\text{Moment of 7 tons} &= 7 * 7 = 49 \text{ ft tons} \\
16x - 60 - 30 - 49 &= 0 \\
16x &= 139 \\
X &= 8.68 \text{ tons} \\
\text{Pressure on } A &= 8.68 \text{ tons} \\
\text{So the pressure on } B &= 15 - 8.68 = 6.31 \text{ tons}
\end{aligned}$$

Example 5: Four calendar rollers of a lap are weighted as shown in Fig. 10.8. The top lever FW is 14 in. long, FP is 3 in. The bottom lever FW is 60 is long, FP is 6 in. long. What pressure is exerted at P, and what is the pressure on the cotton as it passes between each pair of rollers if the top roller weighs 60 lb, the second roller 65 lb, and the third roller 72 lb? The weight of the bottom lever is 12 lb. Both sides of the machine are provided with levers and weights.

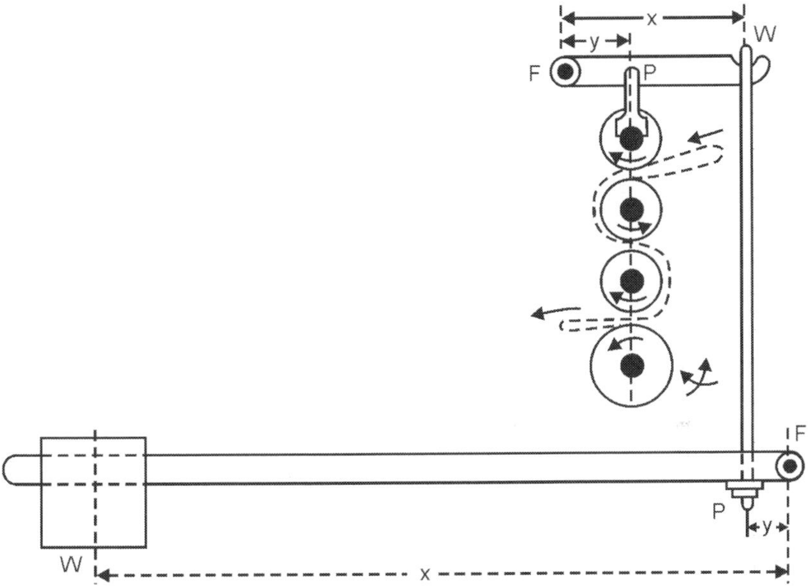

Fig. 10.8

Solution:

In the bottom lever, the pressure P can be found as follows.

$$P \times y = W \times x$$
$$P = W \times x/y$$
$$= 12 \times 60/\ 6$$
$$P = 120 \text{ lb}$$

In the top lever, the weight W is the same as the pressure P on the bottom lever.

Hence,

$$W \times x = P \times y$$
$$P = W \times x/y$$
$$= 120 \times 14/3$$
$$= 560 \text{ lb}$$

So the total pressure is 560 × 2 = 1120 lb

The pressure on the top roller is 1120 lb

The pressure between top pair of rollers = 1120 + 60 = 1180 lb

The pressure between second pair of rollers = 1180 + 65 = 1245 lb

The pressure between bottom pair of rollers = 1245 + 72 = 1317 lb

Example 6: The calendar rollers of the lap end of an opener are weighted as shown in Fig. 10.9. The lever is 72 in. long and the pressure is applied 3 in. from the fulcrum. If a weight of 22 lb is placed 66 in. from the fulcrum, what pressure is put on the cotton between each pair of rollers if the top roller weighs 70 lb and the second roller 78 lb and the third roller 82 lb? Separate levers act on each end of the lever.

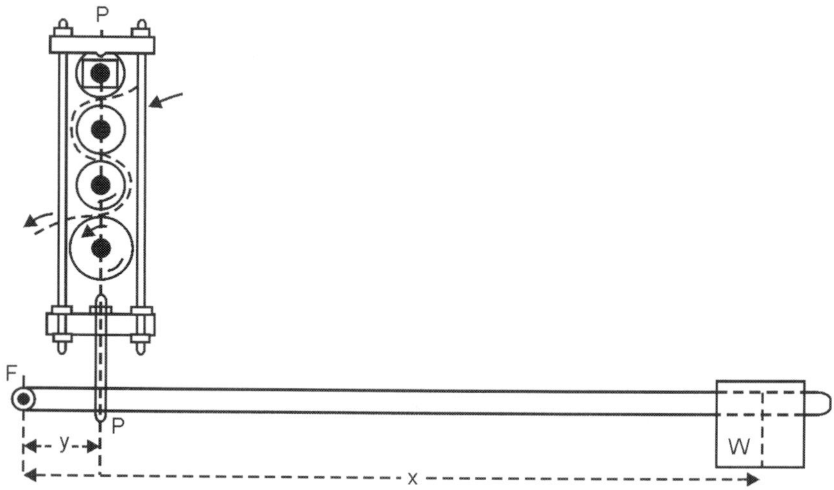

Fig. 10.9

Solution:

$$x = 66 \text{ in.}, y = 3 \text{ in.}$$
$$W = 22 \text{ lb}$$
$$W * x = P * y$$
$$P = W * x/y$$
$$= 22 * 66/3$$
$$P = 484 \text{ lb}$$

Therefore $P = 2 \times 484$ lb $= 968$ lb, which is the total pressure on the roller.

The pressure between the top pair of rollers = 968 + 70 = 1038 lb

The pressure between the second pair of rollers = 1038 + 78 = 1116 lb

The pressure between the third pair of rollers = 1116 + 82 = 1198 lb

This type of application of pressures on Scutcher blowroom line is common in modern machineries. The purpose of this lever arrangement is to consolidate the opened and fluffy cotton in to a lap form and can readily be unrolled without the surfaces of the cotton adhering to each other.

10.7 Moments on bent lever

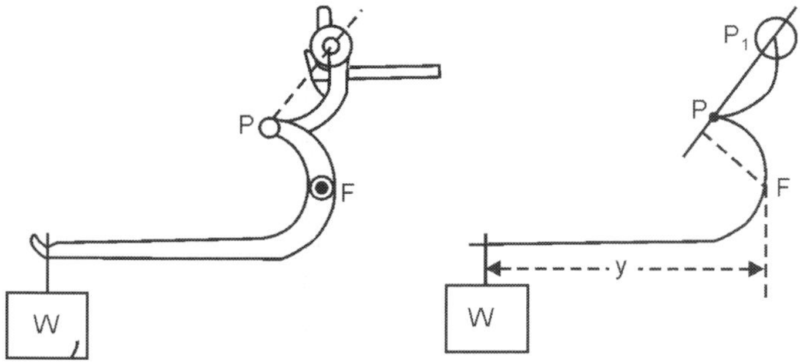

Fig. 10.10

Example 7: The dish feed roller on the carding machine is weighted as shown in Fig. 10.10. The distance $x = 12$ in. and the distance $y = 2$ in. The weight $W = 13$ lb. Calculate the pressure exerted on the feed roller.

Solution:

This is an interesting practical situation that exists in conventional carding machine. By joining P and P', we get the direction in which the pressure acts. The actual connection is made by a curved link.

The distance (x) is the perpendicular distance to the fulcrum in which the weight acts. The distance (y) is the perpendicular distance of the fulcrum in which the pressure acts.

$$W * x = P * y$$
$$P = W * x/y$$
$$= 13 * 12/2$$
$$P = 78 \text{ lb}$$

10.8 Moments of bent lever in weighting arrangement

Figure 10.11 is an illustration of dead weighting arrangement common in some textile machines.

From Fig. 10.11 it is clear that the pressure on the cotton between the pair of rollers is the weight W. On the other hand, this may not be practically possible and the arrangement of dead weight is shown in Fig. 10.11. The pressure on the material between A and B will be practically equal to the weight W as the tension in the connecting chain from W to E is uniform.

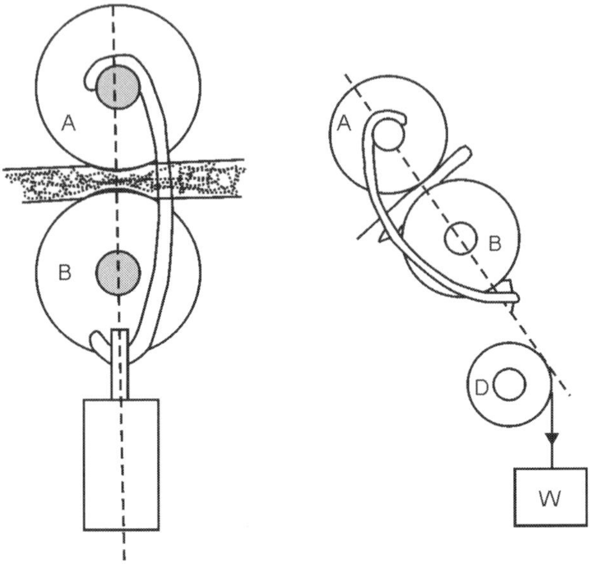

Fig. 10.11

Another type of weighting arrangement is shown in Fig. 10.12. The weight *W* acts vertically, but the pressure along the line joining the centers of *A* and *B*. Figure 10.12 will show the method of finding the pressure. Draw *W* to scale equal to half the weight of W as in Fig. 10.12 and from one end draw *P* parallel to the pressure line. From the other end of *W*, draw a line at right angles to *P*, which will equal the pressure between the rollers. This type of weighting is common in textile spinning machines.

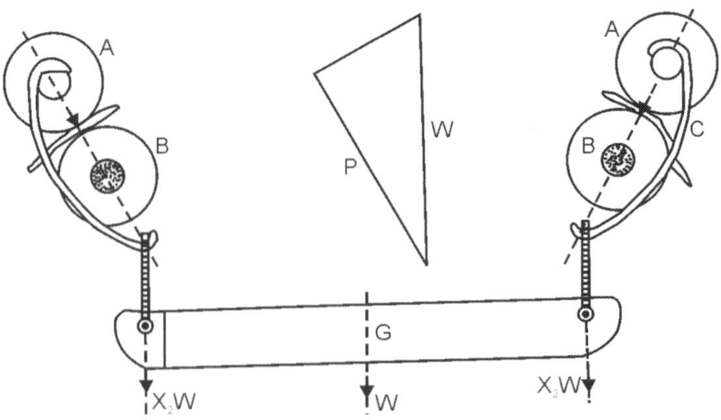

Fig. 10.12

10.9 Bent lever in weighting of single yarns

Yarns from the woven fabrics are unraveled one by one and the weight of the yarns are determined by this method.

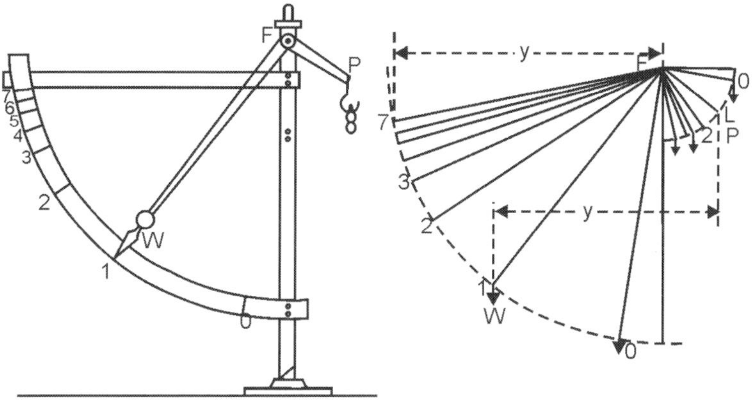

Fig. 10.13

It consists of a bent lever fulcrumed at *F*, one end of arm carrying a pivoted hook or pan on which the yarn is placed; the other arm is weighted and acts as a pointer on a circular scale. This scale is usually a quadrant of a circle, and is divided so that the weight of the yarn can be read off.

When the apparatus is in a state of equilibrium, the long arm pointer is at zero on the scale; this point being some little distance to the left of the strand. A weight in the form of yarn is placed on the hook. This depresses the arm *FP* and raises the pointer *FW*. The effect has been to alter the leverages or moments of the arms; for when the arms come to rest, it is seen that the weight in *P* is nearer to the vertical line through *F*, at the same time *W* is farther away. As *P* increases, its moment decreases and whilst the moment of *W* increase. *W* is a constant quantity. The result of these changes in the moments of arms is to cause the scale to have unequal divisions; the readings get contracting as they ascend the scale. Such scales are graduated by using trial standard weights being used for this purpose.

10.10 Bent lever on Lea strength tester

The Lea strength testing machine is most commonly used testing equipment in spinning and weaving mills to determine the strength of group of threads. A lea, termed as 120 yards, which comprises 80 threads is in the form of an endless loop. It is hung on the fixed top jaw and the other end of the loop is hung on the movable lower jaw. A force or pressure P is applied at the lower movable jaw and strength of the lea is determined when the lea breaks. The strength can be noted in the circular dial graduated in pounds (lb).

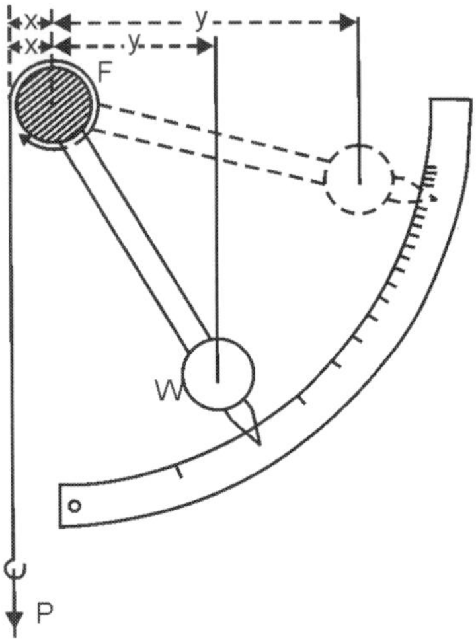

Fig. 10.14

The Lea Strength Tester is also an example of bent lever type. A chain or cord is attached to a drum; the other end of this carries a hook to which the yarn is attached. On the axle of the drum is fixed a weighted arm or pointer W. As the yarn is pulled at P, the drum F is turned and the pointer rises along a quadrant scale which is graduated to indicate the force P in lbs.

In this example, the pull of the yarn has a constant leverage or moment, whilst the weight W has a varying leverage as it moves upwards.

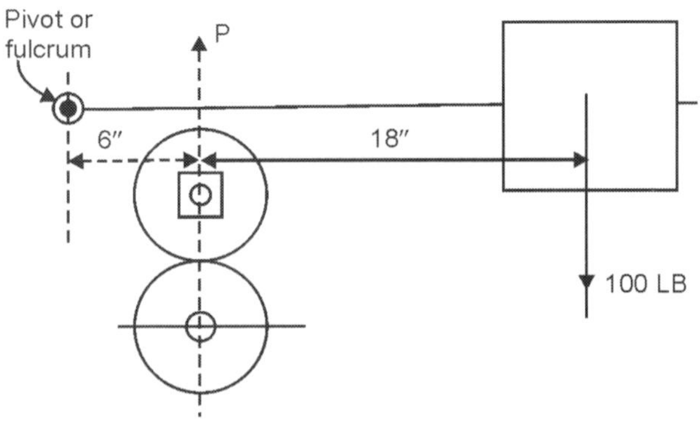

Fig. 10.15

Example 8: Figure 10.15 shows the method of applying pressure to the pair of squeezing rollers in a sizing machine. If similar arrangement is applied to each end of the lever which itself weighs 56 lb, what is the pressure applied to the material passing between the rollers? Ignore the weight of the lever.

Solution:

Taking moments about the fulcrum, we have

$$P \times 6 = 100 \times 24$$

$$P = 2400/6 = 400 \text{ lb}$$

$$\text{Total pressure } P = (2 \times 400) + 56 = 856 \text{ lb}$$

Total pressure applied on the material = 856 lb

Example 9: Figure 10.16 shows the method of applying weighting to a pair of drawing rollers. Ignoring the weight of the lever, links and saddles, find the pressure applied at *A* and *B*.

Solution:

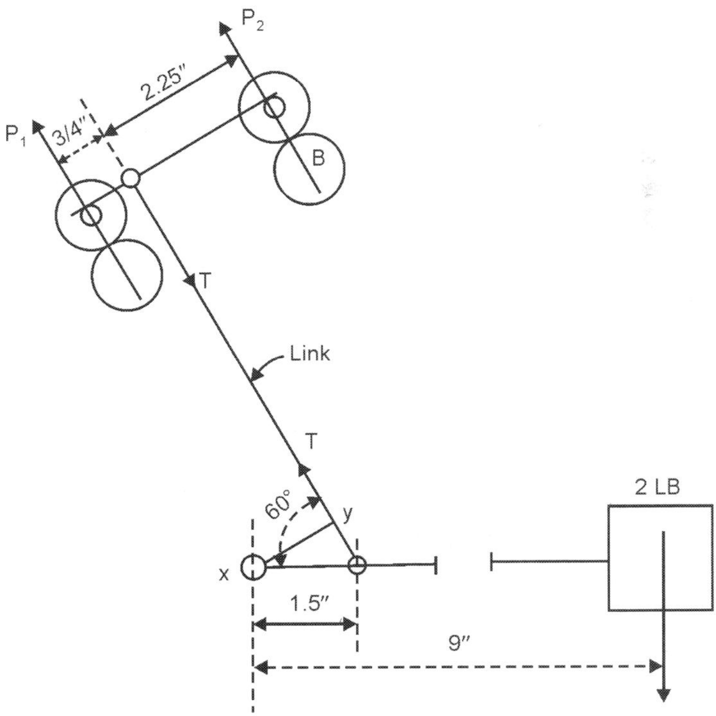

Fig. 10.16

First consider the lever. It is kept in equilibrium condition by the weight, an upward supporting force T in the link and the force at the fulcrum.

Taking moments about lever fulcrum,

$$T * xy = 2 * 9$$

xy is perpendicular distance from the fulcrum to the centre line of the link.

$$xy = 1.5 \sin 60°$$

$$T \times 1.3 = 18$$

$$\text{From which } T = 18/1.3$$

$$= 13.8 \text{ lb}$$

Now the equilibrium of the saddle is to be considered. The saddle is kept in equilibrium by the downward force of T applied by the link and supporting forces and from the roller acting at right angles to the saddle.

Taking moments about the centre of the top roller,

$$P_2 \times 3 = \mathrm{T} \times 3/4$$

$$P_2 = 13.8 \times 3/4/3$$

$$P_2 = 3.45 \text{ lb}$$

Taking moments about

$$P_1 \times 3 = T \times 2.25$$

$$P_1 = 13.8 \times 2.25/3$$

$$P_1 = 10.34 \text{ lb}$$

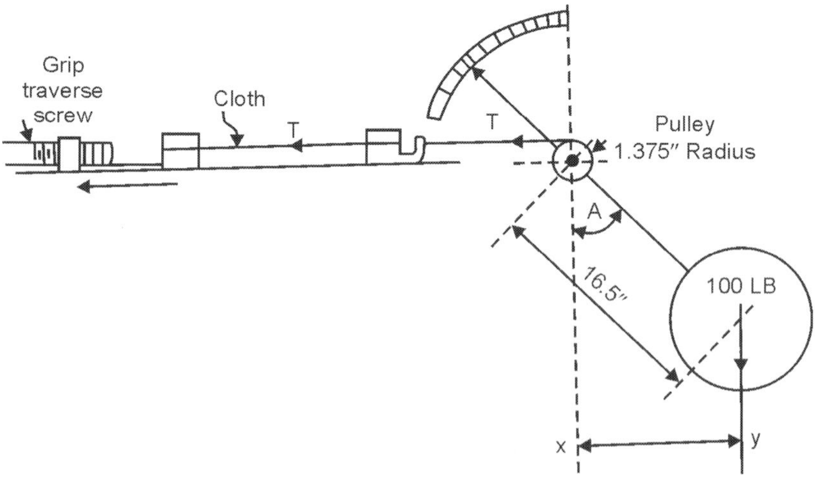

Fig. 10.17

Alternately, if one of the pressures is known, the other pressure can be found out by subtraction.

$$P_1 + P_2 = T\,(13.8 \text{ lb})$$

If $$P_1 = 0.34 \text{ lb, then } P_2 = 13.8 - 10.34 = 3.4 \text{ lb}$$

Example 10: Figure 10.17 is an illustration of applying tension to a specimen in a cloth testing machine. Find the tension on the machine when the angle A is (a) 30 degrees and (b) 60 degrees.

Solution:

Taking moments about the lever fulcrum, we get

(a) 30 degrees

$$T \times 1.375 = W \times xy$$
$$xy = 16.5 \times \sin 30°$$
$$T \times 1.375 = 100 \times 16.5 \sin 30°$$
$$T = 100 \times 8.25/1.375$$
$$T = 600 \text{ lb}$$

(b) 60 degrees

$$T \times 1.375 = W \times 16.5 \sin 60°$$
$$T = 100 \times 16.5 \sin 30/1.375$$
$$T = 100 \times 14.28/1.375$$
$$T = 1039.23 \text{ lb}$$

Example 11: A horizontal weighing bar for a loom has its ends attached to the two beam weighting ropes, 40 in. apart. The bar weighs 10 lb and carries a weight of 50 lb, 6 in. from the left hand rope and a 30 lb weight 10 in. from the right hand rope. Find out the force applied on each of the weighting ropes (Fig. 10.18).

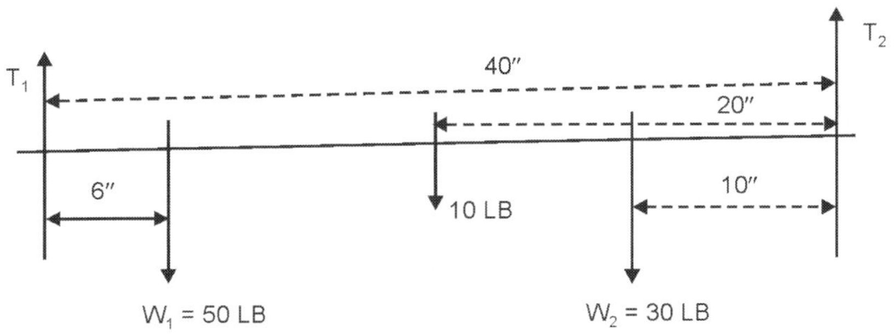

Fig. 10.18

Let T_1 and T_2 be the tensions of the ropes on right and left hand side of the beam weighting mechanism.

Taking moments on the right hand rope, we get

$$T_1 = 2200/40$$
$$= 55 \text{ lb}$$
$$T_1 + T_2 = 50 + 10 + 30$$
$$= 90 \text{ lb}$$
$$T_2 = 90 - 55$$
$$= 35 \text{ lb}$$

T_2 can also be got by taking moments about the left hand rope. It should be remembered that in a system of parallel forces the total downward force must be equal to the total upward force.

It should also be noted that the weight of the bar is 10 lb and is acting at the centre of the bar. If the distance from the fulcrum point is given, then centre of gravity and weight of the bar must also be considered.

Example 12: In a loom taking-up motion, find the force required at the rim of the ratchet wheel to overcome a cloth tension T of the take-up roller.

Solution:

(i) Let us consider first the take-up roller. It is kept in equilibrium by the cloth tension T and the force P applied by the teeth of the swing pinion 15 teeth. Taking moments about the centre of the beam wheel,

$$P \times 6 = 200 \times 2.5$$
$$P = 83.33 \text{ lb}$$

(ii) Next we will consider the swing wheel and the pinion. These are kept in equilibrium by the force P between the teeth of the swing pinion and the beam wheel, and between change wheel and swing wheel.

Taking moments about the centre of the swing wheel,

$$P_1 \times 7.5/2 = P \times 1.5/2$$
$$P_1 \times 3.75 = 83.33 \times 0.75$$
$$P_1 = 16.66 \text{ lb}$$

(iii) Finally, taking moments about the centre of the ratchet wheel,

$$F \times 6/2 = P_1 \times 3/2$$
$$F \times 3 = 16.66 \times 1.5$$
$$F = 16.66 \times 1.5/3$$
$$F = 8.33 \text{ lb}$$

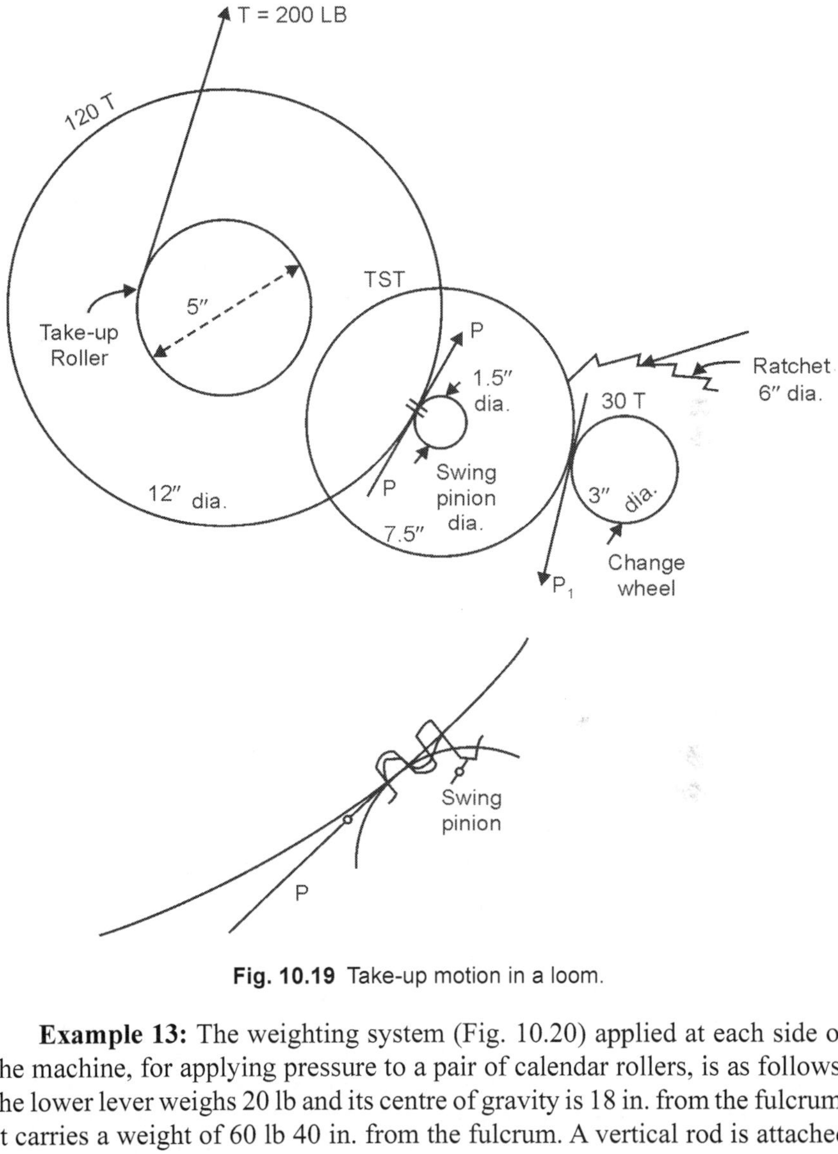

Fig. 10.19 Take-up motion in a loom.

Example 13: The weighting system (Fig. 10.20) applied at each side of the machine, for applying pressure to a pair of calendar rollers, is as follows: the lower lever weighs 20 lb and its centre of gravity is 18 in. from the fulcrum. It carries a weight of 60 lb 40 in. from the fulcrum. A vertical rod is attached to the lower lever, 3 in. from the fulcrum, and its top end is attached to one end of the top lever, 8 in. from the fulcrum at the other end. The top lever, also horizontal, rests on the bush of the top roller. The distance from the centre of the bush to lever fulcrum is 2.5 in. If the weight of the roller is 200 lb, what pressure is applied by the top roller to the cotton material passing between the rollers. Neglect the weight of the top lever.

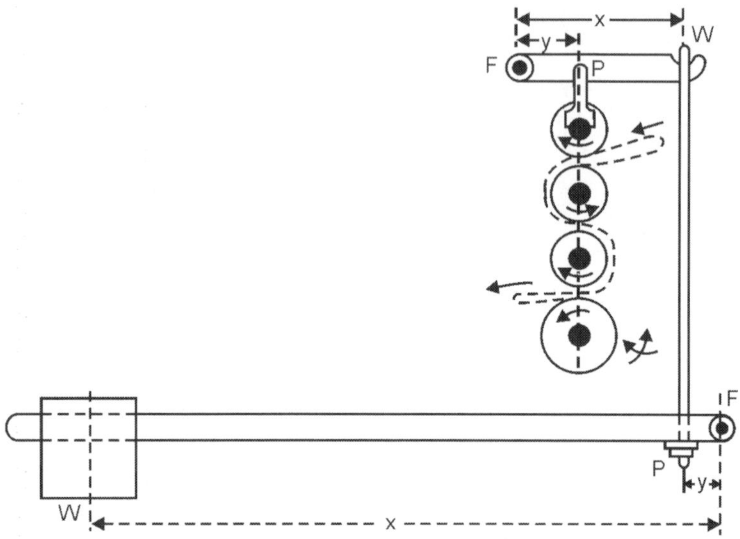

Fig. 10.20

Solution:

In the bottom lever, taking moments about the fulcrum, the pressure P is found as follows.

$P \times y = W * x + G \times z$ where G is the weight of the lever and z is the centre of gravity of the lever from the fulcrum.

$P \times 3 = 60 \times 40 + 20 \times 18; P = 2400 + 360/3$

$$P = 920 \text{ lb}$$

In the top lever, the weight W is the same as the pressure P on the bottom lever, so that pressure $P = 920$ lb which is the weight (W) at the top lever.

Again, taking moments about the lever fulcrum of the top lever, we get

$$W * x = P \times y$$

$$W \times 8 = \text{P} \times 2.5$$

$$920 \times 8 = \text{P} \times 2.5$$

From which, $P = 2944$ lb

Total pressure P on the top lever = 2 × 2944 lb = 5888 lb

Weight of the top roller = 200 lb

Total load acting on the cotton material passing between the rollers is

$$5888 + 200 = 6088 \text{ lb}$$

Example 14: Figure 10.21 shows the method of weighting the front and middle lines of rollers in a ring frame, the roller stand being inclined at an angle of 30° to the horizontal with the link at right angles to the saddle. The weight of the lever is 0.5 lb and its centre of gravity is 3 in. from the fulcrum. Ignore the weight of the saddle and the link. Find the pressure applied by the saddle to the front and middle line of rollers.

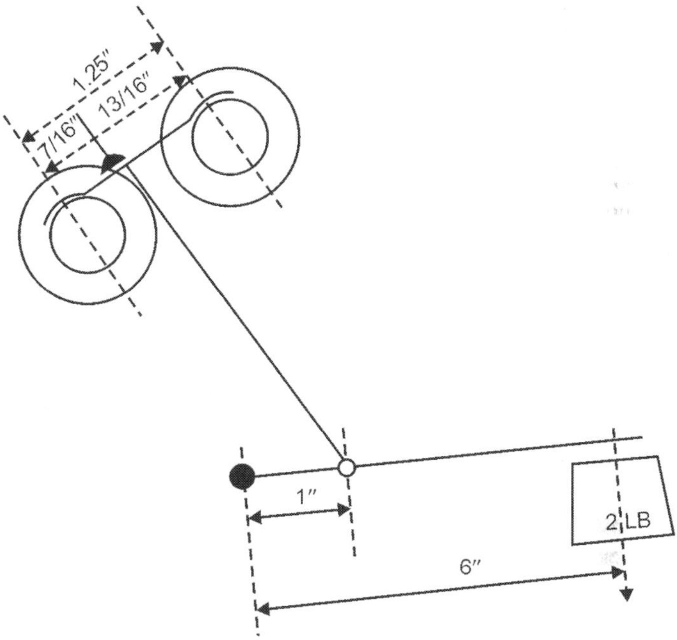

Fig. 10.21

Solution:

First consider the bottom lever. It is kept in equilibrium conditions by the weight, an upward supporting force t in the link, and the force at the fulcrum. The force T in the link, which has pin joints at its ends, must act along the length of the link.

Please note that the moment of t is not $T * x\,y$, but $T * xy$ sin 60°

Taking moments about the lever fulcrum O, we get

$$T \times xy \sin 60° = W \times 6 + G \times 3$$

$T \times ¾ \sin 60° = (2 \times 6) + (0.5 \times 3)$ since the weight of the bottom lever is 0.5 lb and its centre of gravity acts 3 in. from the fulcrum O.

$$T \times 0.6495 = 12 + 1.5$$

$$T = 13.5/0.6495$$
$$T = 20.78 \text{ lb}$$

T is the force acting on the link which applies load to the saddle.

Next, consider the saddle. It is kept in equilibrium conditions by the downward force T applied by the link and supporting forces and from the rollers acting at right angles to the saddle.

Taking moments about the centre of the front top roller,

Let the front roller be and the middle roller be .

$$P_A \times 1.25 = T \times 13/16$$
$$P_A = 20.78 \times 0.8125/1.25$$
$$P_A = 13.50 \text{ lb}$$

Taking moments about PB

$$P_B \times 1.25 = T \times 7/16$$
$$P_B = 20.78 \times 0.4375/1.25$$
$$P_B = 7.27 \text{ lb}$$

or
$$P_A + P_B = T$$
$$P_A = 13.50 \text{ lb}$$
$$P_B = T - P_B$$
$$P_B = 20.78 - 13.50$$
$$= 7.27 \text{ lb}$$

Example 15: Figure 10.22 is an illustration of saddle weighting system applied on the three lines of rollers in a conventional spinning machine. Ignoring the weights of levers, links and saddle, find the pressures applied to the rollers A, B and C.

Solution:

First consider the bottom lever. It is kept in equilibrium conditions by the weight, an upward supporting force t in the link, and the force at the fulcrum. The force T in the link, which has pin joints at its ends, must act along the length of the link.

Now taking moments about the lever fulcrum O,

$$W \times y = T \times x$$
$$4 \times 7.5 = T \times 0.75$$
$$T = 30/0.75$$
$$T = 40 \text{ lb}$$

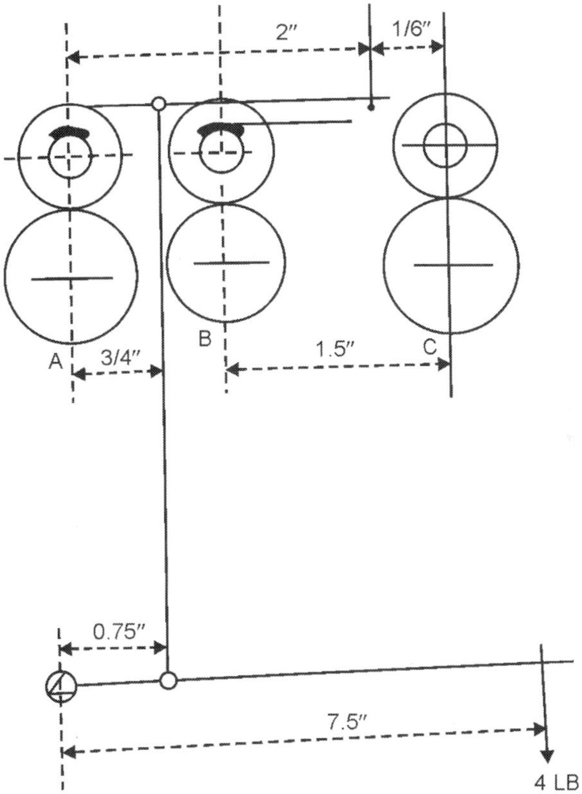

Fig. 10.22

The force T is acting on the saddle. Taking moments about A,

$$P_A \times 2.166 = T \times 1.5$$
$$P_A = 40 \times 1.5/2.166$$
$$P_A = 27.6 \text{ lb}$$

Taking moments about B,

$$P_B \times 1.5 = T \times 0.166$$
$$P_B = 40 \times 0.166/1.5$$
$$P_B = 4.44 \text{ lb}$$

We know that $P_A + P_B + P_C = T$

$$27.6 + 4.44 + P_C = 40$$
$$P_C = T - P_A + P_B$$
$$= 40 - 27.6 + 4.44 = 7.96 \text{ lb}$$

Example 16: In Example 15 shown above, a vertical lifting force of 3.6 lb applied at the end of the lever 10 in. from the fulcrum is just sufficient to lift the weighted lever and the link. If each saddle weighs 0.5 lb and has its centre of gravity at its centre, find the pressures applied to the rollers *A*, *B* and *C*.

Solution:

Vertical lifting force = 3.6 lb

Saddle weight = 0.5 lb for each saddle

Taking moments about the lever fulcrum *O*, we get

$$F * x = 3.6 \times 10 + (0.5 \text{ lb} \times 2)$$
$$= T \times 0.75$$
$$T = 37 / 0.75$$
$$T = 49.33 \text{ lb}$$

This tension (*T*) of 49.33 is applied to the rollers.

Now we have to find the pressures P_A, P_B and P_C

Taking moments about P_A, we get

$$P_A \times 2.166 \text{ in.} = T \times 1.5 \text{ in.}$$
$$P_A = 49.33 \times 1.5/2.166$$
$$P_A = 34.10 \text{ lb}$$

Now we have to calculate the pressure P_B,

Taking moments about we get

$$P_B \times 1.5 \text{ in.} = T \times 0.166 \text{ in.}$$
$$P_B = 49.33 \times 0.166 \text{ in.}/ 1.5$$
$$P_B = 5.45 \text{ lb}$$
$$P_A + P_B + P_C = T$$
$$P_C = T - (P_A + P_B)$$
$$P_C = 49.33 - (34.10 + 5.45)$$
$$P_C = 9.77 \text{ lb}$$

Example 17: A loom beam 15 in. diameter with 5 in. diameter ruffles at the point of turning to let-off the warp is in equilibrium conditions under the action of forces as shown in Fig. 10.23., the weight 120 lb, warp tension *T*, weighting chain tensions, tight side 400 lb, slack side 100 lb, and the reaction at the gudgeon pins. Find the warp tension. All the forces are acting in one plane. If the beam diameter is reduced to 5 in. and its weight 20 lb, what chain tensions are required to keep the warp tension same as before if the tight side tension remains 4 times greater than the slack side tension?

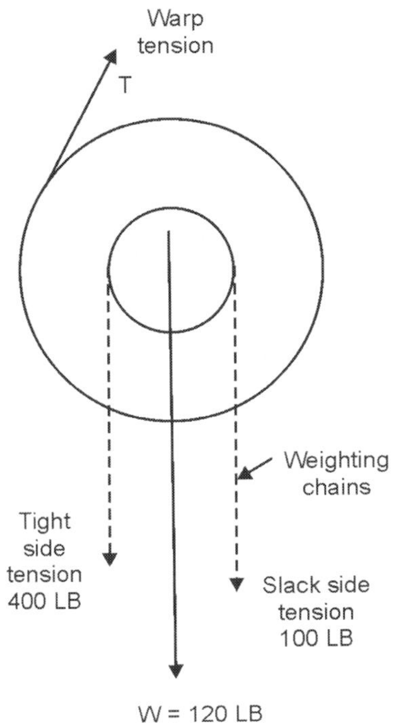

Fig. 10.23

Solution:

Weight (W) = 120 lb

Tight side chain tension (T_t) = 400 lb

Slack side chain tension (T_s) = 100 lb

Beam diameter = 15 in.

Ruffle diameter = 5 in.

Now we have to calculate the warp tension (T).

Taking moments about the beam centre, we have

$$T \times 15 = (T_s + W) \times 5 - (T_s + W) \times 5$$

$$T \times 15 = (400 + 120) \times 5 - (100 + 120) \times 5$$

$$T = 1500 / 15$$

$$= 100 \text{ lb}$$

We know that tight side tension (T_t) is 4 times the slack side tension (T_s)

Therefore $$(T_t - T_s) = T$$

$$T_t = 4T_s$$

$$(4T_s - T_s) = T$$
$$3T_s = 100 \text{ lb}$$
$$T_s = 100/3$$
$$= 33.3 \text{ lb}$$
$$T_t = 4 \times T_s$$
$$= 4 \times 33.3 = 133.3 \text{ lb}$$

If the beam diameter is reduced to 5 in. and the weight is reduced to 20 lb, it is necessary to maintain same chain tensions as before.

Therefore, taking moments about beam centre

$$T \times 5 = W \times 5$$
$$T = 20 \times 5/5$$
$$T = 20 \text{ lb}$$
$$(T_t - T_s) = T \times 5$$
$$(4T_s - T_s) = 100 \text{ lb}$$
$$3T_s = 100 \text{ lb}$$
$$T_s = 100/3$$
$$= 33.3 \text{ lb}$$
$$T_t = 4T_s$$
$$= 4 \times 33.3$$
$$= 133.3 \text{ lb}$$

From the above, the chain tensions are maintained same as before when the beam diameter and the weight is reduced to 5 in. and 20 lb.

Numerical questions

1. In the lever-weighting arrangement of the calendar rollers in a lap end of a scutcher, a weight of 75 lb is placed 5 ft from the fulcrum and the pressure is applied 3 in. from the fulcrum on the same side of the fulcrum as the weight. The weighting arrangement is on each side of the machine. Calculate the pressure between the top and second roller.

   Ans. Total pressure $P = 3000$ lb
2. A bent lever *AFB* is pivoted at *C*. One arm is horizontal and carries a weight of 10 lb placed 11 in. from *C*; the other arm is inclined to the first arm and inclined at an angle of 60 degrees and 4 in. long. Find the pressure exerted at *B*.

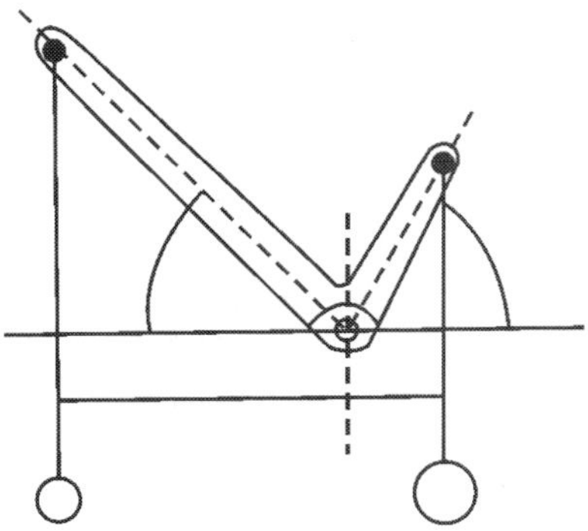

Fig. 10.24

Ans. Pressure at $B = 72.72$ lb

3. Two levers are pivoted on the same end and each is 16 in. long. A cam or tappet with a 3 in. throw acts at one lever 13 in. from the stud and depresses it. What must be the throw of the second cam or tappet placed 9 in. from the stud in order that the second lever is depressed the same amount as the first lever.

4. A lever weighs 5 lb. A vertical spring balance applied 12 in. from the fulcrum and supporting the lever in a horizontal position reads 2 lb. Find the centre of gravity of the lever from the fulcrum.

   Ans. 4.8 in.

5. The underside of the heald shaft of a loom is attached to a horizontal lever. It is at a distance of 16 in. from the lever fulcrum. A reversing spring is attached 4 in. from the fulcrum. The lever must exert a minimum resistance of 10 lb to the lifting of the shaft. What pull must the spring exert when the lever is horizontal?

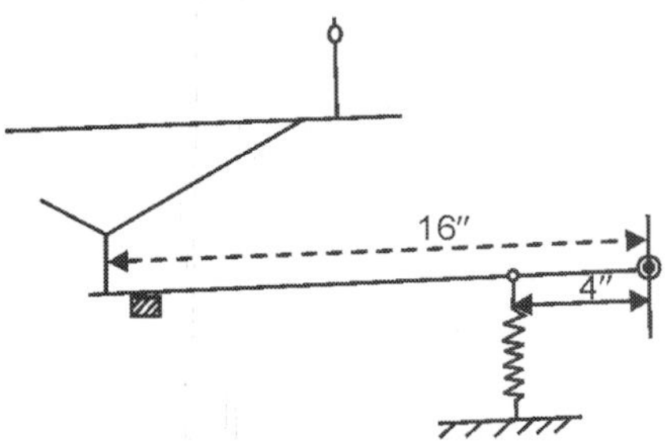

Fig. 10.25

Ans. 40 lb

6. Figure 10.26 shows the weighting lever for the loom let-off motion. Calculate the tension *T* in the chain.

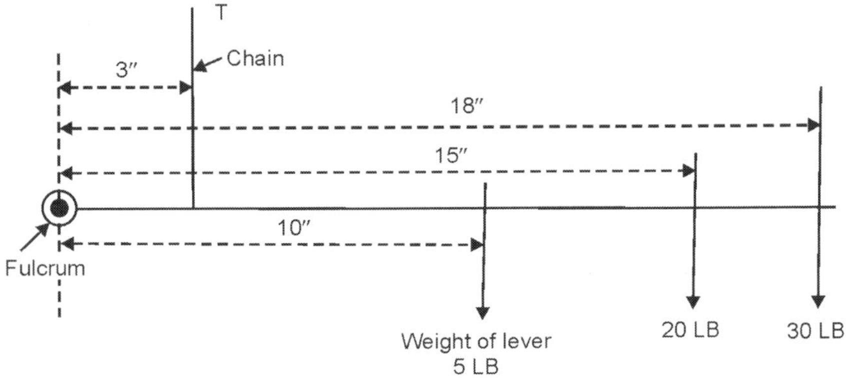

Fig. 10.26

7. A beam of 18 ft long supports number of bales of cotton, each bale weighing 500 lb. One bale is 2 ft from one end, another bale is 5 ft from the end, a third bale is 10 ft from the end and fourth bale is 13 ft. from the end, all the measurements being from the same end. Find the pressure on each support.

   Ans. *PA* = 833.3 lb; *PB* = 1166.6 lb

8. A safety valve is 3 in. diameter. The weight on the lever is 57 lb. The fulcrum is 4 in. from the centre of the valve. At what distance from the fulcrum must the weight be placed in order that 80 lb per sq in. pressure in the boiler will just lift the valve?

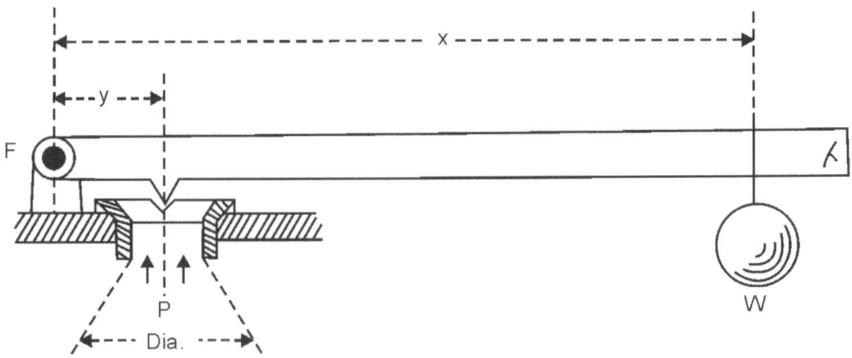

Fig. 10.27

Ans. Total pressure on the valve = 565.20 lb per sq in.

Distance from the fulcrum for the weight to be placed = 39.66 in.

9. A beam 20 ft. long is supported at both the ends and a weight of 5 tons is being raised by a pulley tackle hooked on it at a point 8 ft from one end. Find the reactions on the supports.

   Ans. *PA* = 2 tons; *PB* = 3 tons

10. A uniform rod 4.90 ft and mass weight of 35.2 lb has masses 4.40, 6.6, 8.8 and 11 lb suspended at points 3.93, 5.90, 7.87 and 9.84 in. respectively from one end. It is supported in a horizontal position by means of vertical strings attached to its ends. Find the tensions in the strings.

    Ans. 21.59 lb and 44.52 lb

11
Friction

11.1 Introduction

It has been established since long that the surfaces of the bodies are never perfectly smooth. When even a very smooth surface is viewed under a microscope, it is found to have roughness and irregularities, which is not possible to get detected by normal touch. If a block of one substance is placed over the level surface of the same or of different material, a certain degree of interlocking of the minutely projecting particle takes place. This does not involve any force, so long as the block does not move or tends to move. But if one block moves or tends to move tangentially with respect to the surface on which it rests, the interlocking property of the projecting particles opposes the motion. This opposing force, which acts in the opposite direction of the movement of the upper block, is called the force of friction, or simply friction.

It is thus evident that at every joint in a machine, force of friction arises due to the relative motion between the two parts and hence some energy is wasted in overcoming the friction. Although friction is considered to be undesirable, it plays important role both in nature and in engineering applications, for e.g. walking on road, playing a carom board, movement of shuttle in a loom.

11.2 Types of friction

In practical applications, the friction is of two types.

1. Static friction
2. Dynamic friction

Static friction– It is the friction experienced by a body when it is at rest.

Dynamic friction– It is the friction experienced by a body, when it is in motion. The dynamic friction is also called as kinetic friction and is less than the static friction. The dynamic friction is classified into three types:

(a) Sliding friction: It is the friction experienced by a body when it slides over another body.

(b) Rolling friction: It is the friction experienced between the surfaces which have rollers interposed between them.

(c) Pivot friction: It is the friction, experienced by a body, due to the motion of rotation as in the case of foot step bearings.

The friction may be further classified as friction between lubricated and unlubricated surfaces.

11.3 Limiting friction

Consider a body *A* of weight *W* is lying on a rough horizontal body *B* as shown in Fig. 11.1.

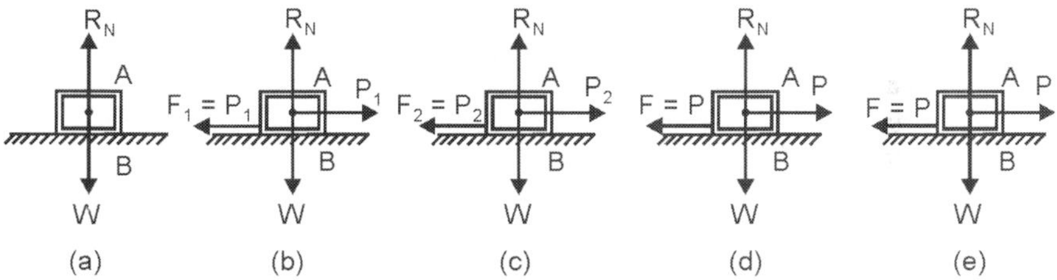

Figure 11.1 Limiting friction

In Fig. 11.1(a), the body *A* is in equilibrium condition under the action of its own weight *W*, and the normal reaction R_N (equal to *W*) of *B* on *A*.

In Fig. 11.1(b), if a small horizontal force *P*1 is applied to the body *A* acting through its centre of gravity, it does not move because of the frictional force which prevents the motion. This means that the applied force P_1 is exactly balanced by the force of friction F_1 acting on the opposite direction.

In Fig. 11.1(c), the applied force is increased to P_2 is still found to be in equilibrium. This means that the force of friction has also increased to a value $F_2 = P_2$. Thus every time the applied force is increased, the force of friction also increases so as to exactly balance the applied force.

In Fig. 11.1(d), any further increase in applied force will not increase the frictional force.

In Fig. 11.1(e), the body begins to move in the direction of the applied force. The maximum value of the frictional force, which comes into play when a body just begins to slide over the surface of the other body, is known as limiting friction.

It may be noted that when the applied force is less than the limiting friction, the body remains at rest, and the friction comes into play, which is called static friction; it may have any value between zero and limiting friction.

11.4 Coefficient of friction (Fig. 11.2)

It is defined as the ratio of limiting friction (F) to the normal reaction (R_N), between the two bodies. It is generally denoted mathematically as coefficient of friction.

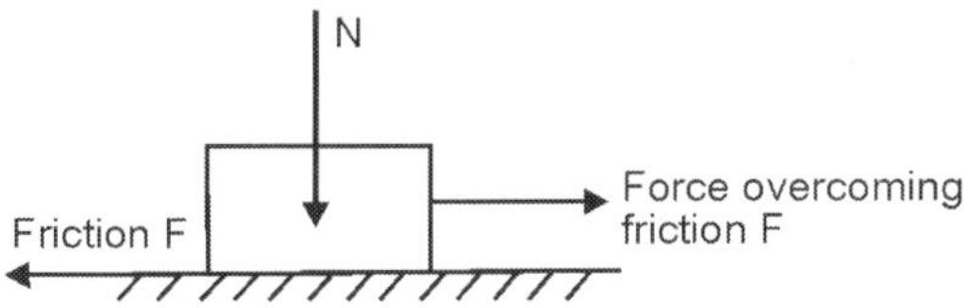

Figure 11.2 Co-efficient of friction

$\mu = F/N$ if is the frictional resistance between two surfaces, one of which is sliding over the other as shown in Fig. 11.2., and N is the total force pressing the two surfaces together, then $F/N = \mu$, is termed as coefficient of friction, and friction

$$F = \mu * N$$

N is the force acting at right angles or normal to the surface. If N acts other than at right angles to the surface, only the components at right angles to the surface causes friction and must be considered in calculating the friction force.

The values of μ depend on the material and the condition of the surfaces. It can only be found experimentally and is approximately constant for any two given materials.

Dry smooth wool on wool: 0.75 to 0.5
Dry smooth metal on metal: 0.15 to 0.3
Lubricated smooth metal on metal: 0.07 to 0.08
Leather on smooth dry metal: 0.56
Cotton yarn on smooth steel: 0.26 to 0.28

The value of μ can be found experimentally in different ways. Two usual methods are shown in Fig. 11.3(a) and (b).

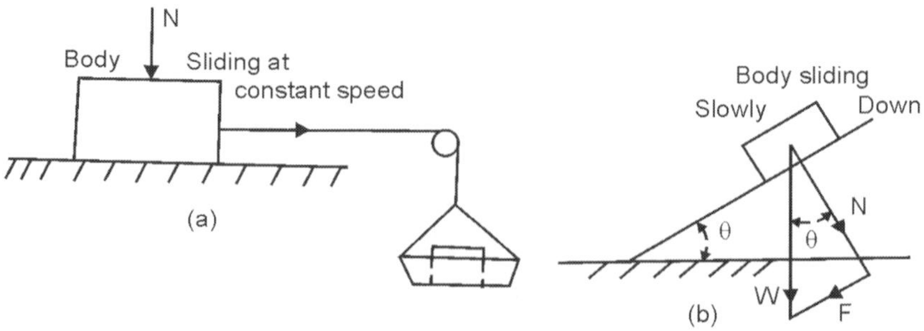

Figure 11.3 (a) and (b)

In Fig. 11.3(a), *F* is obviously equal to the friction and $\mu = F/N$, and at Fig. 11.3(b) the plane is tilted until the body slides slowly down at uniform speed. The weight *W* of the body acts vertically downwards and can be resolved into its components. N is at right angles to the surface, and *R* is parallel to the surface. *N* is the force normal to the surface. *F* is the force pushing the body along the plane. If the body is moving slowly down, *F* is equal to the frictional resistance between the surface and

$F/N = \mu + \tan\theta$, where θ is termed as angle of repose or the angle of friction between two surfaces.

11.5 Laws of solid friction

(a) Friction is directly proportional to the total pressure between the surfaces.
(b) Friction is independent of the areas of the surfaces in contact.
(c) Friction is independent of the speed of sliding.
(d) Friction depends on the nature of the two surfaces.
(e) Static friction is greater than the kinetic friction.

The laws of solid friction mentioned above are only approximately true. For very low and very high pressures, the coefficient of friction is higher than at moderate pressures. Friction is only roughly independent on speed. In general, as speed increases, μ value decreases. This is not always true. For example, the friction between the thread and the metal surface generally increases as speed increases.

Example 1: A body weighing 100 lb. is moved along a horizontal plane by a force P, against friction $\mu = 0.4$. Find the force P required (a) if applied horizontally, (b) if applied as pushing force at an angle of 30 degrees with the horizontal, (c) if applied as a pulling force at 30 degrees with the horizontal.

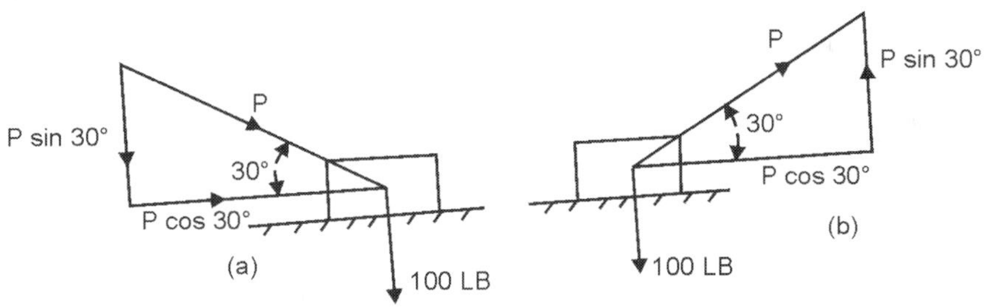

Figure 11.4 (a) and (b)

Solution:

$$\mu = 0.4$$

(a) Force applied horizontally is

$$F = \mu * R$$
$$= 0.4 * 100 = 40 \text{ lb}$$

(b) Resolving the forces horizontally and vertically,
When a push of P is applied on the block

$$F = P \cos 30° = P * 0.866 = 0.866\, P$$
$$R_1 = W + P \sin 30°$$
$$= 100 + 0.866\, P$$
$$F = \mu * R_1$$
$$= 0.866\, P = 0.4 * (100 + 0.866\, P)$$
$$P = 60.06 \text{ lb.}$$

(c) When a pull of P is applied on the block
Resolving the forces horizontally and vertically,

$$F = P \cos 30° \pi$$
$$= 0.866\, P$$
$$R_1 = W - P \sin 30°$$
$$= 100 - 0.866\, P$$
$$F = \mu * R_1$$
$$0.866\, P = 0.4 * (100 - 0.866P)$$
$$P = 37.52 \text{ lb}$$

Angle of friction is

$$\text{Tan } \theta = F/R = 40/100 = 0.4$$
$$\theta = 21.80°$$

Example 2: A leather brake block is pressed against the rim of the brake pulley with a normal force of 30 lb. If μ is 0.5, what is the frictional resistance at the pulley rim? If the pulley is 1 ft diameter, how much work is done per revolution in overcoming brake friction?

Solution:

$$F = \mu * R$$
$$= 0.5 * 30 = 15 \text{ lb.}$$

Frictional resistance of the pulley rim is 15 lb.

$$\text{Work done} = \text{Resistance} * \text{displacement}$$
$$= 15 * 1 * 3.14 = 47.1 \text{ lb.}$$

Example 3: A loom swell applies a force of 10 lb to the back edge of the shuttle, as shown in Fig. 11.5 ignoring the friction between the shuttle and the shuttle box. What force P must be applied to the rear end of the shuttle to overcome swell friction?

μ between shuttle and swell and shuttle and shuttle box is 0.3.

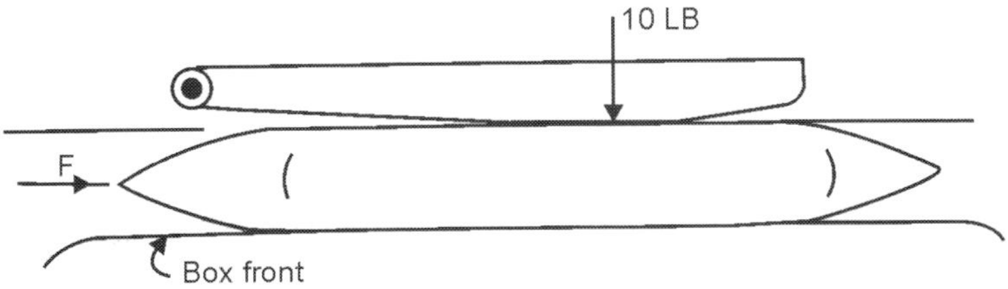

Figure 11.5

Solution: There is a frictional resistance to the shuttle movement both at the front and back edges of the shuttle. Hence, the total resistance is the sum of the two. If the friction at the bottom of the shuttle is negligible, then the force applied by the front edge of the shuttle to the box front is 10 lb, the shuttle merely monitoring the swell pressure to the box front.

Hence total frictional resistance = 2 * 0.3 * 10 = 6 lb.

11.6 Friction of shaft in its bearings

If L is the total load on a shaft bearing and μ is the coefficient of friction between shaft and bearing surfaces, then the resistance F at the shaft surface due to bearing friction is approximately μ * L.

If L is in lb, the shaft diameter is d in. and N is the number of revolutions per minute, then the work done to overcome bearing friction is given by μ * L * * d/12*N ft lb per minute. The resistance acts tangentially to the shaft surface, opposing its movement as shown in Fig. 11.6.

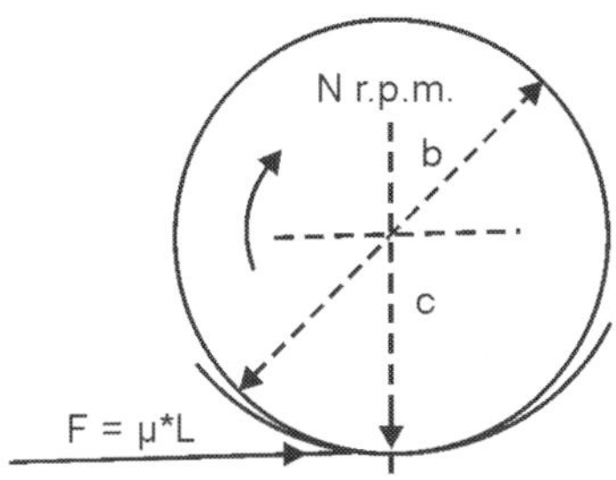

Figure 11.6

Friction clutch is used to connect the engine to the driven shaft below the bearing centre.

Example 4: A carding engine cylinder weighs 1200 lb. and is supported by two bearings of 3 in. diameter. When the cylinder is disconnected from the other parts of the machine, it requires 0.25 HP to run it idly at 165 revolutions per minute. What is the approximate value of μ in its bearings?

Solution:

Work done against friction per revolution

$$= 0.25 * 33000/165 = 50 \text{ ft lb}$$

$$\mu * 1200 * \pi * 3/12 = 50 \text{ ft lb}$$

11.7 Friction clutches

A friction clutch has its principal applications in the transmission of power of shafts and machines which must be started and started frequently. Its applications are also found in cases in which power is to be delivered to machines partially or fully loaded. The force of friction is used to start the driven shaft from rest and gradually brings it up to the proper speed without excessive slipping of the friction surfaces. In automobiles, friction clutch is used to connect the engine to the driven shaft. In operating such a clutch, care should be taken so that the friction surfaces engage easily and gradually bring the driven shaft up to proper speed. The proper alignment of the bearing surfaces must be maintained and it should be located as close as possible to the clutch.

The following points are valid:

1. The contact surfaces should develop a frictional force that may pick up and hold the load with reasonably low pressure between the contact surfaces.
2. The heat of friction should be rapidly dissipated and tendency to grab should be at minimum.
3. The surfaces should be backed by a material stiff enough to ensure a reasonably uniform distribution of pressure.

   The friction clutches are classified into many types.

   (a) Disc or plate clutches (single disc or multiple disc clutch)
   (b) Cone clutches
   (c) Centrifugal clutches

11.7.1 Single disc or plate clutch

A single disc or plate clutch is shown in Fig. 11.7. It consists of a clutch plate whose both sides are faced with a friction material. It is mounted on the hub which is free to move axially along the splines of the driven shaft. The pressure plate is mounted inside the clutch body bolted to the flywheel.

Both the pressure plate and the flywheel rotate with the engine crankshaft or the driving shaft. The pressure plate pushes the clutch plate towards the flywheel by a set of strong springs which are arranged radially inside the body. The three levers also known as release levers or fingers are carried on pivots suspended from the case of the body. These are arranged in such a manner that the pressure plate moves away from the flywheel by the inward movement of the thrust bearing. The bearing is mounted upon a forked shaft and moves forward when the clutch pedal is pressed.

When the clutch pedal is pressed down, its linkage forces the thrust release bearing to move in towards the flywheel and pressing the longer ends of the levers inward. The levers are forced to turn on their suspended pivot and the pressure plate moves away from the flywheel by the knife edges, thereby compressing the clutch springs. This action removes the pressure from the clutch plate and thus moves back from the flywheel and the driven shaft becomes stationary. On the other hand, when the foot is taken off from the springs to extend and thus the pressure plate pushes the clutch plate back towards the fly wheel.

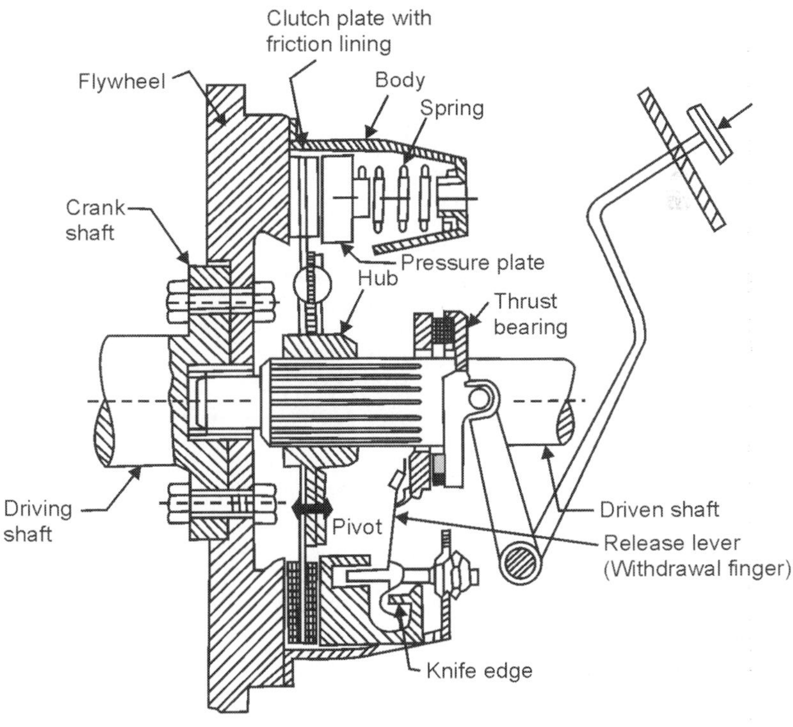

Figure 11.7 Single disc or plate clutch

11.7.2 Multi-plate clutch

The diagram of multi-plate clutch is shown in Fig.11.8. Multi-plate clutch with friction plates have linings on both sides except the first plate which is adjacent to the flywheel. This plate is having friction lining on one side. The friction plates are connected on the top to the flywheel. Hence the friction plates rotate with the flywheel and hence with the driving shaft. The friction plates are also free to move axially.

The discs or plates are also supported on splines of the driven shaft. Hence these plates rotate with the driven shaft. These plates are located in between the friction plates and can also slide axially as shown in Fig. 11.8, which shows the position of the friction plates and disc plates in disengaged position.

In the engaged position (when the foot is taken off from the clutch pedal), the set of strong springs will press the discs into contact with the friction plates. Hence the power will be transmitted from driving shaft to the driven shaft.

Multiplate clutches are normally used when large torque is to be transmitted as in motor cars and in machine tools.

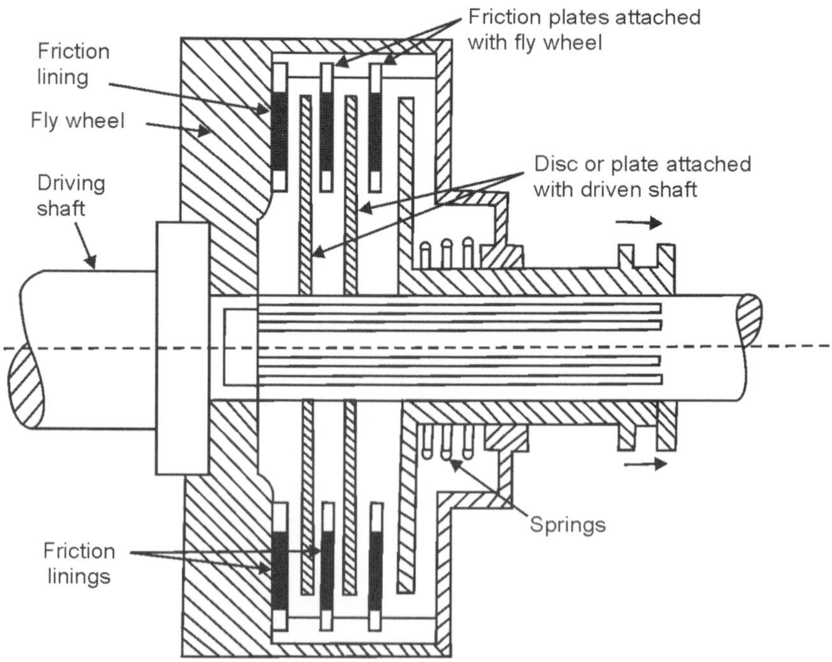

Figure 11.8 Multi-plate clutch

11.7.3 Cone clutch

Figure 11.9 shows the diagram of a cone clutch, in which the contact surfaces are in the form of cones. The driver cone is keyed to the driving shaft whereas the driven cone is keyed to the driven shaft. In the engaged position, the friction surfaces of the two cones are in complete contact due to the spring pressure. In this engaged position, the torque is transmitted from driving shaft to the driven shaft. For disengaging the clutch, the force of the spring pulls back the driven cone through a lever system.

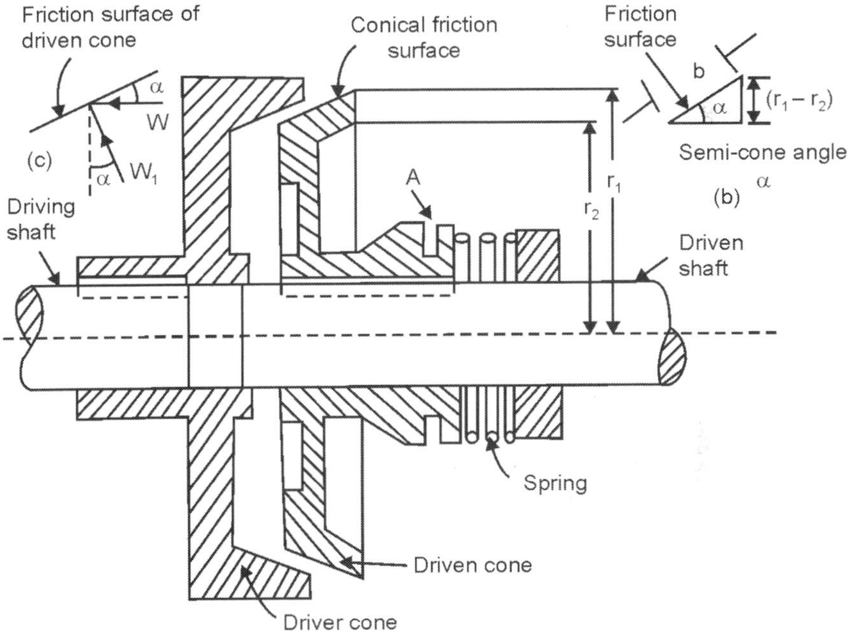

Figure 11.9 Cone clutch

The contact surfaces of the clutch may be metal to metal contact, but more often the driven cone surface is lined with some friction material.

11.8 Power transmission

In the types shown in Fig. 11.10(a) and (b), the total force normal to the friction surfaces is equal to the end force P, and the maximum available friction force $F = \mu * P$, where μ is the coefficient of friction at the driving faces. The maximum available driving torque is $F * r = P * r * \mu$, and the maximum horse power that is transmitted is given by

$\mu * P * r * 2 * \pi * N/33000$

where N is the revolutions per minute, P being in *lb*, and r in *ft*. If the difference between the outer and inner radii of the friction faces is considerable, the friction force $\mu * P$ can be taken as acting at the mean radius, on the assumption that the rate of wear is constant over the friction surface.

In type (c) the conical faces give increased friction for a given force P, as compared with the flat faces. The normal force R between the friction faces must have a horizontal component equal to P, where $P/R = \sin\alpha$. Hence $R = P/\sin\alpha$ and the friction force $\mu R = \mu * P/\sin\alpha$. The smaller the angle α, the greater the friction force available for a given force P, but α must not be too small, or the clutch may stick. The angle is usually from 4 to 10 degrees. The friction force can be assumed to act in the mean radius

$r! + r2/2$

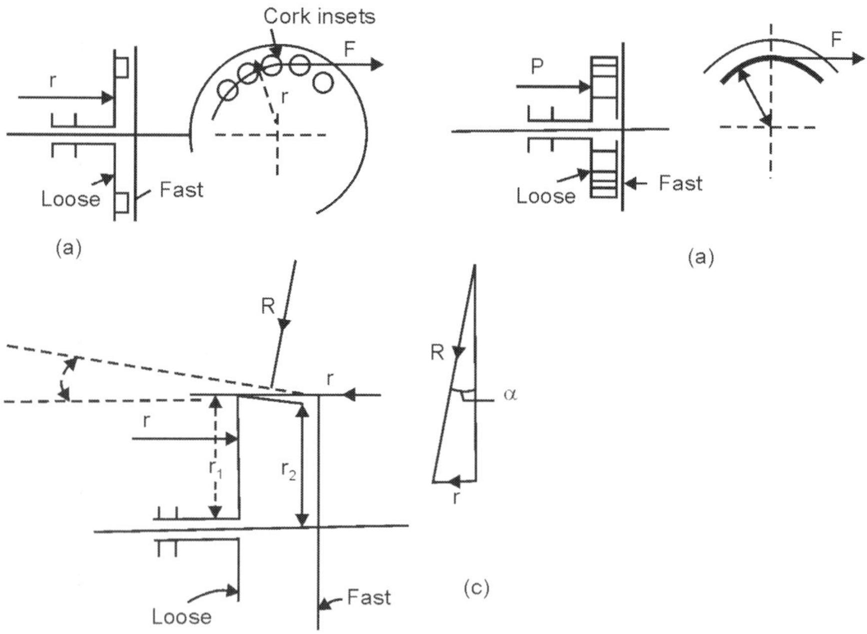

Figure 11.10

Example 5: A beam warping machine is driven by a face type clutch type (b) in Fig. 11.10. The mean radius of the friction faces being 6 in. if μ between the faces is 0.15, what force must be applied axially to the loose half to transmit 0.5 HP at 40 rpm of the driving shaft?

Solution:

Work done per minute = 0.5 * 33000 = Torque * 2 * * 40

Torque required = 0.5 * 33000/2 * π * 40 = 66 lb ft

Friction force required at the driving face = 66/mean radii

$$66/6 * 12 = 132 \text{ lb}$$

$$\text{Axial force } P = 132/\mu$$

$$= 132/0.5 = 880 \text{ lb}$$

Example 6: Figure 11.11 shows the arrangement of the starting handle and clutch for a loom drive. It is found experimentally that a force of 30 lb is required at the starting handle. In addition to the force required to overcome the spring in the handle, to put the handle in to its "on" position. Calculate the maximum horse power the clutch can transmit at 180 rpm, if μ at the friction faces is 0.5?

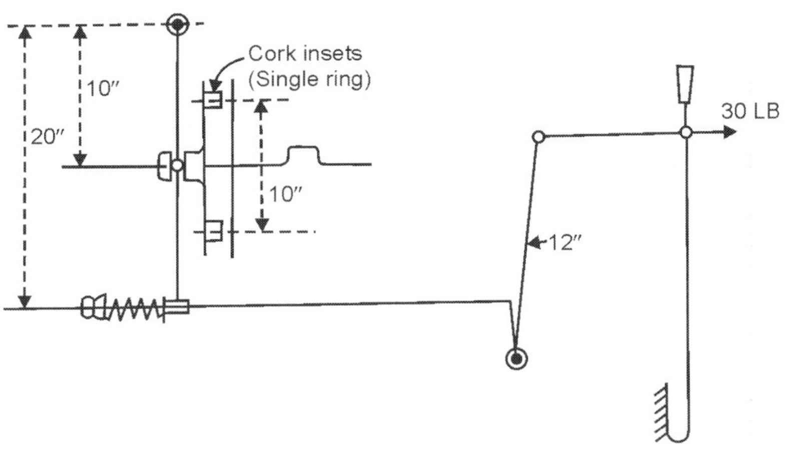

Figure 11.11

Solution:

Force between clutch faces = 30 * 12/4 * 2/1

$$= 180 \text{ lb}$$

Maximum friction force available = 0.5 * 180

$$= 90 \text{ lb}$$

Maximum torque available = 90 * 5.12

$$= 37.5 \text{ lb ft}$$

Maximum horse power clutch can transmit

$$= T * 2 * \pi * N/33000$$
$$= 37.5 * 2 * \pi * 180/33000$$
$$= 1.28$$

Example 7: A flanged warper's bobbin weighs 1.15 lb when full, 0.4 lb when empty; bobbin diameter is 4 in. full, 1.25 in. empty. It rests on and is driven by a friction disc on an upright revolving spindle as shown in Fig. 11.12. What yarn tension during winding will cause slip at the driving disc (a) when the bobbin is full, (b) when it is almost empty? μ between disc and flange is 0.3.

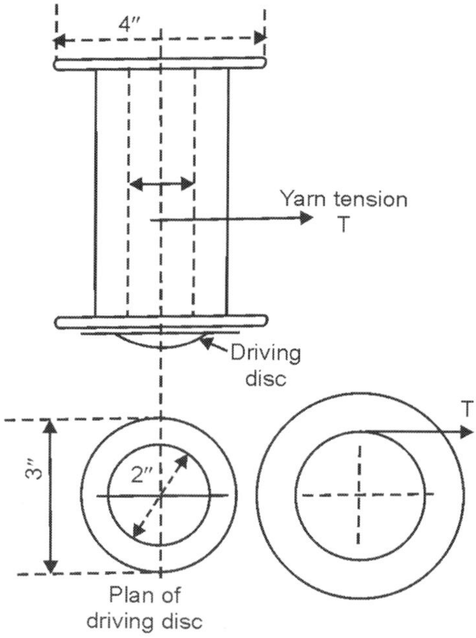

Figure 11.12

Solution:

Frictional driving force acts at the mean radius of the disc 1.25 in., the frictional driving torque is $\mu * W * 1.25 = 0.3 * 1.25\ W = 0.375$ lb in.

When the slip occurs, this will be equal to the retarding torque due to yarn tension T.

(a) At full bobbin,

$$W = 1.15 \text{ lb and } T \text{ acts at 2 in. radius}$$

Then $$T * 2 = 0.375 * 1.15$$
$$T = 0.216 \text{ lb.}$$

(b) At empty bobbin,

$$W = 0.4 \text{ lb and } T \text{ acts at } 0.625 \text{ in.}$$
$$T * 0.635 = 0.375 * 0.4 \text{ lb}$$
$$T = 0.24 \text{ lb}$$

Numerical questions

1. Find the minimum force required to move a body weighing 100 lb, slowly down a horizontal surface if $\mu = 0.25$.

   Ans. Force = 25 lb. Tan $\theta = 14°$

2. A leather brake block is pressed against the rim of a pulley through the medium load carried by block. If μ is 0.5, what leverage is required to provide a braking force of 25 lb at the pulley rim?

   Ans. 5 to 1

3. A ring frame traveler moving round a ring of 1.5 in. diameter at 8000 rpm presses against the ring with a normal force of 2000 grains. What horse power is required in overcoming the friction if $\mu = 0.1$ and there are 400 travellers in the frame? Assume the friction force acts at 0.75 in. from the centre of the ring.

   Ans. H.P. = 1.09

4. The bearing of a mill line shaft carries a load of 250 lb due to weight of shafts and pulleys. In addition, belt pulls of 80 lb each act on either side of the bearing as shown in Fig. 11.13. Find the total load on the bearing causing friction and the HP lost in 10 such bearings, if the shaft is 2.5 ft diameter and the speed is 250 rpm, μ at bearings = 0.05.

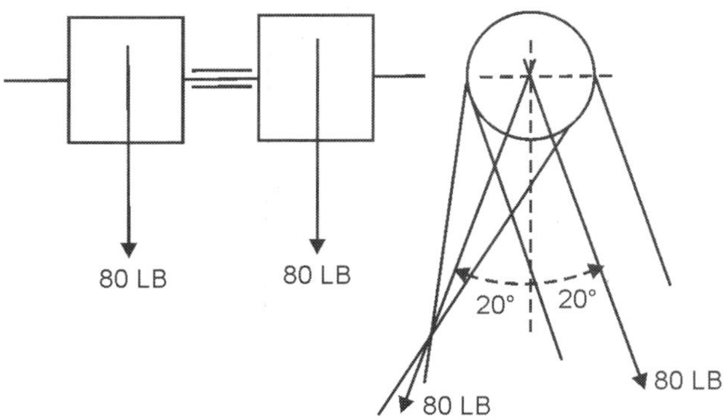

Figure 11.13

Ans. Total load on the bearings = 552 lb; Horse power lost = 1.09.

5. The backing of clutch on a mule spinning frame rim shaft is a leather-faced cone type. The angle of the cone is 10° and μ for the friction faces is 0.5. Mean diameter at the friction faces is 16 in. During backing off, the clutch is required to exert a retarding torque of 65 lb ft. What axial force must be applied to the clutch to produce this torque?

   Ans. 34 lb

6. Figure 11.14 shows a war per's bobbin, winding on a drum winding machine. Ignoring friction at the bobbin spindle, find the weight W that must be used to enable winding to be done without bobbin slip, against a yarn tension t of 4 oz, μ = 0.75. Assume the cradle lever carrying the bobbin to be vertical throughout filling of bobbin.

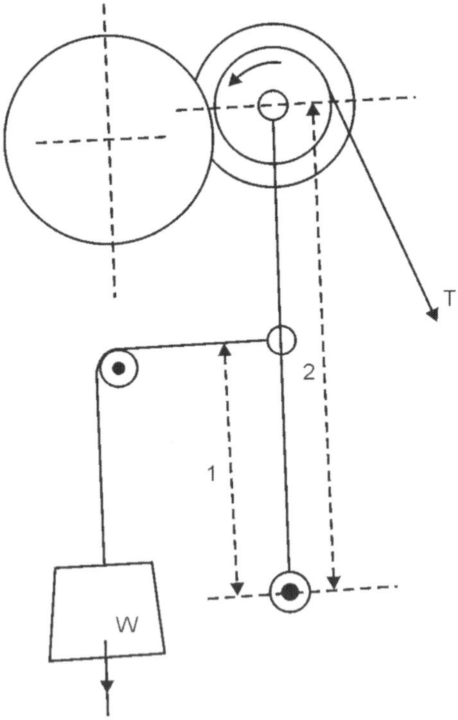

Figure 11.14

   Ans. Weight (*W*) = 2 lb

7. A friction drive for the section shaft of a section warping machine is shown in Fig. 11.15. The driving radius r at which the leather

covered driving bowl acts on the disc being changed automatically as the section fills, so that $r = R$ the section radius changes from 4 to 8 in. from start to the end of section. Assuming that the drive has to transmit a maximum of 0.25 HP, what is the minimum force P pressing the driving bowl against the disc, which will enable this to be done without slip at the beginning and at the end of the section? Warping speed is 100 yards/min, $\mu = 0.5$.

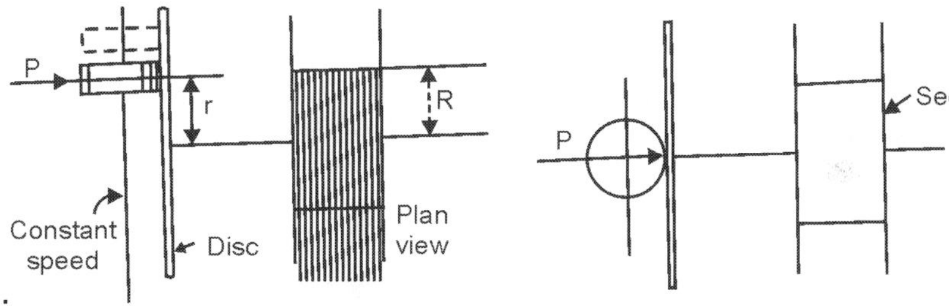

Ans. (a) 55 lb, (b) 110 lb

12
Coil friction

12.1 Introduction

Coil friction is the friction between a flexible band such as a belt, rope, chain or thread and a curved surface. It is particularly extreme importance in textile machinery and processes, as it not only in such common applications as belt and rope driving, band brakes and in loom beam weighing motions and yarn tensioning devices and also a thread passes over a curved surface.

Coil friction is often usually applied, as in belt driving, but on the other hand, it may be undesirable and have to be reduced as far as possible. This generally occurs when a thread passes round a curved surface in any process. At the first instant, coil friction does not appear to obey the laws on the first inch = μ solid friction, particularly which states that friction is independent of the area of surfaces in contact. Coil friction does depend on the area of the contact between the band and the curved surfaces, or on the angle of lap, but it is entirely due to the effect of the curved surface on the pressure between the surfaces, and not on the area of the surfaces. Thus, an increase in the angle of lap increases the pressure between the band and surface.

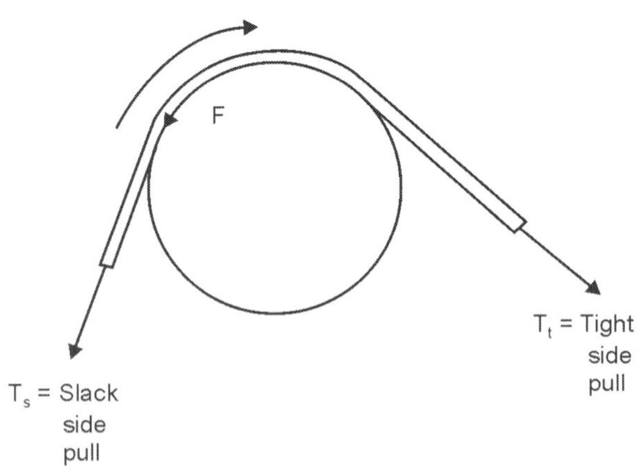

Figure 12.1

Consider a flexible band or thread of negligible thickness, sliding slowly over a fixed circular pulley as shown in Fig. 12.1, if the band is moving clockwise, friction F between band and pulley is opposing the movement as shown in Fig. 12.1.

The pull T_t required at the tight side of the band must obviously be greater than any pull T_s at the on-going slack end, since T_s must overcome any pull T_t and also the total friction force F opposing the movement, $T_t = T_s + F$. Generally, the only force causing the band to press against the pulley face and causing friction is due to the tension at the end of the band. At the same time, there must be some tension at the slack side (T_s).

Example 1: A thin leather band (Fig. 12.2) makes contact for 10 in. with the rim of a fixed pulley of 10 in. radius, i.e. angle of lap is 1 radian. $T_s = 100$ units: $\mu = 0.5$. The force triangle pqr for the first inch of the band is similar to the sector *a, b*, and *c*.

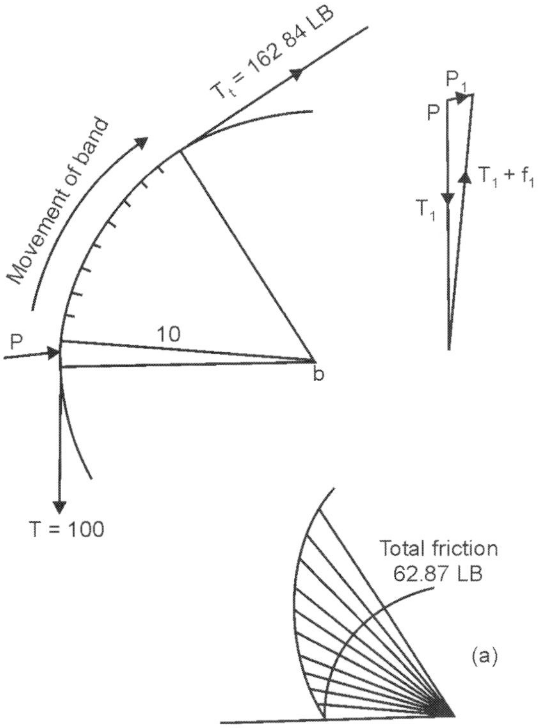

Figure 12.2

Hence $P1 = 1/10$ (approx.), and the friction f1 on the first inch $= \mu P1 = 0.5$ $1/10\ T_s - 1/20\ T_s$. The same applies for each successive inch of lap, the friction on any inch being approximately 1/20 * tension at the lower end of that inch.

Mathematical formula: As already stated, it can be proved mathematically, that when slip occurs, T_1 bears a definite relationship T_s. The relation is

$T_t / T_s = e^{\mu\theta}$ where $e = 2.718$, the base of Naperian logarithm, μ = coefficient of friction, θ = angle of lap in radians. Applying this to the above example,

$$
\begin{aligned}
Ts &= 100 \text{ units}, \mu = 0.5, \theta = 1,\\
T_t / T_s &= e^{\mu\theta} = 2.718^{(0.5 \times 1)}\\
&= 1.6486\\
T_t / T_s &= 1.6486\\
T_s &= 100 \text{ units}\\
\text{Therefore} \qquad T_t &= 1.6486*100\\
&= 164.86 \text{ units}\\
F &= (T_t - T_s)\\
&= 164.86 - 100\\
&= 64.86 \text{ units}
\end{aligned}
$$

If the angle of lap is increased to 2 radians,

$$
\begin{aligned}
T_t / Ts &= 2.718^{0.1} = 2.718\\
T_t &= 2.718^{100} = 271.8 \text{ units}\\
F &= 271.8 - 100 = 171.8 \text{ units}
\end{aligned}
$$

Effect of coil friction of the tension conditions at the ends of the band (Fig. 12.3).

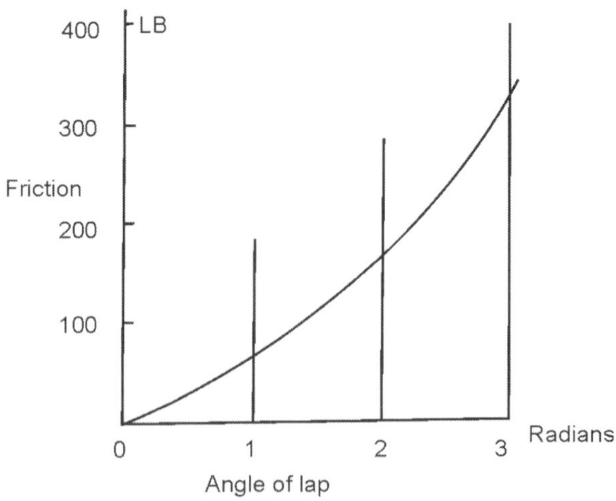

Figure 12.3

The above example shows how greatly the magnitude of coil friction may depend on the angle of lap. However, the friction increases with increase of lap if *Ts* remains constant. The familiar example is found in some applications of band brakes and loom beam weighing motions.

Example 2: Figure 12.4 shows a friction let-off for a loom beam, a rope fixed at *A* to the loom frame, coiled round the weighing ruffle and attached to the weighting lever at *B*. If $\mu = 0.15$, what warp tension *T* will cause slip and let-off warp? There is a similar braking arrangement at each end of the beam.

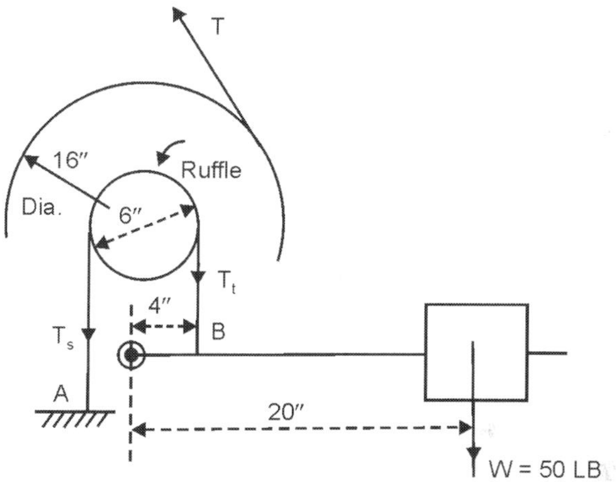

Figure 12.4

Solution:

In this let-off motion, the brake band is stationary and the ruffle moves, when slip occurs, $T_t / T_s = e^{\mu\theta}$.

Here, however, T_t is constant, since it must always support lever and weight. Now taking moments about the lever fulcrum,

We have

$$T_t * 4 = W * 20$$
$$T_t = 50 * 20/4 = 250 \text{ lb}$$
$$T_t / T_s = 2.718^{(0.15 \times 3.14)}$$
$$T_t / T_s = 1.603$$
$$T_t = 250 \text{ lb}; \; T_s = 250/1.603 = 155.95 \text{ lb.}$$
$$F = T_t - T_s$$
$$250 - 155.95 = 94.04 \text{ lb}$$

Friction at one ruffle = 94.04 lb

Friction at two ruffles = 2 * 94.04 = 188.08 lb

Now, taking moments at the beam centre, we have

$$T * 8 = F * 3$$
$$T = 188.08 * 3/8 = 70.53 \text{ lb}$$

Warp tension *T* to overcome friction is 70.5 lb.

Example 3: In the same problem, if the lap is increased to 1.5, what will be the effect of warp tension *T*?

Solution:

$$T_t / T_s = 2.718^{(0.15 \times 3.14)} \ (1.5 \text{ laps} = 3\pi)$$
$$T_t / T_s = 4.103$$
$$T_s = 250/4.103 = 60.93$$
$$\text{Friction } F = (T_t - T_s)$$
$$= 250 - 60.93 = 189 \text{ lb at each ruffle}$$

Friction at two ruffles is 2 * 189 = 378.13 lb

Now, taking moments about beam centre,

$$T * 8 = F * 3$$
$$T = 378.13 * 3/8 = 141.79 \text{ lb}$$

Warp tension T to overcome friction is 141.79 lb, when the angle of lap is increased to 1.5 laps.

Example 4: If, instead of arranging the lever as shown, the end *A* of the rope was attached to the lever, and end *B* fixed to the frame, then T_s would be constant instead of T_t and a much smaller weight would be required to give the same friction and let-off tension as before.

In the above Example 3, what weight on the lever would be required to give the warp tension of 142 lb with the angle of lap 1.5?

Solution:

$T_t = 250$ lb; $T_s = 61$ lb. But *Ts* supports the lever and by taking moments about the lever fulcrum,

$$T_s * 4 = W * 20$$
$$W = 61*4/20 = 12.2 \text{ lb}$$

Example 5: In the above problem, if μ rises from 0.15 to 0.17, what is the effect of warp tension with 1.5 laps?

Solution:

With T_t is constant, the slipping ratio $T_t / T_s = 2.718^{(0.17 \times 0.3 \times \pi)} = 4.95$

Therefore, $T_t = 250$ lb; $T_s = T_t /4.95$

$$= 250/4.95 = 50.5 \text{ lb}$$

$$\text{Friction } (F) = (T_t - T_s) = 250 - 50.5 = 199.49 \text{ lb.}$$

$$\text{Friction at one ruffle} = 199.9 \text{ lb}$$

Friction at two ruffles = 2 * 199.9 = 399.8 lb

Now taking moments about the beam centre, we have

$$T * 8 = F * 3$$

$$T = 399.8 * 3/8 = 150 \text{ lb}$$

Rise in warp tension, due to rise of μ from 0.15 to 0.17, from 142 lb to 150 lb is 5.6%.

With T_s constant, $T_s = 61$ lb; $T_t = 61 * 4.95 = 302$ lb

$$F = (302 - 61) = 241 \text{ lb}$$

For two ruffles, the friction is 2 * 241 = 482 lb.

Now taking moments about beam centre,

$$T * 8 = F * 3$$

$$T = 482 * 3/8 = 180.75 \text{ lb.}$$

Therefore, rise in warp tension from 142 lb to 180 lb is 27.28%.

An important thing to note in coil friction is that coil friction multiplies tension.

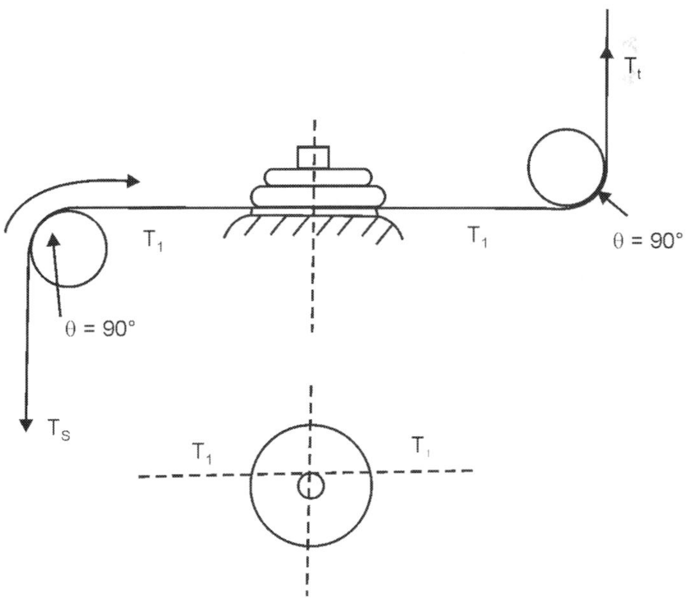

Figure 12.5

Example 6: Figure 12.5 shows a thread passing round two guide bars with a drag washer tensioner resting on the yarn. If μ at the guide bars and also between thread and drag washer is 0.26, what is the leaving tension T_l, when the drag washer weight is 2 oz and the entering tension T_1 is (a) 1 oz, (b) 2 oz, (c) if the drag washer weight is increased to 4 oz when *T*1 is 1 oz?

Solution:

Slipping ratio at each guide bars is$2.718^{(0.26 \times 180/2)}$ radians.

Slipping ratio = 1.50

The friction at the drag washer will be 2 * μ * *W*, since friction acts at both the top and bottom of the thread. It means that when *W* = 2 oz, drag washer tension = 2 * 0.26 * 2 = 1.04 oz.

Now the tensions at various points are shown in Table 12.1.

Table 12.1 Tensions at various points

T_S	T_1	T_2	T_t
1	1/5	2.54	3.8
2	3.0	4.04	6.15
1	1.5	3.58	5.65

In the above problem, the greater increase in final tension by increasing the T_s by 1 oz, than by increasing the drag washer friction by 1 oz. Since, the latter simply adds 1 oz to the tensions, whereby increasing T_s by 1 oz increases the tension by 1.5 oz at the first guide bar alone, due to the multiplying effect of coil friction.

12.2 Friction on curved surface

Assume a band passes round a succession of curved surfaces, all with the same value of μ; it can be easily proved that the ratio T_t / T_s between the tensions at the end and at the start is eμθ, where θ = sum of successive angles of lap.

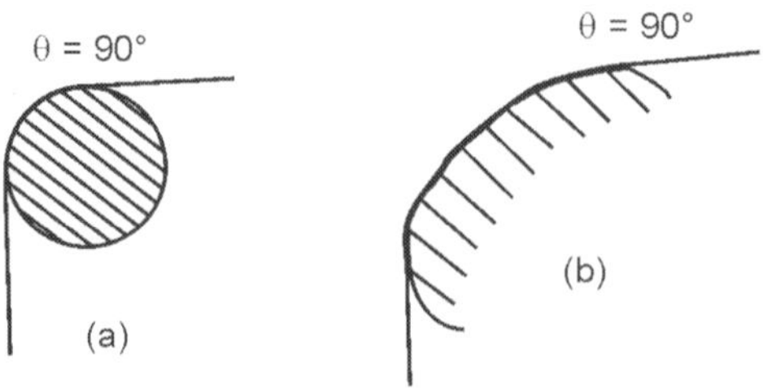

Figure 12.6

It is realized that the slipping ratio does not depend on the diameter or radius of the curved surface. This is not quite the case if the diameter is very small, but it can be assumed generally true. Hence, the surface need not be of constant radius of curvature, flexible material against a solid surface the friction, for a given value of μ depending upon the angle by which the direction of the band is changed.

Thus, for the two curved surfaces (a) and (b) shown in Fig. 12.6, the slipping ratio would be the same, if μ is the same for both.

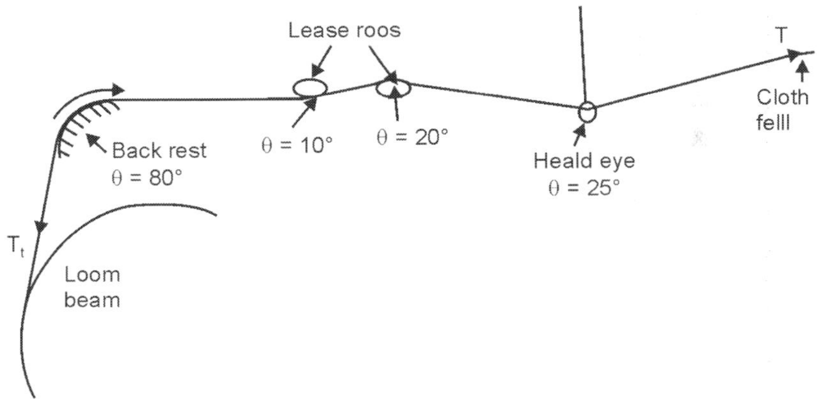

Figure 12.7

Example 7: Figure 12.7 shows the path of the warp thread passing through a loom from loom beam to cloth fell. Assuming μ to be 0.25 at each of the friction surfaces, find the tension in the thread at the cloth fell to pull the yarn slowly forward against a let-off tension at the loom beam of 1 oz.

Solution:

Since the μ is same at all the curved surfaces, the lap may be considered continuous.

Hence, total angle of lap = 80 + 10 + 20 + 25 = 135/57.3 = 2.36 radians.

$$T_t / T_s = 2.718^{(0.25 \times 2.36)} = 1.8$$

Therefore, when $T_s = 1$ oz, $T_t = 1.8$ oz, it means the friction increases the tension by 80%.

12.3 Coefficient of friction on solid surface

It can be found experimentally the coefficient of friction (μ) of a solid surface by passing a cotton thread on a steel surface with the help of the coil friction formula. A simple method is indicated in Fig. 12.8., the thread being hung over a bar of metal, and the thread is loaded as shown until it slides slowly and at uniform speed over the bar.

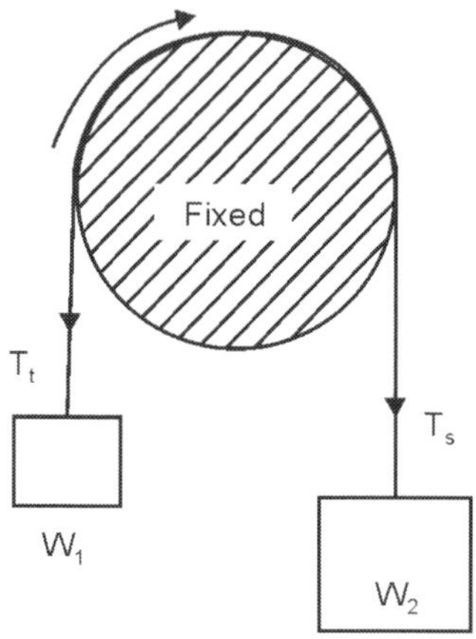

Figure 12.8

Then $T_t / T_s = W_1 / W_2 = 2.718^{\mu\theta}$ from which μ can be easily calculated.
Then $T_t / T_s = 2 = 2.718^{\mu\theta}$.
Taking logarithm on both sides,

$$\log 2 = \mu * \pi * \log 2.718$$
$$0.301 = \mu * \pi * 0.4343$$
$$\mu = 0.221$$

This method will often found to be useful for finding experimentally the slipping ratio for any angle of lap other than 180 °. The slipping ratio explained in the above problem is for 180°. For a lap of 90°, *Tt* / *Ts* = 2*90/180 = 2*1/2 = 1.41, while for 360°

$$T_t / T_s = 2*360/180 = 2*2 = 4.$$

12.4 Transmission of power by belts and rope drives

Power is transmitted by means of belts and ropes which transmit through the friction between belt or rope and the pulley rim. The maximum friction that can be applied by a belt on a flat pulley rim occurs

When the belt is at the point of slipping, i.e. when $T_t / T_s = e^{\mu\theta}$, where T_t is the tension on the tight side or driving side of the belt and T_s is the tension on the slack side.

Friction = $(T_t - T_s)$. The belt drive is designed, however, on the assumption that when transmitting the maximum power required, the belt will be near the point of slipping.

This reduces the size of the belt to a minimum. In many machine belt drives, the maximum power is required when turning up the machine the power then often being two or three times the normal running power, if the machine is to start up in reasonably short time. The drive must be designed to start with this maximum power requirement.

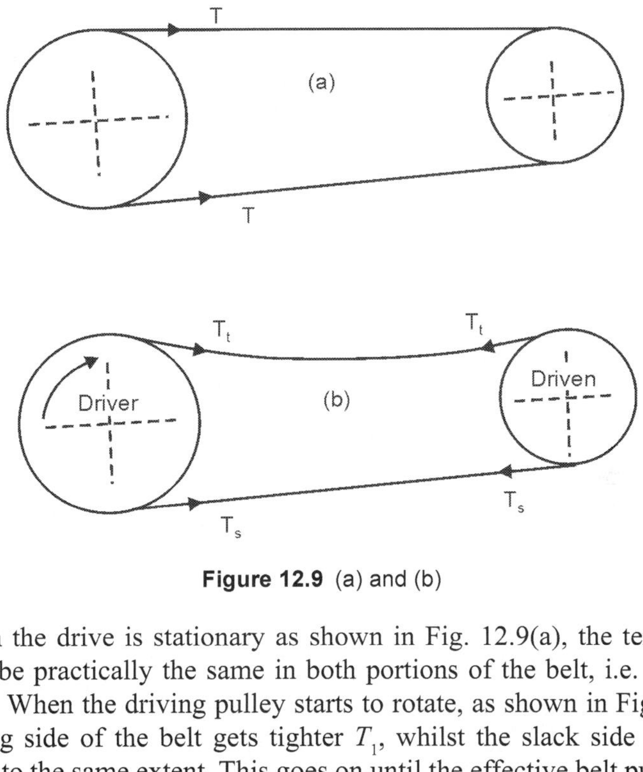

Figure 12.9 (a) and (b)

When the drive is stationary as shown in Fig. 12.9(a), the tension will normally be practically the same in both portions of the belt, i.e. the initial tension T. When the driving pulley starts to rotate, as shown in Fig. 12.9(b), the driving side of the belt gets tighter T_1, whilst the slack side tension T_s decreases to the same extent. This goes on until the effective belt pull applied to the driver pulley $(T_t - T_s)$ is sufficient to overcome the resistance offered by this pulley, when it begins to turn. As long as the resistance remains constant $(T_t - T_s)$ will now remain constant.

12.4.1 Conditions for transmission of maximum power

Let T_t = Tension on the tight side of the belt in lb

T_s = Tension on the slack side of the belt in lb

v = Speed of the speed of the belt in feet per minute

$$\text{Work done} = (T_t - T_s) * v$$
$$= (T_t - T_s)\ \pi * DN,$$

where D is the diameter in feet and N is the speed in rpm of either pulley.

Horse power transmitted = $(T_t - T_s) * v/33000$.

Maximum tension in the belt

A little consideration will show that the maximum tension in the belt (T) is equal to the total tension in the tight side of the belt (T_t).

The initial tension $T = T_t + T_s/2$, as stated above, T_s decreases at the same extent T_t increases. This is based on the assumption that the belt material is perfectly elastic.

Maximum tension in the belt is T_t and its value depends on the width of the belt required.

Working strength of the belt

The working load must always be lower than the breaking strength. For leather belts of single layer thickness, the working strength may be taken as 50 to 60 lb per inch width of the belt, the lower value being safe for most of the machine drives. The material will stand a much greater load than this, but the belt joint will reduce the strength of the belt.

Example 8: A belt drive from a motor to a line shaft has to transmit a maximum of 10 HP. Motor pulley has diameter of 5 in. and the speed is 960 rpm. Angle of lap at motor pulley is 172°; $\mu = 0.35$. Maximum working load on the belt is 80 lb per inch width of the belt.

Solution:

$$\text{Belt speed} = \text{Motor pulley rpm} * \pi * D$$
$$= 960 * \pi * 5/12 = 1260 \text{ ft per min}$$
$$\text{Effective belt tension} = \text{HP} * 33000/\text{speed of the belt in ft min}$$
$$= 10 * 33000/1260 = 262 \text{ lb}$$

Angle of lap at the pulley = 172/57.3 = 3 radians

$$\text{Slipping ratio} = T_t / T_s = 2.718^{(0.25 \times 3)}$$
$$= 2.86$$
$$T_s = T_t/2.86$$

Substituting the value of T_s in $(T_t - T_s)$

We have, $$T_s = \frac{T_t}{2.86} = 262$$

$$T_t - 0.35Tt = 262$$
$$T_t = 400 \text{ lb}$$

Belt width required = 400/80 = 5 in.

Initial tension required = $T_t + T_s/2$ = 400 + 140/2 = 270 lb.

12.4.2 Factors affect the power transmission by a belt

There are certain factors that affect the transmission of power by a belt drive. With horizontal drives, the tight side of the belt should be underneath, if possible, to get maximum angle of contact. The belt must be kept flexible.

One of the chief factors affecting the power transmitted is the speed of the belt. According to the horse power formula, power is directly proportional to the speed of the belt V. Hence, for a given amount of power, the higher the speed V, lower the belt tensions required and tighter the belt required. This condition is true for low and moderate speeds. When the speed goes beyond 5000 ft/min, instead of power increasing when belt speed is increased, the

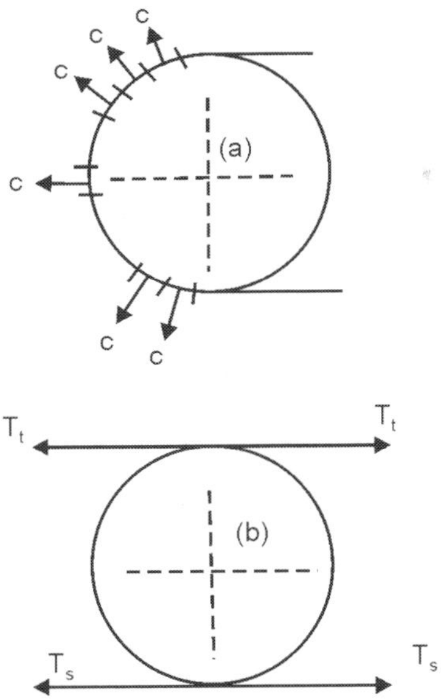

Figure 12.10 (a) and (b)

power decreases. This is attributed to the effect of centripetal force. It is the force needed to change the direction of motion of the belt as it goes round the pulley. The force must come from the tension in the belt, and this reduces the tension available for pressing the belt against the pulley on which the friction depends. The effect is equivalent to each short element of the belt round the pulley being acted on by a force C as shown in Fig. 12.10(a), pulling it away from the pulley rim and so reducing the pressure.

The net effect of all these forces C is equivalent to tensions T_c acting as shown in Fig. 12.10(b) opposing T_t and T_s. It simply means that without centripetal tension the full values of the forces T_t and T_s are pulling the belt against the pulley rim and are balanced by the reaction from rim to belt, when the forces T_c exist. The forces pulling the belt against the pulley rim are reduced to $(T_t - T_c)$ and $(T_s - T_c)$.

The effective belt tension is still $(T_t - T_s)$. But the belt will now slip at $(T_t - T_c) / (T_s - T_c) = e^{\mu\theta}$.

It is proved that centripetal tension $T_c = Wv^2/g$ where W = weight of 1 ft of belt in lb, V = belt speed in ft/s, g = acceleration due to gravity which is 32.2 ft/s.

Example 9: In a spinning machine, each of the two driving belts is 3 in. wide and will withstand a maximum working load of 300 lb. Pulley diameter = 16 in.; speed = 960 rpm; angle of lap = 3 radians; $\mu = 0.35$. Calculate the maximum horse power that can be transmitted (a) ignoring the centripetal tension, (b) allowing for centripetal tension, if the weight of the belt is 0.32 lb/ft.

Solution:

$$\text{Belt speed} = 960 * \pi * 16/12 = 4020 \text{ ft/min and } 67 \text{ ft/s}$$

$$T_c = Wv^2/g = 0.32 * 67^2/32.2$$

$$= 45 \text{ lb}$$

Slipping ratio of the belt = $2.718^{(0.25 \times 3)} = 2.86$.

(a)

$$T_t = \text{Maximum working load} = 300 \text{ lb}$$

$$T_t / T_s = 2.86$$

$$300 / T_s = 2.86$$

$$T_s = 300/2.86 = 105 \text{ lb}$$

$$\text{Horse power} = (T_t - T_s) * v/33000$$

$$(300 - 105) * 4020/33000 = 23.8 \text{ HP}$$

(b)

$$T_c = 45 \text{ lb}$$

$$T_t - T_c / T_s - T_c = 2.86$$

$$300 - 45 / T_s - 45 = 2.86$$

$$\text{From which } T_s = 134 \text{ lb.}$$

$$\text{Horse power transmitted} = (300 - 134) * 4020/33000 = 20.2 \text{ HP}$$

12.5 Transmission of power by V-belts and rope drives

A rope or V-belt in a V-shaped groove exerts a greater frictional resistance to slipping, for a given tension, as compared with a belt on a flat pulley rim. The rope or V-belt does not rest on the bottom of the groove but wedges itself into the groove as shown in Fig. 12.11(b) and (c). Hence, friction acts on both sides of the groove and it is due to the normal force *P*.

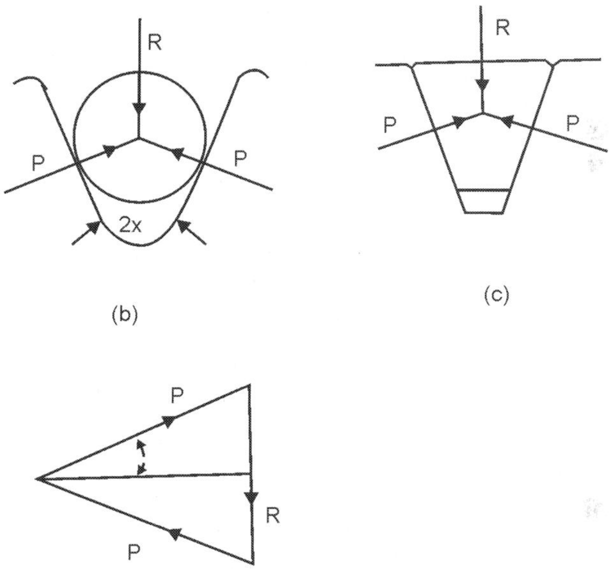

12.11 (b) and (c)

The total friction = 2 μP

From the triangle of forces, $R/2P = \sin \alpha$.

$2P = R / \sin \alpha = R \operatorname{cosec} \alpha$

Slip now takes place when $T_t / T_s = 2.718^{(\mu\theta \operatorname{cosec} \alpha)}$

In a flat belt shown in Fig. 10.4(a), the frictional force available = μR, for rope or v-belt the friction = $2\mu P = 2\mu P \operatorname{cosec} \alpha$.

For a rope in a 45° groove (α = 22.5°), the slipping ratio becomes $2.718^{(0.3 \times \pi \times \operatorname{cosec} 22.5^\circ)}$ where $\alpha = 0.3$, $O = 180°$

Example 10: A tex rope drive for a loom has 2 ropes in grooves. It must be able to transmit 2 HP at 180 rpm. Driving pulley diameter is 14 in. angle of lap at the motor pulley = 3 radians; μ by experiment = 0.5. What maximum pull does the rope can withstand?

Solution:

$$\text{Cosec } \alpha = \frac{9}{3} = 3$$

$$\text{Slipping ratio} = 2.718^{(0.3 \times 3 \times 3)} = 89.$$

Speed of the rope = 180 * π * 14/12 = 660 ft/min.

Maximum pull of the rope = $(T_t - T_s)$ = 2 * 33000/660 = 100 lb.

50 lb per rope

$$T_t - T_t/89 = 50$$

$$T_t = 50.5 \text{ lb}$$

It is known that that the slack side tension is negligible and the maximum tension will be at tight side of the rope.

12.6 Brakes

A brake is a device used either to bring a body which is in motion or to hold a body in a state of rest or of uniform motion against the action of external forces or couples. Actually the brake offers frictional resistance to the moving body and this frictional resistance retards the motion and the body comes to rest. During the process, the kinetic energy is absorbed by the brakes.

12.6.1 Types of brakes

The brakes are classifies as follows:

(a) Hydraulic brakes
(b) Electric brakes
(c) Mechanical brakes

In textile applications, mechanical brakes are commonly used.

The important mechanical brakes are as follows:

(a) Simple block or shoe brake
(b) Band brake
(c) Band and block brake
(d) Internal expanding shoe brake

12.6.2 Simple block or shoe brake

A simple arrangement for applying a braking force is shown in Fig. 12.12(a) and (b). The face of the brake is made up of special material which possesses high coefficient of friction.

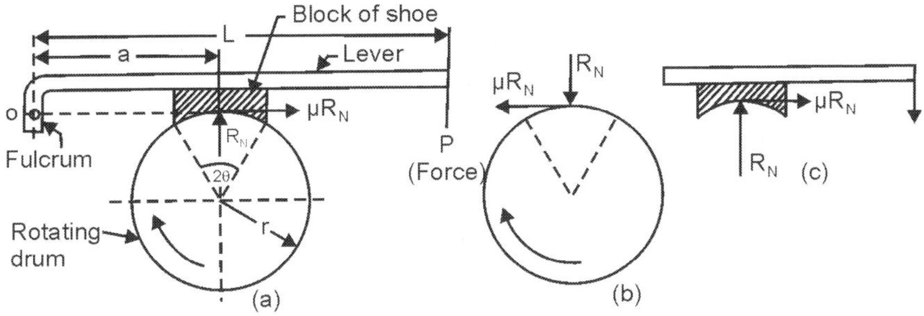

Figure 12.12 (a), (b) and (c)

A single block or shoe brake which is pressed against a rotating drum is shown in Fig. 12.12. The block is rigidly fixed to the lever. The force is applied at one end of the lever and the other end of the lever is pivoted on a fixed fulcrum *O*. As the force is applied to the lever, the block is pressed against the rotating drum.

The block is made of a softer material than that of the drum so that the block can be replaced easily in the event of wearing. For light loads, wood and rubber are used; whereas for heavy loads, cast steel is used.

Let P = Force applied at the lever end

r = Radius of the drum

μ = Coefficient of friction

R_N = Normal reaction on the block

2θ = Angle made by the contact surface of the block at the centre of the drum

F = Frictional resistance acting on the block = μR_N

T_B = Braking torque

When force P is applied at the lever end, the block is pressed against the rotating drum. The block exerts a radial force on the drum or in other words, this force passes through the centre of the drum. At the same time, the drum will exert a normal reaction (R_N) on the block. Hence the radial force on the drum will be equal to the normal reaction (R_N) on the block.

Assuming that the normal reaction (R_N) and the frictional force $F = (\mu R_N)$ act at the mid point of the block, we have

Braking torque on the drum = Frictional force * radius

$$T_B = F * r$$

$$= \mu R_N * r \quad \text{...(12.1)}$$

The braking torque can be calculated if the value of R_N is known in Eq. (12.1). The value of R_N is obtained by considering the equilibrium of the block.

In Fig. 12.12, the drum is rotating clockwise. Hence the frictional force on the drum will be acting in the opposite direction, i.e. in the anticlockwise direction as shown in Fig. 12.12(b). The frictional force on the drum will be opposite to the frictional force on the drum. Hence the frictional force on the block will be in the clockwise direction as shown in Fig. 12.12(c) (i.e. in the same direction in which drum is rotating).

Let the line of action of this frictional force (μR_N) passes through the fulcrum(O) of the lever. The forces acting on the block are:

(a) Normal reaction (R_N)
(b) Frictional force (μR_N)
(c) Applied force (P)

Taking moments about the point O, we get,

$$R_N * a = P * L$$
$$R_N = P * L/a$$

The frictional force (μR_N) passes through O and hence its moment is zero.

Substituting the value of R_N in Eq. (12.1), we have the braking torque

$$T_B = \mu * P * L/a * r \qquad \text{... (12.2)}$$

It is not necessary that the line of action of the frictional force (μR_N) should pass through the fulcrum of the lever O. The line of action of the force may be at a distance b below or above the fulcrum O of the lever.

Let us consider these two cases.

Case 1: When the line of action of the frictional force (μR_N) is at a distance "b" below the fulcrum O and the drum rotates clockwise as shown in Fig. 12.13. The forces acting on the block are:

R_N, μR_N and P

Taking moments about the fulcrum O, we get

$$R_N * a + \mu R_N * b = P * L$$
$$R_N (a + \mu * b) = P * L$$
$$R_N = P * L/(a + \mu b)$$

Substituting the value of R_N in Eq. (12.1), we get braking torque (TB) as

$$T_B = \mu * P \times L/(a + \mu b) * r$$
$$= \mu \times P \times L \times r / (a + \mu b) \qquad \text{... (12.3)}$$

If the drum is rotating in anticlockwise direction as shown in Fig. 12.14, then the frictional force (μR_N) will also be acting in anti-clockwise direction. The moment of all the forces acting on the block (R_N, μR_N and P) will give

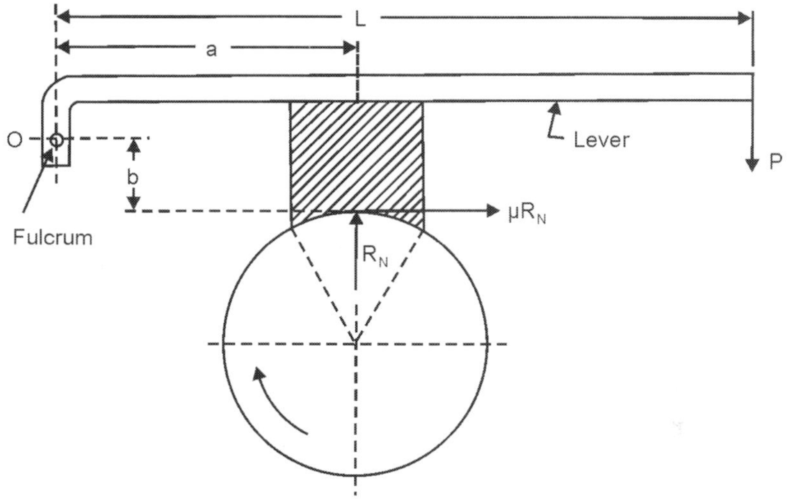

Figure 12.13

$$R_N \times a = P \times L + \mu R\_N \times b$$
$$R_N \times a - \mu R_N \times b = P \times L$$

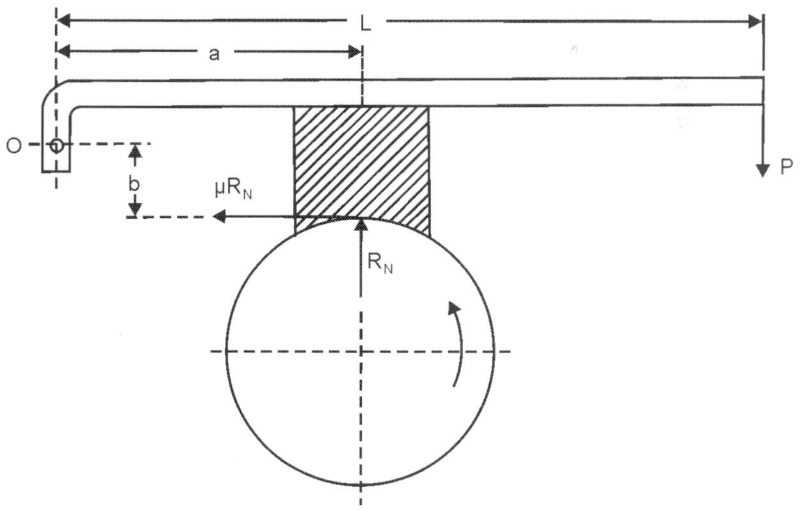

Figure 12.14

$$R_N = P \times L/(a - \mu b)$$

Substituting the value of *RN* in Eq. (12.1), we get the braking torque (*TB*) as

$$TB = \mu \times P \times L/(a - \mu b) *r$$
$$= \mu \times P \times L \times r/(a - \mu b) \qquad \dots (12.4)$$

Case 2: If the line of action of the braking force (μR_N) is at a distance of 'b' above the fulcrum O and the drum rotates clockwise as shown in Fig. 12.15. The forces acting are:

(a) R_N

(b) μR_N

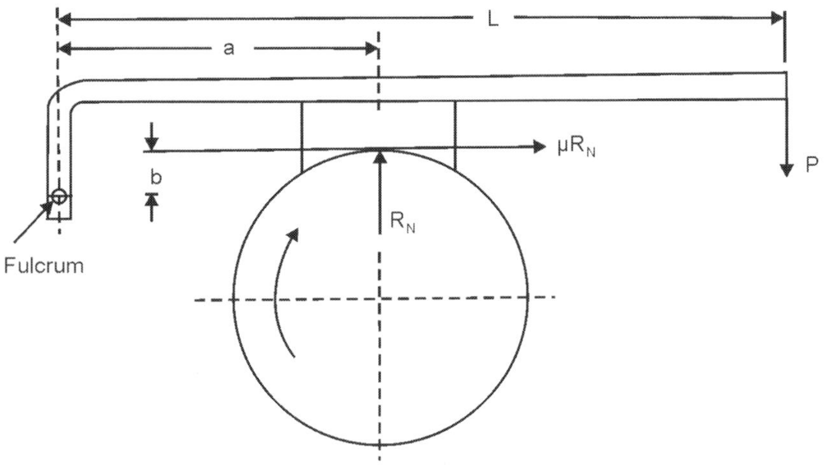

Figure 12.15

Taking moments about fulcrum O, we get

$$R_N \times a = P \times L + \mu R_N \times b;\ R_N \times a - \mu R_N \times b = P \times L$$
$$R_N (a - \mu \times b) = P \times L$$
$$R_N = P \times L/(a - \mu b)$$

Substituting the value of R_N in Equation 1, we get the braking torque (T_B) as

$$T_B = \mu \times P \times L /(a - \mu b) * r$$
$$= \mu \times P \times L \times r/(a - \mu b) \qquad \dots (12.5)$$

12.6.3 Double block or shoe brake

When single block brake is pressed against a rotating drum, a side thrust on the bearing of the shaft supporting the drum will act due to normal reaction (R_N). This results in bending of the shaft. This situation can be prevented by using two blocks on the two sides of the drum as shown in Fig. 12.16. Hence the braking torque required is twice as that required for single block brake. The braking torque is given by

$$T_B = \mu RN_1 \times r + \mu RN_2 \times r$$
$$= (\mu RN_1 + \mu\, RN_2) \times r$$

The value of RN_1 is obtained by taking moments of the forces RN_1, μRN_1 and P about fulcrum O_1. Likewise, the value of RN_2 is obtained by taking moments of the forces RN_2, μRN_2 and P about fulcrum O_2.

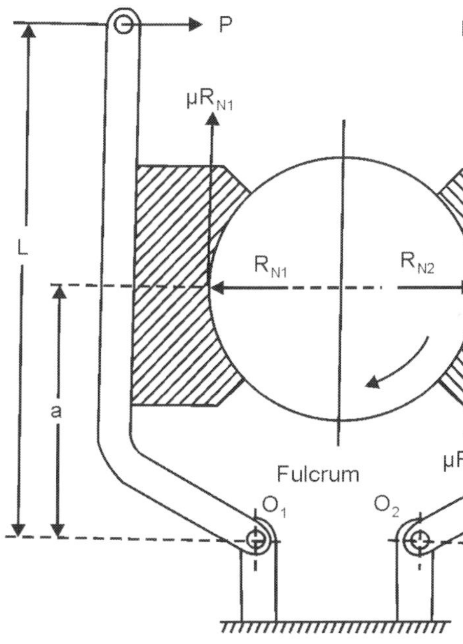

Figure 12.16

12.6.4 Band brake

A band brake may be a simple band brake or differential brake. In beam warping machines, simple band brakes are commonly used to produce the necessary braking force.

(a) Simple band brake: Simple band brake consists of one or more ropes, belt or flexible steel band lined with friction material which covers a part of the circumference of the rotating drum. Figure 12.17 shows a simple band brake in which one end of the band is attached to the fulcrum or fixed pin of the lever while the other end is attached to the lever at a distance 'a' from the fulcrum.

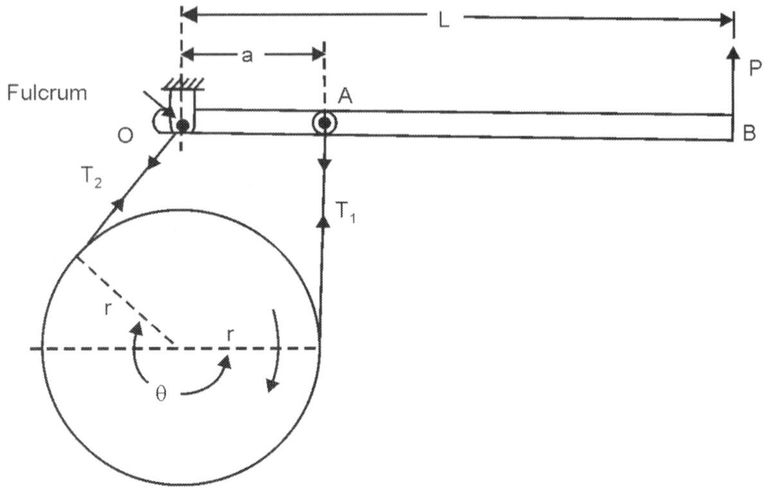

Figure 12.17

The force P is applied at the free end of the lever which turns about the fulcrum O. This tightens the band on the drum and hence the brakes are applied. The braking force is provided by the friction between the band and the drum. The force P at the end of the lever for clockwise and anti-clockwise rotation of the drum is explained as follows:

Let θ = angle of lap on the drum

T_1 = Tension in the tight side of the band

T_2 = Tension on the slack side of the drum

r = Radius of the drum

μ = Coefficient of friction between band and the drum

t = Thickness of the band

r = Effective radius of the drum = $(r + t/2)$

P = Force at the end of the lever

Slipping ratio or limiting ratio of tensions is given by

$$T_1/T_2 = e^{\mu\theta}$$

Braking torque on the drum is given by

$T_B = (T_1 - T_2) \times r$, if the belt thickness is neglected

$(T_1 - T_2) \times r_c$, if belt thickness is considered. We will consider two cases in which the values of P for clockwise and anti-clockwise rotation of the drum are

Case 1: Value of P for clockwise rotation of the drum

For clockwise rotation of the drum as shown in Fig. 12.17, the end of the band connected to the fulcrum O will be slack side with tension T_2 and the end of the band attached to A will be tight side with tension T_1. Taking moments about the fulcrum O, we get

$P \times L = T_1 \times a$ (Since T_2 passes through a) where L = Distance OB and a = perpendicular distance from O to the line of action of T_1.

Case 2: Value of P for anti-clockwise rotation of the drum (Fig. 12.18)

For anti-clockwise rotation of the drum as shown in Fig. 12.18, the end of the band connected to the end of the fulcrum O will be the tight side with tension T_1 and the end of the band attached to A will be the slack side with tension T_2. Taking moments about the fulcrum O, we get

$P \times L = T_2 \times a$ (since T_1 passes through O).

Where L = Distance OB, length of lever from fulcrum

a = Perpendicular distance from O to the line of action of Tension T_2.

Note: In simple band brake, one end of band is always connected to the fulcrum.

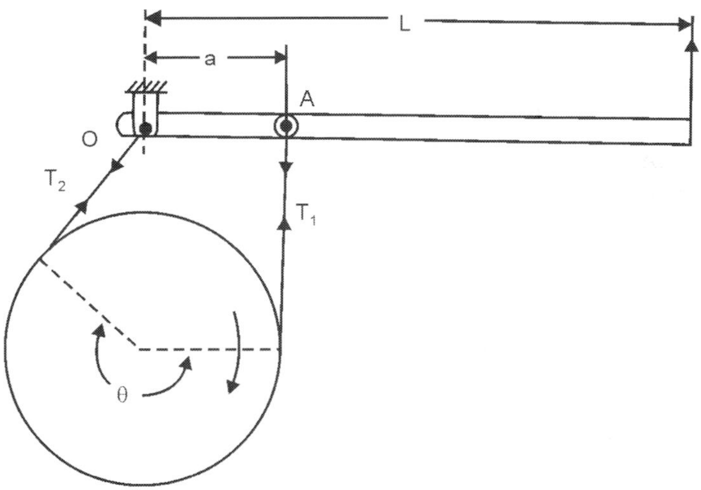

Figure 12.18

Example 11 (In SI units): The bake drum of a single block brake is running at 500 rpm in the clockwise direction. The diameter of the drum is 400 mm. The force required at the end of the lever is 300 N. The angle of lap is 30° and length (L) is 1 m, a = 333 mm and b = 25 mm. Determine the braking torque if the coefficient of friction is 0.3.

Solution:

Refer Fig. 12.13,

Speed (N) = 500 rpm

Radius of the drum = 400/2 = 200 mm or 0.2 m

Force at the end of the lever (P) = 300 N

Angle of contact $2\theta = 30°$

Length of lever from fulcrum (L) = 1 m

Distance of centre of the block from fulcrum (a) = 300 mm = 0.3 m

Perpendicular distance between line of action of frictional force and fulcrum (b) = 25 mm = 0.025 m

Rotation of the drum clockwise

Coefficient of friction (μ) = 0.3

Taking moments of all the forces (R_N, μR_N and P) about fulcrum, we get

$$R_N \times a + \mu R_N \times b = P \times L$$

$$R_N \times 0.3 + 0.3\ RN \times 0.025 = 300 \times 1$$

$$R_N = 300/\ 0.3 + 0.3 \times 0.0025$$

$$= 975.6\ N$$

Braking torque (T_B) is given by

$$T_B = \mu R_N \times r$$

$$= 0.3 \times 975.6 \times 0.2$$

$$= 58.536 \text{ Nm}$$

Example 12 (In SI units): A simple band brake is applied to a rotating drum of diameter 500 mm. The angle of lap on the drum is 270°. One end of the band is attached to a fulcrum pin of the lever and the other end is to a pin 100 mm from the fulcrum. If the coefficient of friction is 0.25 and a braking force of 90 N is applied at a distance 600 mm from the fulcrum, find the braking torque when the drum rotates (i) clockwise and (ii) anti-clockwise directions.

Solution:

In simple band brake, one end of the band is connected to fulcrum.

d = 500 mm = 0.5 m, r = 500/2 mm = 250 mm = 0.25 m.

$\theta = 270°$, distance a = 100 mm = 0.1 m; L = 600 mm = 0.6 m

$\mu = 0.25$, $P = 90\ N$

Let T_B = braking torque

(i) Drum rotates in clockwise direction

The braking torque T_B is given by $(T_1 - T_2) \times r$ ... (12.6)

Let us first find the values of T_1 and T_2. Taking moments of all forces as shown in Fig. 12.19 about O, we get

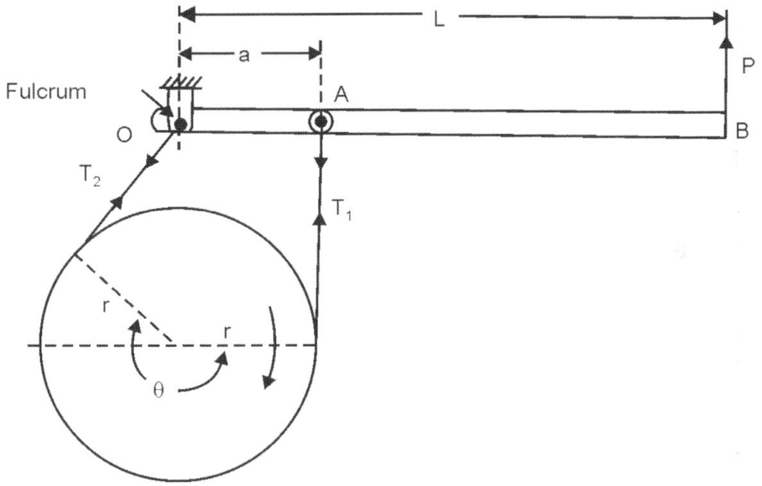

Figure 12.19

$$T_1 \times a = \text{P} \times \text{L}$$
$$T_1 = 90 \times 0.6/0.1$$
$$= 540 \text{ N}$$

Slipping ratio or limiting ratio of tension is given by

$$T_1/T_2 = e^{\mu\theta}$$
$$T_1/T_2 = 2.718^{0.25 \times 4.712} \ (270/57.3 = 4.712)$$
$$T_1/T_2 = 3.247$$
$$T_1 = 540 \ N$$
$$540/T_2 = 3.247$$
$$T_2 = 540/3.247$$
$$= 166\text{N}$$
$$\text{Braking Torque } (T_B) = (T_1 - T_2) \times r$$
$$T_B = (540 - 166) \times 0.25$$
$$= 93.5 \text{ Nm}$$

(ii) Drum rotates anti-clockwise

Refer Fig. 12.20

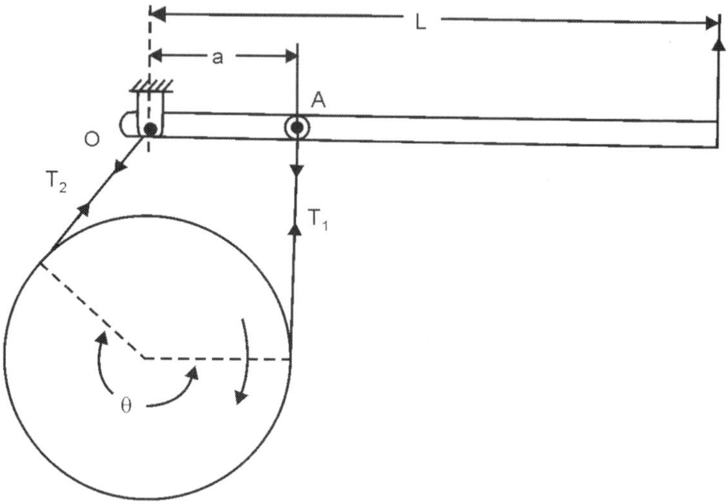

Figure 12.20

Taking moments about all the forces about fulcrum O, we get

$$T_2 \times a = P \times L$$
$$T_2 = 90 \times 0.6/0.1$$
$$T_2 = 540 \text{ N}$$
$$\text{Slipping ratio} = 3.247$$
$$T_1/\mathrm{T}_2 = 3.247$$
$$T_1 = T_2 \times 3.247$$
$$T_1 = 540 \times 3.247$$
$$= 1753.28\text{N}$$

Substituting the values of T_1 and T_2 in Eq. (12.6), we get

$T_B = (1753.28 - 540) \times 0.25 = 303.34$ Nm

(iii) Band or block brake

The band brake may be lined with blocks of wood or other material as shown in Fig. 12.21. The braking action is provided by the friction between the blocks and the drum. Let there are 'n' number of blocks each subtending at an angle 2θ at the centre and the drum rotates in anti-clockwise direction.

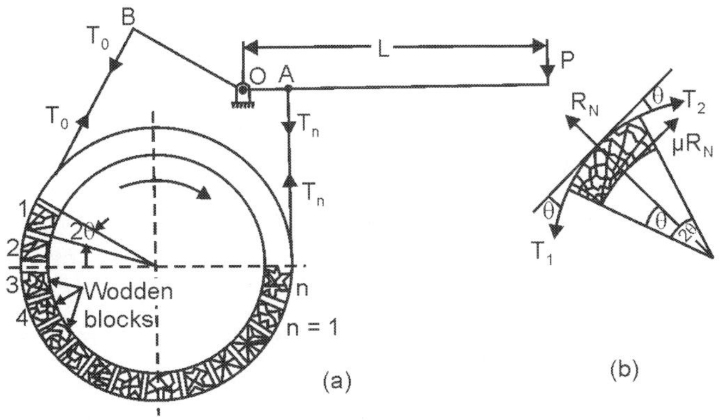

Figure 12.21

Let

T_1 = Tension on the tight side

T_2 = Tension on the slack side

μ = Coefficient of friction between blocks and drum

T_1' = Tension in the band between first and second block

T_2', T_3', etc. be Tension in the band between second and third block, between the third and fourth block and so on.

We will consider one of the block as shown in Fig. 12.21(b).

This is in equilibrium conditions under the action of the following forces.

1. Tension in the tight side (T_1)
2. Tension (T_1') in the band between the first and second block
3. Normal reaction (R_N) of the drum on the block
4. Force of friction (μR_N)

Resolving the forces radially, we get

$$(T_1 + T_1') \sin \theta = R_N \qquad \ldots (12.7)$$

Resolving the forces tangentially, we get

$$(T_1 - T_1') \cos \theta = \mu R_N \qquad \ldots (12.8)$$

Dividing Eq. (12.8) by Eq. (12.7),

$(T_1 - T_1') \cos \theta/(T_1 + T_1') \sin \theta = \mu R_N/R_N$

$$(T_1 - T_1') = \mu \tan \theta \, (T_1 + T_1')$$

$$T_1/T_1' = 1 + \mu \tan \theta/1 - \mu \tan \theta.$$

Similarly, for other blocks also it can be proved that

$$T_1'/T_2' = T_2'/T_3' = T_{n-1}/T_2 = 1 + \mu \tan \theta \,/\, 1 - \mu \tan \theta$$

$$T_1/T_2 = T_1/T_1' \times T_1'/T_2' \times T_2'/T_3' \times \ldots T_{n-1}/T_2$$

$$= \left(\frac{1 + \mu \tan \theta}{1 - \tan \theta}\right) n$$

Therefore, braking torque on the drum of effective radius rc,

$$T_B = (T_1 - T_2) * r_c$$

$$= (T_1 - T_2) * r, \text{ if the thickness of the band is ignored}$$

Point to be noted: For the first block, the tension in the tight side is T_1 and in the slack side is T_1' and for the second block the tension in the tight side is T_1" and the tension in the slack side is T_2". Similarly, for the third block the tension in the tight side is T_2" and in the slack side is T_3" and so on. For the last block, the tension in the tight side is T_{N-1} and in slack side is T_2".

12.6.5 Internal expanding brake

Internal expanding brake is shown in Fig. 12.22. It consists of two shoes S_1 and S_2. The outer surface of the shoes are lined with some friction material which have high coefficient of friction to increase the friction and prevent premature wearing away of the metal. Each shoe is pivoted at one end about a fixed fulcrum O_1 and O_2 and made to contact a cam at the other end. When the cam rotates, the shoes are pushed outwards against the rim of the drum. The braking torque is obtained by the friction between the shoes and the drum and in due course the speed of the drum get reduced. The shoes are normally held in off position by a spring as shown in Fig. 12.22. The drum prevents the entry of dust and moisture. This type of brake is commonly used in motor cars and light trucks.

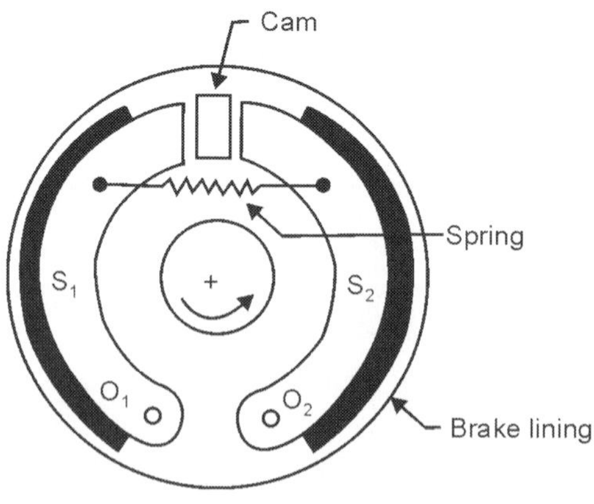

Figure 12.22

Example 13: A band brake for a beam warping machine is arranged as shown in Fig. 106. If $\mu = 0.25$ and the angle of lap is 240 degrees, what braking torque the brake can apply if the pulley rotates (i) clockwise and (ii) anti-clockwise.

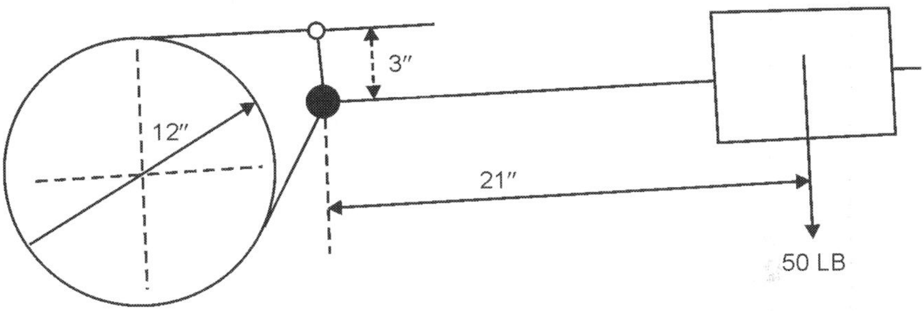

Figure 12.23

Solution:

$$\text{Load } (P) = 50 \text{ lb.}$$
$$\text{Distance } (L) = 21 \text{ in.}$$

Perpendicular distance (a) from the line of action of Tension (T_1) = 3 in.
Radius of the drum (r) = 6 in.

Coefficient of friction (μ) = 0.25

Angle of lap = 240 degrees; T_1= Tight side tension and T_2 = Slack side tension of the belt.

(i) Drum rotates clockwise

Taking moments about fulcrum O, we have

$$T \times a = P \times L \text{ (Since } T_1 \text{ passes through } O)$$
$$T_2 = P \times L/a$$
$$T_2 = 50 \times 21/3$$
$$= 350 \text{ lb.}$$

Slipping ratio

$$T_1/T_2 = e^{\mu\theta}$$
$$T_1/T_2 = 2.718^{0.25 \times 4.188} \ (240/57.3)$$
$$= 4.188$$
$$T_1/T_2 = 8.11$$
$$T_1 = T_2 * 8.11$$
$$T_1 = 350 * 8.11$$
$$T_1 = 2838.5 \text{ lb.}$$

$$\text{Braking torque } (T_B) = (T_1 - T_2) \times r$$
$$= (2838.5 - 350) \times 6$$
$$= 14931 \text{ in. lb}$$
$$T_B = 14931/12 = 1244.25 \text{ ft. lb.}$$

(ii) Drum rotates anti-clockwise

Taking moments about fulcrum O for all the forces, we have

$$T \times a = P \times L \text{ (since } T_2 \text{ passes through } O)$$
$$T_1 = 50 \times 21/3 = 350 \text{ lb.}$$
$$T_1 = 350 \text{ lb.}$$
$$\text{Slipping ratio} = 8.11$$
$$T_1/T_2 = 8.11$$
$$T_2 = 350/8.11$$
$$= 43.15 \text{ lb}$$
$$\text{Braking torque } (T_B) = (T_1 - T_2) \times r$$
$$= (350 - 43.15) \times 6$$
$$= 1841.06 \text{ in. lb}$$
$$T_B = 1841.06/12$$
$$= 153.42 \text{ ft lb}$$

Example 14: A warp is tensioned by passing it over and under a series of four round tension bars, the warp making an angle of lap of 60 degrees with each bar. It is found by experiment that when the angle is 180 degrees, the warp slips when the tight side tension is 3 times the slack side tension. If the warp tension is 20 lb before reaching the bars, how much is the tension increased by the bars?

Solution:

$$\text{Input warp tension} = 20 \text{ lb}$$
$$\text{Tight side tension } (T_t) = 3T_s$$

Angle of lap on each bars = 60°

Slip occurs at 180 degrees

Total angle of lap in all the 4 round bars = 240°

Slip occurs at 180°

$$T_t / T_s = 240 - 180 = 60° \ (60/57.3) = 1.047$$
$$T_t / T_s = 2.718^{1.047}$$
$$= 2.848$$

Slip occurs at third bar, hence

$$
\begin{aligned}
2.848/3 &= 0.948 \text{ at each bar} \\
\text{Input tension} &= 20 \text{ lb} \\
\text{Slipping ratio} &= 65.86/23.11 = 2.848 \\
T_t / T_s &= 2.848 \\
T_t &= 3T_s \\
3\,T_s / \mathrm{T_s} &= 2.848 \\
T_s &= 2.848/3 \\
&= 0.949 \\
\text{Input tension} &= 20 \text{ lb} \\
T_s &= 20/0.949 = 21.07 \text{ lb} \\
T_t &= 3T_s \\
&= 3 \times 21.07 = 63.22 \text{ lb} \\
T_t &= 63.22/0.949 = 66.6 \text{ lb}
\end{aligned}
$$

The warp tension is increased by 66.6 lb

Numerical questions

1. A thread is passing through a ring frame traveler is changed in its direction by 80 degrees. If μ between the thread and the traveler is 0.26, by what percentage the yarn tension is increased by the traveler?

   Ans. Tension has increased by 43%.

2. The weighing chains of a loom friction let-off motion each make 1.5 laps round the ruffle and μ = 0.15. The ruffle diameter is 6 in. and the tight side tension on each chain is 300 lb.

   (a) Find the warp tension required to turn the beam against friction, when the beam diameter is 16 in.

   (b) If the angle of lap of the chain is reduced to 0.5 lap, what effect will this have on warp tension required to turn the beam?

   Ans. (a) 170 lb; (b) 85 lb

3. A carding engine requires 1 HP to keep it running steadily at 180 rpm. If the pulley diameter is 18 in., what is the effective tension on the belt? A 3 in. wide belt is used, loaded to 225 lb. maximum on the tight side. What maximum HP is the belt capable of transmitting, if the slipping ration T_t to T_s is 3. Also, when will be the belt loaded to the maximum?

   Ans. Effective tension = 38.9 lb; maximum horse power at the start-up = 3.87.

4. A V- belt drive with three belts connects a motor to a machine, the motor pulley being 4 in. diameter and its speed 960 rpm. If each of the belts will stand a maximum working tension of 60 lb. and the belt slips when the tight side tension is 1.5 times the slack side tension, what maximum horse power can the drive transmit? With what initial tension must the belt be put on? What will be the effect on power the belts can consume if wear reduces the initial tension to half its correct value?

   Ans. HP = 5.1; initial tension = 32 lb per inch; power is reduced by 50 percent.

5. A rope drive from motor to machine has 2 ropes. The motor pulley is 3 in. diameter at the rope centre and has grooves of 40 degrees. The machine pulley has flat face and 18 in. in diameter. Find the tension ratio at which the slip can occur and the maximum horse power the drive can transmit if the maximum tension in the rope does not exceed 50 lb. Motor speed = 960 rpm; $\mu = 0.5$; angle of lap is 146° at the motor pulley and 214° at the machine pulley.

   Ans. Slipping ratio = 6.4; Horse power = 1.93

6. A ring frame spindle has to be driven at 10000 rpm by a tape making an angle of lap of 90° with the 7/8 in. wharve diameter on the spindle. μ between tape and wharve is 0.26. If the maximum resisting torque to the spindle is 1000 grain-inches, calculate the frictional driving force the tape must apply to the wharve and the minimum initial tension required in the tape ignoring the effect of centripetal force. If the tape weighs 0.1 oz per ft, what is the centripetal tension and how will it affect the required initial tension?

   Ans. Driving force = 2289 grains

   Initial tension = 5725 grains

   Centripetal tension = 2000 grains

   Initial tension should increase by 2000 grains.

7. A rope drive from an engine flywheel to a line shaft has 5 ropes of 1.75 in. diameter. Rope speed is 4800 ft per min. Maximum working load per rope (lb) is 160 D_2. Weight of rope is 0.3 D_2 lb per ft; angle of rope groove is 45°; $\mu = 0.25$; angle of lap on smaller pulley is 3 radians; what maximum horse power the rope can transmit at the point of slipping? Allow for centripetal force.

   Ans. Maximum horse power = 219

13

Lifting machines and velocity ratio

13.1 Introduction

A machine is a contrivance which receives energy in some available form and uses it for doing a particular useful work. For example, the steam engine converts the energy provided by steam in to motion of translation.

In the similar way, lifting machines have been dealt with. A lifting machine may be defined as a contrivance to overcome a force or load (W) applied at one point by means of another force called effort (P) applied at another point. In most of the cases, P is smaller than W but sometimes it may be greater than W.

13.2 Definitions

Let an effort P be applied through a distance y to lift a load W through a distance x, the ratio

Distance moved by y / Distance moved by x

$= y/x$ is called the velocity ratio (VR) of the machine

The ratio weight lifted / effort applied

$= W/P$ is called the mechanical advantage (MA) of the machine.

In an ideal machine, in which friction is negligible, the work put in by the effort is equal to the work done on the load, i.e.

$$Py = Wx$$

or $$y/x = W/P \qquad \dots (13.1)$$

Hence, Velocity ratio = Mechanical advantage

However, no machine is perfectly ideal, since some amount of work done is lost in overcoming the friction and in lifting parts of the machine. So the work applied is always greater than the useful work obtained.

$$Py > Wx$$

or $$Y/x > W/P$$

The ratio "useful work obtained / Actual work applied" is called the efficiency of the machine. Thus

$$\text{Efficiency } (\eta) = Wx / Py = W/P\,(y/x) = W/PV$$

We can also write $\eta = W/PV = (W/P)/V$

or

$$\text{Efficiency} = \text{Mechanical advantage / Velocity ratio}$$

$$\text{Again, } \eta = W\,x/P\,y = (W\,x/y)/P \qquad \text{... (13.2)}$$

But from Eq. (13.2), we know that in an ideal machine, the effort P required to lift a load W is given by $P = W\,x/y$. Hence Eq. (13.3) can be written as

η = Effort required to lift load W in an ideal machine/Actual effort required to lift load W

= ideal effort/Actual effort

Efficiency is a fraction less than one. It is usually expressed in percentage multiplied by 100.

Condition for irreversibility of a machine

Let P denote the actual effort to lift a load W. Now the ideal effort P required to lift a load W is given by

$$y/x = W/P$$

Hence effort lost in friction

$$= P - P' = P - W/V$$

When the effort P is removed, the force resisting the falling down of weight W will be the force of friction, i.e. $P - W/V$. The machine will be reversible, i.e. W will not fall down on removal of P, provided the force $P = W/V$ is greater than P which is the force necessary to balance W. That is

$P - W/V > W/V$

or $P > 2W/V$

$\frac{1}{2} > W/PV$

$W/PV > \frac{1}{2}$

Therefore, the machine is said to be irreversible or self-locking, if the efficiency is less than 50%.

13.3 Velocity ratio

The velocity ratio may be defined in various forms, all meaning the same thing. For instance,

Movement of the first driver/Movement of the last driver = velocity ratio

or

The first movement in a given time/the last movement in a given time = velocity ratio

or

Space moved over at driving end/space moved at finishing end = velocity ratio

It will be seen that in any given arrangement of driving mechanism, it is simple matter to find out how much faster or how much slower the resulting speed or movement is than the starting speed.

or

Starting movement/resulting movement = velocity ratio

This may be illustrated in the case of a simple lever as shown in Fig. 13.1. If the load is 8 lb and the weight of 32 lb is required to balance it on the other arm in the position shown. Then

Load lifted/Load applied = 32/8

= 4 is the mechanical advantage

If instead of weight being used, we applied other forms of force such as a driving effort, then there would be exerted also at the other end a resultant effort or load, so that the effect can be expressed in the form

Load/Driving effort = 32/8

= 4 is the mechanical advantage

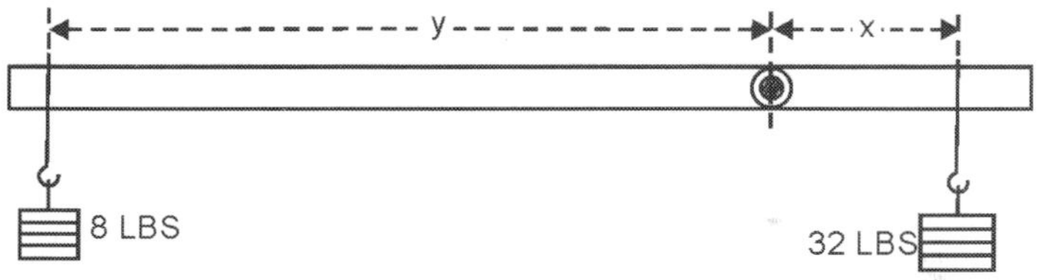

Figure 13.1

We will now suppose that the lever moves round its fulcrum as in Fig. 13.2.

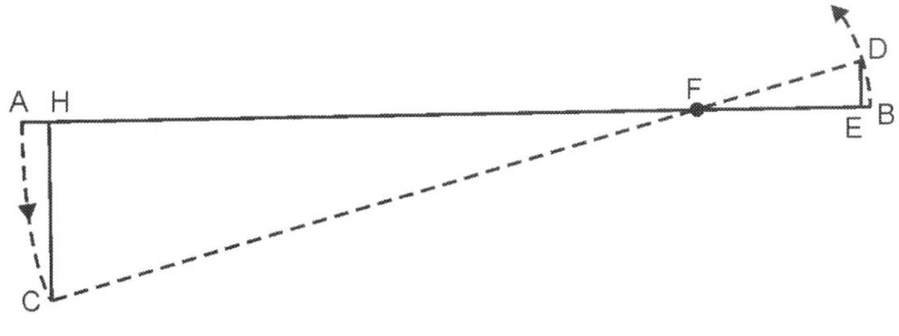

Figure 13.2

When the lever has moved from position *AB* to *CD*, the end *A* has traversed a portion of a circle *AC* and the end *B* has moved to the circular path *BD*, so that

AC/*BD* = velocity ratio

It is quite easy to say that since *AF* is four times greater than *FB*, the arc *AC* is four times longer than *BD*, and hence the velocity ratio is four. This shows that the mechanical advantage is equal to the velocity ratio, and as a rule, the mechanical advantage of most of the simple machines is based on finding the velocity ratio of the mechanism. Another important factor which must not be overlooked is friction, for it will be found on making tests on various kinds of apparatus, the friction destroys some of the advantage, so this factor must always be taken in to account.

13.4 Velocity ratio on simple pulleys

Figure 13.3 represents an ordinary simple pulley. In this arrangement, a cord *B* passes over the pulley *A* to one end of which is attached a weight *W*. It has to be raised by pulling at the other end *P*. If the force applied at *P* were equal to the weight *W*, there would be a balancing effect, for the weight will exert its force equally along the whole length of the cord *B*. To raise *W*, we must apply at *P* a pressure in excess of *W*. In addition, an extra pressure must be added to *P* to overcome the friction on the axle or pin that carries the pulley *A*. This extra pressure will vary according to the load lifted. In this arrangement, it will be seen that the pulley offers no mechanical advantage, for the pressure is equal to the weight and the movements of each end of the cord are equal.

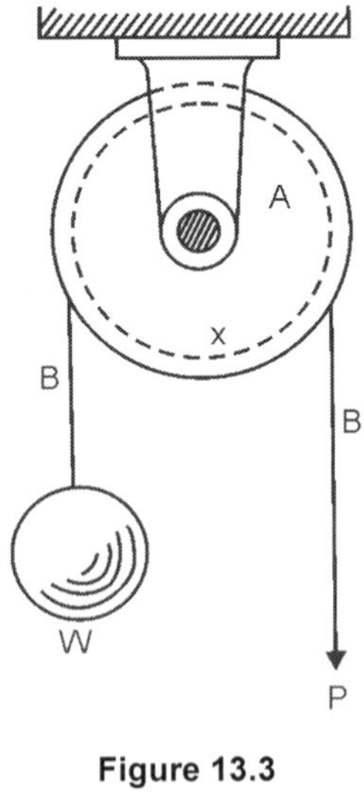

Figure 13.3

In another arrangement of the pulleys as represented in Fig. 13.4, pulleys are arranged to afford great mechanical advantage and are extensively used for this purpose.

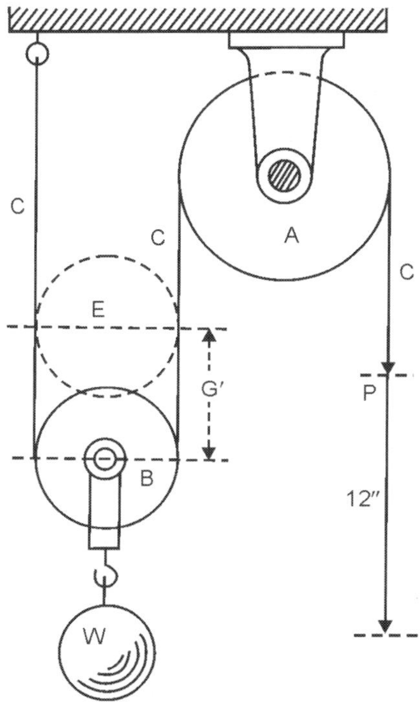

Figure 13.4

In Fig. 13.4, a cord *C* is fastened to the frame work and is passed over a movable pulley *B* and over a fixed pulley *A*. A pressure at *P* will raise the weight *W* which is attached to the movable pulley *B*.

The pressure *P* will be equal along the whole cord *C*, so that the pulley *B* is being supported by two cords, each transmitting a pressure *P*: the weight therefore will be equal to 2*P* and the mechanical advantage is two.

If we assume that the pressure *P* pulls the cord downwards a distance of 12 in., then the pulley *B* and weight would be raised to 6 in., because it would require 6 in. of each supporting cord to obtain 12 in. of *P*'s movement. The space ratio or velocity ratio in this case is two.

In another system of arrangement as represented in Fig. 13.5, it is an extended application of previous Fig. 13.4 by using more pulleys.

If *P* moves downwards 12 in., then pulley *B* is raised by 6 in.

If pulley *B* is raised 6 in., then pulley *C* is raised by 3 in.

If pulley *C* is raised by 3 in., then pulley *D* is raised by 1.5 in.

If pulley *D* is raised by 1.5 in., then pulley *E* is raised by 0.75 in.

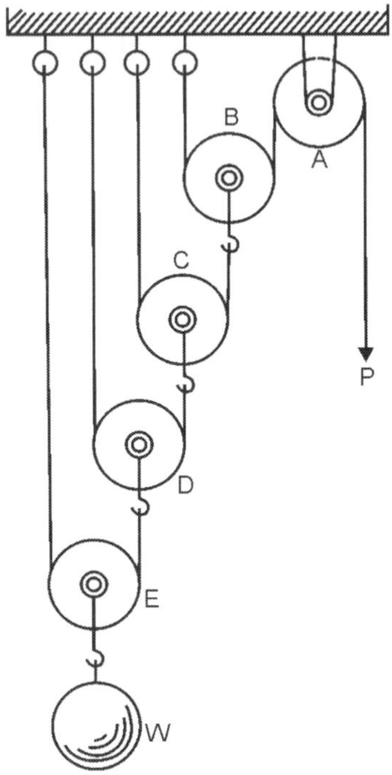

Figure 13.5

For 12 in. movement of *P*, we therefore obtain 0.75 in. movement of the weight *W*, so that

= First movement / Last movement = *P/E*

= 12/0.75

= 16 is the velocity ratio and mechanical advantage

Suppose if the pull of *P* is 10 lb, then the tension of the whole of the cord round *B* = 10 lb, and *B* will support 2*P* = 20 lb

The cord round *C* has a tension of 20 lb and then the pulley *C* will support 40 lb.

The cord round *d* has a tension of 40 lb, then the pulley *D* will support 80 lb.

The cord round e has a tension of 80 lb then the pulley *E* will support 160 lb.

A pull of 10 lb at *P* will support a weight of 160 lb hanging from the pulley E, so that

160/10 = 16 is the mechanical advantage

The beam will support *W* + *P* so that the total load on the beam will be 160 + 10 = 170 lb

Another arrangement of pulleys is shown in Fig. 13.6.

In this arrangement, a continuous cord is used which passes over all the pulleys. The upper pulley is fixed to the framework whilst the lower ones are movable. It will be once seen that the weight is supported by five cords and as the tension throughout the cord is equal to *P*, we consider that P will support a weight equal to five times its own force so that the mechanical advantage equals five. If the force P moves 10 in. downwards, the five cords must each contribute an equal portion to make up the 10 in. movement so that the distance the weight is raised equals 10/5 = 2 in., therefore, the velocity ratio is

= First movement/ Last movement

= 10 in. / 2 in. = 5 in.

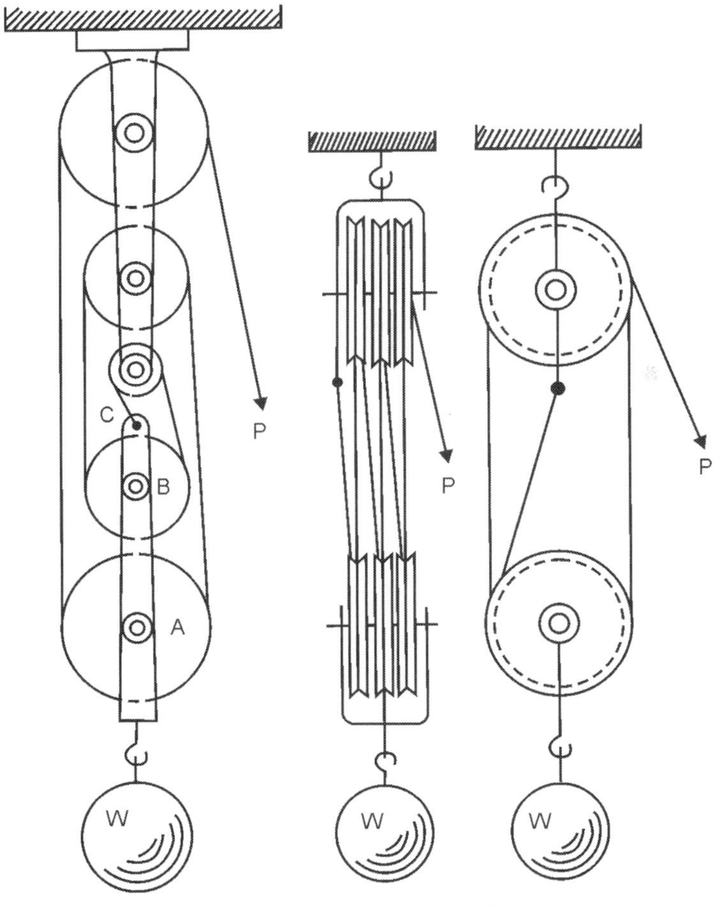

Figure 13.6 (a) and (b)

It is clear from the above illustration that many combinations of pulleys are possible but in practice this arrangement is extensively used in "pulley block tackle" for raising weights. From Fig. 13.6(b), it will be noted that the upper fixed pulleys are side by side on a single axle and the lower pulleys are treated in the same way, all the pulleys are of the same size. When the cord is fixed on the upper block, the number of pulleys in each block will be the same but if fixed on lower block there will be one less pulley in the lower than the upper block. The number of cords supporting the lower block in Fig. 13.6(b) is 6, so the mechanical advantage and velocity ratio will be 6.

One other arrangement of pulleys is shown in Fig. 13.7. Here it is quite simple to track the tensions in various cords and as each cord is attached to the weight, the total weight will be equal to the sum of tensions in the cords.

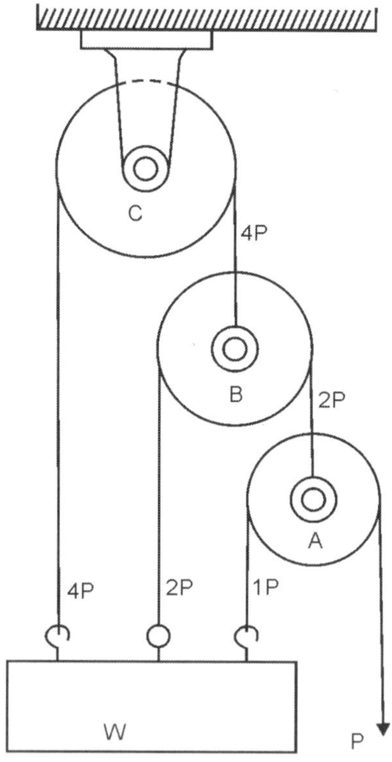

Figure 13.7

The tension in each part of the cord = P total weight = P

The tension in each part of the second cord = 2 P = total weight = $3P$

The tension in each part of the third cord = $4P$ = total weight = $7P$

If more pulleys are used we should continue as –

The tension in each part of the fourth cord = $8P$ total weight = $15\ P$

The tension in each part of the fifth cord = $16P$ total weight = $31P$

The tension in each part of the sixth cord = $32P$ = total weight = $63P$

It will be seen from the above discussion that the total weight in each case is equal to two raised to the number of cords or pulleys minus 1 or 2 power $n - 1 = W$, where n = number of cords or pulleys. In Fig. 13.7 the cords are three in number so that 2 power $3 - 1 = 7$; the mechanical advantage and velocity ratio being 7.

13.5 Weston differential pulley block

Weston pulley block arrangement is extensively used for hoisting purposes. The arrangement is represented in diagrammatic form.

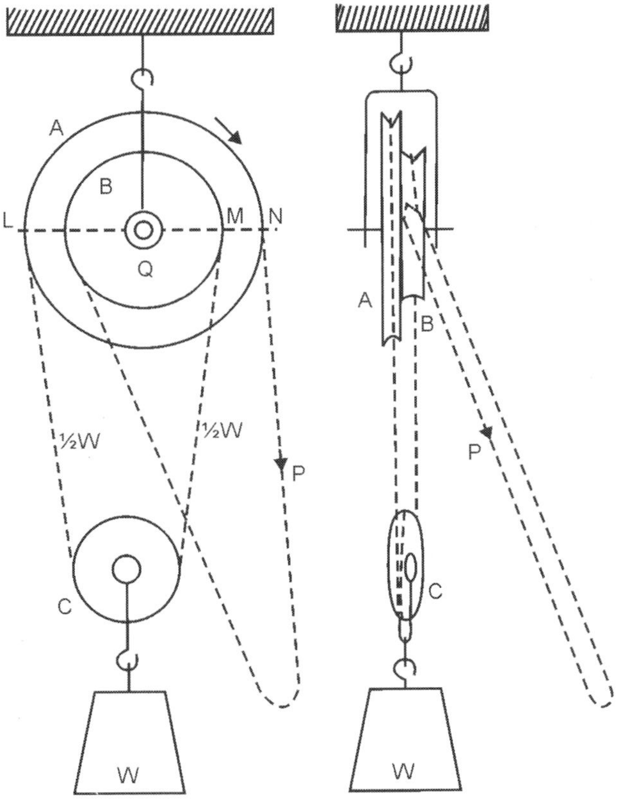

Figure 13.8

Two pulleys or sheaves, A and B of different diameters, are cast together so that both revolve as one on the axle, which is carried by a forked bearing hooked at its upper end. The rims of the sheaves are provided with spaced projections to receive the links of a chain, so that no slippage of the chains occur. The lower pulley C is single and carries the load to be raised.

An endless chain passes over the pulley A down and round the pulley C, then up and round the pulley B, and down again to the starting point where the ends are joined and hang freely. If the chain is pulled at P, two pulleys A and B are revolved in clockwise direction, the effect of which is that the larger pulley A raises the sheave C, whilst the chain from the smaller pulley B lowers the sheave C. The difference between these two movements is the amount the weight will be raised.

The velocity ratio may be found by assuming that the chain is pulled at P until A and B make one complete revolution in clockwise direction.

Then A will wind up a chain of length $= \pi A$

And B will unwind a chain of length $= \pi B$

The sheave C will be raised half the difference between the two circumferences of the two pulleys A and B, so that the weight is raised

$$\pi A - \pi B/2 = \pi(A - B)/2$$

Since P moves the chain a distance of πA, the velocity ratio is

First movement/Last movement or Movement of pressure/Movement of weight

$= \pi A / \pi(A - B)/2 = 2 * \pi A / \pi(A - B)$

$= 2A/A - B$ is the velocity ratio

The velocity ratio is numerically the same as the mechanical advantage.

Alternate method

The mechanical advantage of the Weston Block Pulley can also be found out by principle of moments.

In Fig. 13.8, there are three forces on a line drawn through the centre of the pulleys. These forces are

(a) P tending to turn the pulley clockwise acting at N

(b) $\frac{1}{2}W$ tending to turn the pulley clockwise acting at M

(c) $\frac{1}{2}W$ tending to turn the pulley anti-clockwise acting at point L

In order to obtain equilibrium or balance, the clockwise moments = anti-clockwise moments

$P \times NQ + \frac{1}{2}W \times MQ = \frac{1}{2}W \times LQ$

Where $NQ = \frac{1}{2}$ diameter of A, $MQ = \frac{1}{2}$ diameter of B and $LQ = \frac{1}{2}$ diameter of A.

Then $(P \times A/2) + (W/2 \times B/2) = W/2 \times A/2$

$$P \times A/2 = (W/2 \times A/2) - (W/2 \times B/2)$$

$$P \times A/2 = \tfrac{1}{2} W (A/2 - B/2)$$
$$P = \tfrac{1}{2} W (A/2 - B/2)/A/2$$
$$= \tfrac{1}{2}W (A - B)/A$$
$$= W/P = 2A/A - B$$

The result is the same as the velocity ratio obtained by the previous method.

13.6 Types of machines

All machines, however complicated, are made up of simple machines. They are (a) lever type and (b) inclined plane. Type(a) includes not only levers of different kinds but also pulleys, wheel and axle, and gear driving. Type (b) may be modified in the form of wedge, the screw, worm and worm wheel and cams. Examples of the simple machines are shown in Fig. 13.9.

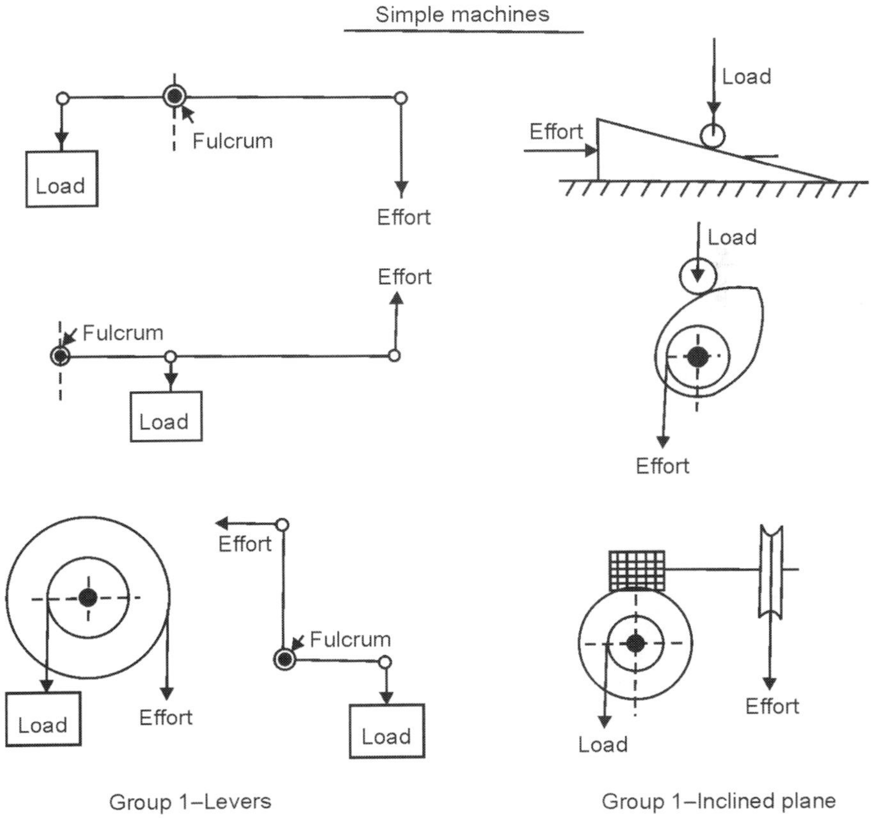

Figure 13.9

The efficiency of the lever-type machines is generally higher than that of the inclined-type machines. This is mainly due to the fact that the friction force required to overcome a distance is generally much greater in inclined plane than in lever arrangements.

13.6.1 Applications in textile machines

It is important to clearly distinguish the term efficiency between mechanics and efficiency of production which is familiar to textile engineers.

Efficiency of the production is the ratio of actual production to calculated production and it has no relevance to the mechanical efficiency of the machine.

13.7 Law of a machine

The term "Law of a Machine" states that if load W is increased, both the theoretical effort P and the actual effort Q must be increased obviously, since $P = W/VR$, P is directly proportional to W. The actual effort Q does not increase to the same extent; however, since the part of Q which overcomes friction, although higher for increased load will not be higher in proportion to the load. The result of this is that the efficiency generally increases with increase of load, rapidly at first, more slowly later.

If a machine is tested at different loads, results are obtained which are plotted, give curves as shown in Fig. 13.10., P and Q against W being straight line curves. The equation to the straight line connecting Q and W is known as the "Law of the machine." In Fig. 13.10, the law is $Q = 0.12W + 3$.

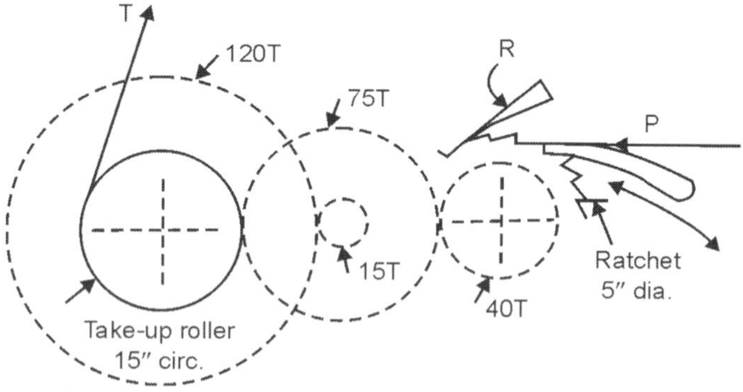

Figure 13.10

Example 1: Figure 13.10 shows the gearing of taking up motion of a loom, the pushing pawl P turning the gearing and taking up the cloth against the cloth tension T. Find the velocity ratio and the actual effort at the ratchet pawl to take up the cloth against a tension of 200 lb, if the efficiency is 75 percent.

Solution:

Velocity ratio = Movement of *p*/Movement of *T*

For 1 revolution of the take up roller, the take up roller (*T*) moves 15 in.

Movement of the ratchet wheel is 1* 120/15 * 75/40 * 3.14 * 5 in.

Velocity ratio = 120/15 * 75/40 * 3.14 * 5/15 = 15.7

Theoretical force (*P*) to overcome the effort (*T*) = 200/15.7 = 12.7 lb

Actual force (*Q*) required if the efficiency is 75 percent = *P*/*Q* = η

= 12.7/.75 = 17 lb

As the efficiency is more than 50 percent, the machine is said to be reversible. Consequently, as the pawl *P* is moving back to take another tooth, *T* should turn the gearing backwards unless prevented by the retaining pawl *R*, on a fixed pivot. *R* retains any forward movement of the gearing given by *P*.

Example 2: The drive to the chain of flats in a carding machine is shown in Fig. 13.11. What is the velocity ratio? If the efficiency is 0.1 percent and a force of 8 lb is required at the rim of the driving rope pulley to move the flats, what resistance do they offer?

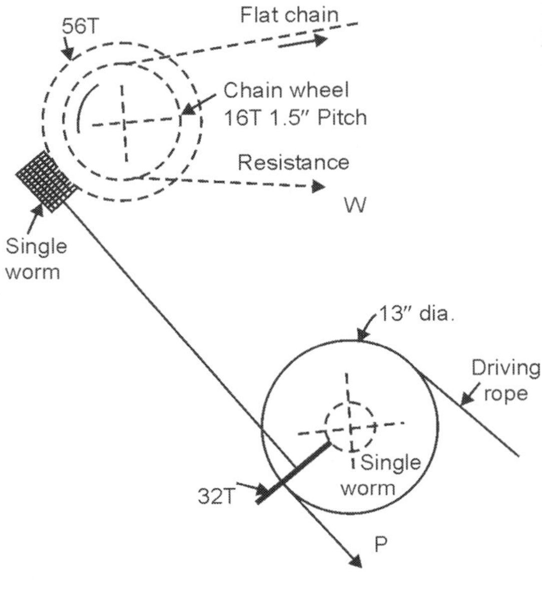

Figure 13.11

Solution:

For 1 revolution of the driving rope pulley, the movement of the rope is 3.14 * 13 and the movement made by the flat chain is:

From the gearing, it is calculated as

1/32 * 1/56 * 16 * 1.5 in.

Velocity ratio = Movement of rope pulley/ Movement of flat chain

$$= 3.14 * 13/1/32*1/56*16*1.5$$

$$= 3048$$

$$\text{Effort available } (Q) = 8 \text{ lb}$$

$$\text{Efficiency} = 0.1 \text{ percent}$$

$$\text{Pull } (P) = Q * \text{Efficiency}$$

$$= 8 * 0.1 = 0.8 \text{ lb}$$

$$W/P = VR$$

Resistance of the flat chain to movement is

$$W = P * VR$$

$$= 3048 * 0.8$$

$$= 2440 \text{ lb}$$

The resistance W is the frictional resistance necessary for doing the useful work so as to move the flat chain. Of the 8 lb effort is needed at the driving rope pulley, only 0.8 lb is effective in overcoming the friction at the flat chain. The remaining 7.2 lb is needed to overcome the friction between the driving pulley and the flat chain itself, mainly the friction at the two worm gears.

Another point to be noted is that due to the nature of the load even if the efficiency is above 50 percent, the machine is not reversible, since the frictional load W disappears as soon as the flat chain movement is stopped.

Example 3: The drive to the griffe of a single lift Jacquard loom is shown in Fig. 13.12. If the resistance of the griffe to being lifted due to its weight and downward pull of the harness, in the position shown is 50 lb, what effort P, at the rim of the hand wheel, would be required to lift the griffe if the drive is frictionless? Calculate the approximate efficiency and the actual effort Q required to lift the griffe slowly by the hand wheel?

Solution:

If P moves 1 in. the griffe will rise by 1* 2/6* 24/20 = 2/5 in.

$$\text{Velocity ratio } (VR) = 1/2/5 = 2.5$$

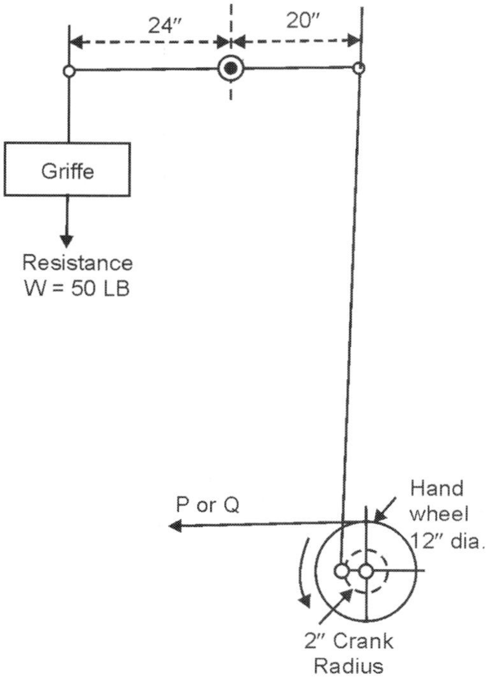

Figure 13.12

$$\text{Theoretical effort } P = W/VR$$
$$= 50/2.5 = 20 \text{ lb}$$

The approximate efficiency, if the machine is just reversible is 0.5. Effort $Q = 20/0.5 = 40$ lb

In this example, velocity ratio will not remain constant. This is because when the crank shaft has turned 60 degrees from the given position, the rate of griffe movement, for the same rate of turning the hand wheel would be reduced to half for the position shown, and hence the velocity ratio would be 5 instead of 2.5. In this position, an actual effort of 20 lb should be sufficient to keep the griffe rising.

Also, it is not essential that the velocity ratio of a machine should be greater than unity. It may be a fraction, in which case the mechanical advantage (MA) also will be less than one and the effort will be greater than the load.

Example 4: A loom picking mechanism was tested with the sley and shedding disconnected. The driving pulley rim was found to move, during the picking period, 3 in. for a 6 in. movement of the picker. The picker offers a resistance of 6 lb, a pull of 18 lb was needed at the rim of the driving pulley to move the picker. Calculate the velocity ratio and approximate efficiency of the parts which transmit motion from the driving pulley to the picker.

Solution:

Velocity ratio = Distance moved by the driving pulley/ distance moved by the picker

3/6 in. = 0.5

Resistance offered by the picker = 6 lb

$$P = 6/VR = 6/0.5 = 12 \text{ lb}$$

Pull (Q) required at the rim of the driving pulley = 18 lb

$$\text{Efficiency } (\eta) = 12/18 = 0.67.$$

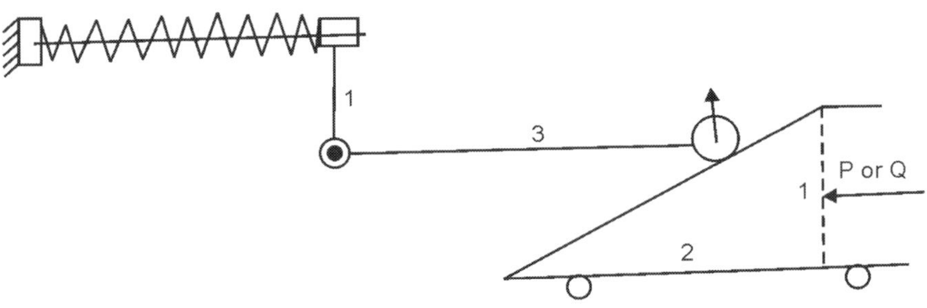

Figure 13.13

Example 5: A spring on a spinning mule is compressed by the horizontal movement of an inclined plane, acting through the "bell crank" lever. The spring has a stiffness of 12 lb per in. and has an initial resistance of 24 lb. If the efficiency is 50%, what horizontal force Q is needed at the inclined plane, and how much work is done by Q in compressing the spring 1 in.? The lever arms are in the proportion 3:1 to be horizontal and vertical throughout (Fig. 13.13).

Solution:

For the compression of the spring to 1 in., the bell crank lever must rise on the inclined plane 3 in. and the incline will move horizontally 6 in.

Velocity ratio = 6

Spring resistance at the start of the compression = 24 lb

Spring resistance at the end of the compression = 24 + 12 = 36 lb

Average resistance = 24 + 36 = 60/2 = 30 lb

Work done on the spring = 30 lb * 1 in. = 30 in. lb

At the start of the compression,

Theoretical effort P = Initial resistance/VR = 24/6 = 4 lb

Actual effort Q = 4/0.5 (efficiency) = 8 lb

At the end of the compression,

Theoretical effort P = Final resistance/VR = 36/6 = 6 lb

Actual effort Q = 6/0.5 = 12 lb

Average actual effort = 8 + 12 = 20/2 = 10 lb

Work done by Q = 10 lb * 6 in. = 60 in. lb

Or work done by Q = work on spring/Efficiency

= 30/0.5 = 60 lb

Example 6: A mule carriage is drawn by a friction clutch, 12 in. diameter, running at 120 rpm near the end of the acceleration of the carriage, the clutch is transmitting 4 horse power and the carriage is moving at 3 ft/s. If the efficiency of the transmission of force from the clutch to the carriage is 90 percent, what driving force is acting at the clutch face, and what resistance is the carriage offering in its movement?

Solution:

Speed of the clutch = π * 12 * 120 = 4521.6 in.

$$HP = T * 2*\pi*N/33000$$

$$T = 4 * 33000/2*\pi$$

$$= 175.15 \text{ in. lb}$$

Friction force available at the driving face of the clutch is

$$T/6/12(\text{radius}) = 175.15/0.5 = 350.12 \text{ lb}$$

$$\text{Work done per revolution} = 4 * 33000/N * 6/12 \text{ (radius)}$$

$$= 2200 \text{ ft. lb}$$

$$\text{Work done} = \text{Carriage speed} * \text{resistance}$$

$$2200 = 3 * \text{Resistance}$$

$$\text{Resistance} = 2200/3$$

$$= 733.3 \text{ lb } (Q)$$

If the efficiency of the transmission is 90 percent,

$$P/Q = \text{Efficiency}$$

$$P = Q * 0.90$$

$$= 733.3 * 0.90 = 660 \text{ lb}$$

Example 7: A pair of double sheaved pulley blocks shown in Fig. 13.14(a) is used to lift a load of 224 lb. If the efficiency is 90 percent, what pull is required on the rope?

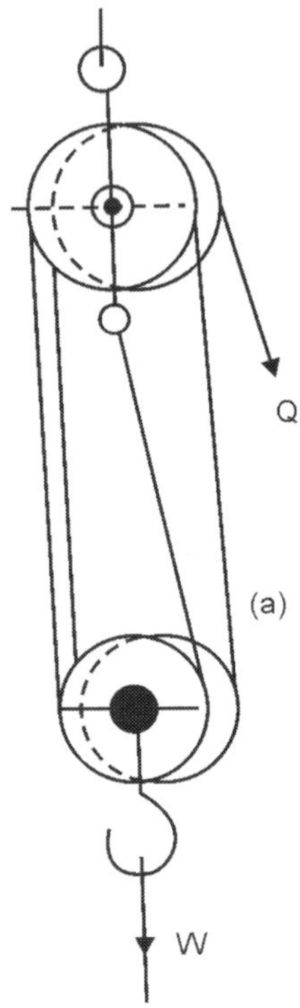

Figure 13.14 (a)

Solution:

Load (W) = 224 lb
Efficiency = 90%
Pull (Q) = ?

If the bigger pulley rotates clockwise, the pull on the rope also moves down clockwise. Since the weight (W) is hung on the smaller pulley tend to move upwards by half the distance moved by the bigger pulley. Hence the weight raised will be 112 lb.

Number of cords passing over the pulleys is two and hence the velocity ratio is also 2.

$$\text{Pull } (Q) \text{ on the rope} = W/VR \times \text{Efficiency} = 112/2 \times 0.90 = 62.2 \text{ lb}$$

Example 8: A Weston differential pulley block shown in Fig. 13.14(b) is used to lift a load of 224 lb. What is the velocity ratio? The pulley spindle is 1 in. diameter, with $\mu = 0.15$ and the total load on the spindle including the weight of the block itself is 280 lb. Calculate the efficiency and also find the pull Q needed on the chain to lift the load.

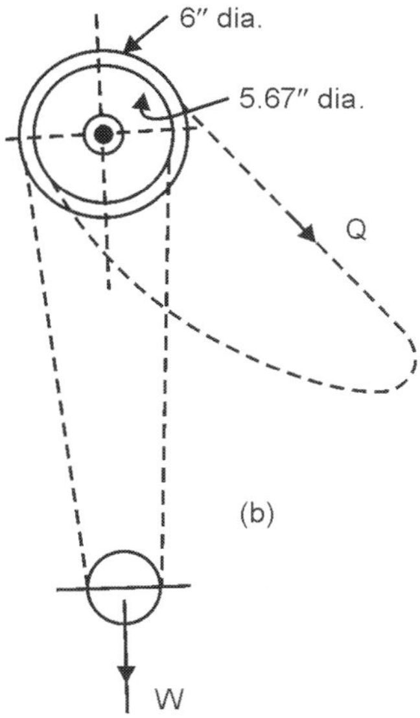

Figure 13.14 (b)

Solution:

$$\text{Load } (W) = 224 \text{ lb}$$

Diameter of the pulley spindle = 1 in.

Coefficient of friction (μ) = 0.15

Total load on the spindle = 280 lb

Let A = Diameter of the larger pulley which is 6 in.

B = Diameter of the smaller pulley which is 5.67 in.

C = Diameter of the lower pulley carrying the weight W

As given in the Weston differential pulley block, the velocity ratio is

First movement / last movement

$$= \left(\frac{\pi A}{\pi(A - B)}\right) = 2A/(A - B)$$

$$= 2 \times 6 / (6 - 5.67) = 36.36$$

Let the total thrust on the spindle is P.

Taking moments about the pulley spindle P, we get

$$P \times 3 = 0.15(P + 280) \times 0.5 + (1/2W \times 2.835) - (\tfrac{1}{2}W \times 3)$$

From which P = 21.39 lb

Mechanical Advantage (MA) = W/P

= 224 + 140 / 21.39

= 17.01 lb

Efficiency = W/PV

= 224 + 140/21.39 * 36.36

= 46.80

Pull (Q) required is W/V = 280 + 224/36.36

= 13.67 lb

Numerical questions

1. The machine in a mill requires 100 horse power measured at the machine driving pulleys. If the efficiency of power transmission by shaft and belts is 0.3, what horse power must the engine give out?

   Ans. HP = 125

2. A carding engine is found to absorb 0.625 horse power when running without material and 0.9 horse power when the machine runs with the material. What is the efficiency?

   Ans. 0.305

3. In a mill with mechanical transmission, power is taken from the engine flywheel by ropes to the line shafts and by belts from line shafts to machines. If the efficiency of the rope drive, shafting and belt drives are 90%, 95% and 95%, calculate the over-all efficiency of the transmission.

   Ans. 81.25 percent

4. A weaving shed use electrical power from outside transformer. If the efficiency of the supply is 97 percent, of the cable 97 percent, of the motors 82 percent, and of the gear drives from motors to looms 97 percent, what proportion of power supplied to the transformer is available at the looms?

   Ans. 74.8 percent

5. In the negative taking up motion shown in Fig. 13.15, the weight Q is normally lifted every pick and then allowed to act as shown in Fig. 13.15 to take up the cloth against tension T. If the efficiency of the gearing is 0.25, at what maximum tension T will cloth be taken up if the beam is 6 in. in diameter and the worm is single?

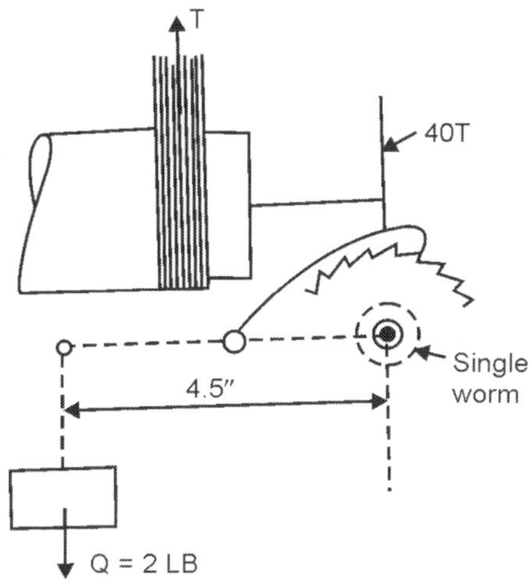

Figure 13.15

Ans. 30 lb

6. The beam in a ball beaming machine is driven by a 14 in. diameter clutch pulley compounded with a 30 tooth wheel driving one of 120 teeth gear on the beam shaft. If the warp tension is 100 lb, what torque is required at the driving pulley when the beam is (a) 4 in. (b) 12 in. diameter, assuming an efficiency of 70 percent for (a) and 75 percent for (b)?

   Ans. (a) 5.95 lb ft; (b) 16.6 lb ft.

7. In a take-up motion for a woolen loom, the driving chain wheel is compounded with a gear wheel of 20 teeth. The latter drives a change wheel compounded with a pinion of 18 teeth driving a wheel of 84 teeth. This wheel is compounded with a wheel of 85 teeth on the take-up roller which is 20 inches circumference. When a change wheel of (a) 40 teeth and (b) 60 teeth is being used, what torque is required at the driving chain wheel to take-up the cloth against a cloth tension of 100 lb, if the efficiency is 0.3?

   Ans. (a) 0.11 lb ft; (b) 0.074 lb ft

14

Stress, strain and elasticity

14.1 Stress

Stress is defined as when a body is subjected to a load P, it is said to be in the state of stress. The "intensity" of stress is mentioned as stress as it is the load per unit area carrying the load. If A is the area of the section carrying the load P, then

$$\text{Stress} = P/A$$

If a rod of metal is 0.5 sq in. sectional area carries a load of 1 ton, the stress can be calculated as 1/0.5 = 2 tons per sq in.

14.1.1 Types of stress

Change in dimensions or distortions of a body may be caused by forces other than a direct pull (tension) or a direct push (compression).
Stresses in textile materials

It is practically difficult, but not impossible to measure the true cross-sectional area carrying the load to which a yarn or thread is subjected. This is because the cross-section of a yarn is made up of large number of fibres, which are of different sizes and densities. Also, it is not practically possible to express the stress in the yarn in usual way, stress = load/unit cross-sectional area.

For this reason, the stress in the yarn is expressed as "specific stress".

14.1.2 Specific stress

Specific stress is defined as the ratio of the force to the mass per unit length. This especially holds good for yarns which are counted on the direct system; for e.g., polyester, nylon, rayon yarns in which the specific stress is generally expressed in grams per denier. If a polyester yarn of 100 denier carries a load of 80 grams, the specific stress is calculated as 80/100 = 0.8 grams per denier.

14.1.3 Strain

Strain is defined as the change of form or dimensions of a body due to the application of load. In engineering applications, nobody is absolutely rigid, and the application of any force or load to it invariably causes some change, however small may be, in dimensions and shape. In other words, it can be explained as when a body is in state of stress it is also strained.

Strain due to tensile or compressive stress is measured by the ratio,

Increase or decrease in length / Original length

Thus if L is the original length and l is the change in length

$$\text{Strain} = L/l$$

14.1.4 Elasticity

A body is said to be in the state of elastic when the strain disappears entirely on the removal of stress which caused it, or elasticity can be defined as the property of the body to recover its original form and dimensions on the removal of straining load.

An elastic material is not necessarily one which stretches much when subjected to a tensile load. For e.g., steel is much more elastic material than rubber.

14.1.5 Permanent stress

Permanent stress is defined as that part of the stress which remains even after the removal of load.

14.1.6 Elastic limit

Elastic limit is defined as the stress at which a body takes a marked permanent set.

14.1.7 Modulus of elasticity

Modulus of elasticity is defined as the stress which would double the length of a bar of the material, or under compression, would reduce its length to zero, assuming the material to remain perfectly elastic throughout. This modulus is known as Young's Modulus of Elasticity.

In an elastic body, the strain is directly proportional to the stress, i.e. doubling the load on the bar will double the resulting stretch, however, for a given load. It obviously differs with different materials and to denote the differences in strain, due to given stress, in different materials the "Modulus" or measure of elasticity is the term used.

14.1.8 Hooke's Law

Hooke's Law states that within elastic limit, stress is directly proportional to strain. Thus,

$$\begin{aligned}\text{Stress} &= \text{Straining load / Area of section}\\ &= P/A\\ \text{Strain} &= \text{Change in length / Original length}\\ &= l/L\\ \text{Stress } \alpha \text{ Strain} &= \text{Strain * Constant } E\\ E &= \text{Stress/Strain}\\ &= P/A * L/1 \text{ or}\\ PL &= A/E, \text{ where } E \text{ is the Young's Modulus of Elasticity}\end{aligned}$$

It is important to note that higher the value of E for a given material, the less it will stretch for a given stress. Some typical examples are shown in the following Table 14.1.

Table 14.1 Young's Modulus values.

Material	E (lb per sq in.)
Cast iron	17,000,000
Mild steel	30,000,000
Wood – Oak	1,500,000
Leather	25,000

Example 1: A leather picking strap, 16 in. long, 1.25 × 0.25 in. in cross section, is subjected to a tensile load of 100 lb. How much will it stretch if E is 25000 lb per sq in.?

Solution:

$$\begin{aligned}\text{Stress} &= P/A\\ P &= 100 \text{ lb}; A = 1.25 \times 0.25 \text{ sq in.}\\ \text{Stress} &= 100/1.25 * 0.25 = 320 \text{ lb per sq in.}\\ \text{Strain} &= l/L\\ &= l/16\\ E &= 25000 = \text{Stress/Strain} = PL = A/E\\ &= 320 * 16/l\\ L &= 320 * 16/25000 = 0.205 \text{ in.}\end{aligned}$$

Energy stored in a strained body.

Consider a load, e.g. a tensile load is gradually applied to a body, the stretch increases uniformly as the load increases. If P is the final load and l is the stretch, the average force acting is $P/2$ and the work done is

$$\text{Average load * stretch} = Pl/2.$$

14.2 Energy

Energy is defined as the capacity of doing a work.

14.2.1 Strain energy

Strain energy is defined as the ability of the body to return to its original length on the removal of the straining load of doing as much work was done in straining it. Thus the energy stored in the strained body is termed as strain energy or resilience.

Springs, whether circular spiral springs or flat springs, are generally elastic if not over loaded. This can be explained by the fact that the stretch or compression in the case of spiral spring, or the deflection of a flat spring, is proportional to the load causing it. Again the strain energy of the spring is $Pl/2$, where P is the load required to stretch, compress, or deflect the spring by an amount (l).

14.2.2 Tenacity

Tenacity is defined as the specific stress at the breaking load.

14.2.3 Breaking length

The other way of expressing the tenacity is "breaking length", which is a length of the yarn of which the weight is equal to the breaking load. The breaking length can be expressed in kilometers (km) or in yards or by any other convenient units.

14.2.4 Count-strength product

The stress or tenacity of the yarns which are counted on the indirect system of length units per unit of weight, e.g. cotton, worsted yarns, is expressed generally as the count × strength product. In direct system, it is expressed in strength/count.

For the indirect system, the specific stress is again load/units of mass per unit length. Consider a hank of cotton yarn, as the unit of length, in a mass of 1 lb, there are number of length units equal to the count, i.e. units of mass per unit length = 1/count and specific stress is load/units of mass per unit of length, which is equal to load/1/count or load × count. If the load is the breaking load, the tenacity, which is the specific stress at the breaking load, is breaking load × count. This is known as count-strength product. The strength is usually the lea strength in lb. or single yarn strength in grams.

For yarns of given material, the count strength product is approximately constant. The cross-sectional area of yarns of a given material of different counts is approximately inversely proportional to the count. This can be best illustrated by an example. 20s cotton yarn is approximately double the cross-sectional area of 40s yarn. Hence for the same breaking stress, 20s count yarn should stand approximately double the breaking load of 40s count yarn, and the product count × strength should be the same for both yarns. This concept is roughly true, since other factors also have to be considered in the yarn. For cotton yarns, the lea strength × count in the neighborhood of 2500 to 3000, whereas for rayon staple fibre yarn it is around 3000. This is true also in the case of single yarn strength in grams. The count strength product is also referred to as breaking length. This is not perfectly correct, although the two values are proportional to each other, and the yarn with high count strength product will have a correspondingly high breaking length.

14.2.5 Load-extension and stress-strain diagrams

The relationship between load and extension or stress and strain is explained clearly by load-extension or stress-strain diagrams. A typical example is shown in Fig. 14.1.

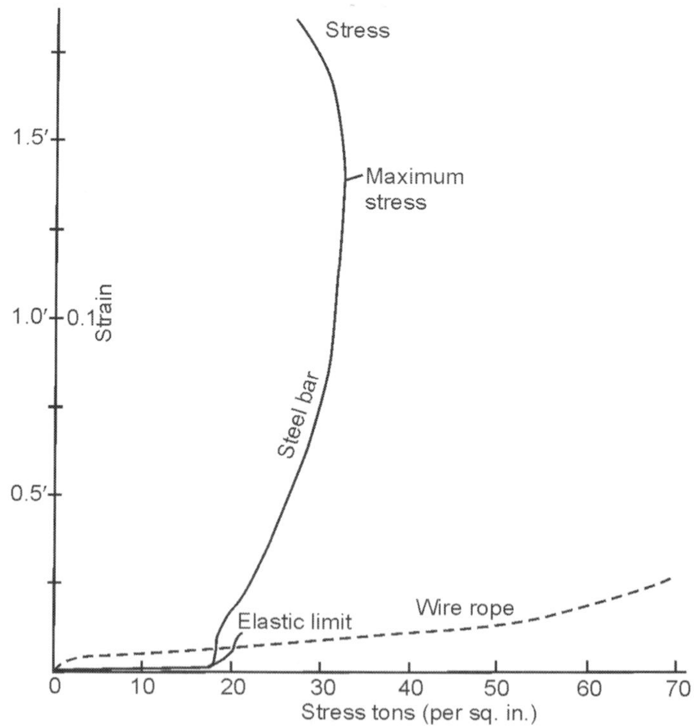

Figure 14.1

The load-extension diagram of a ductile material (mild steel) and also one of steel wire rope is shown in Fig. 14.1. The steel is practically elastic to about half the breaking stress, the stretch then being only about 1/1000 of the length. The material reaches the yield point and thereafter stretches rapidly, in doing so, the diameter decreases uniformly until the maximum stress is reached. It then stretches locally where the break is to occur, decreasing rapidly in diameter there, and finally breaks with a stress less than the maximum. The wire rope stretches rapidly at first due to the tightening up of the strands. Then the rope is nearly elastic up to ¾ of the breaking load after which stress again reaches rapidly to the break.

Load-extension diagrams for three textile yarns are shown in Fig. 14.2.

From the diagram, it is learnt that the rayon yarn behaves in the same manner as steel, whereas the curves for cotton and woolen yarns are similar to the wire rope. This is mainly due to the fact that the rayon filament is more homogeneous material than the cotton and woollen yarns and also there is little twist in the rayon filaments.

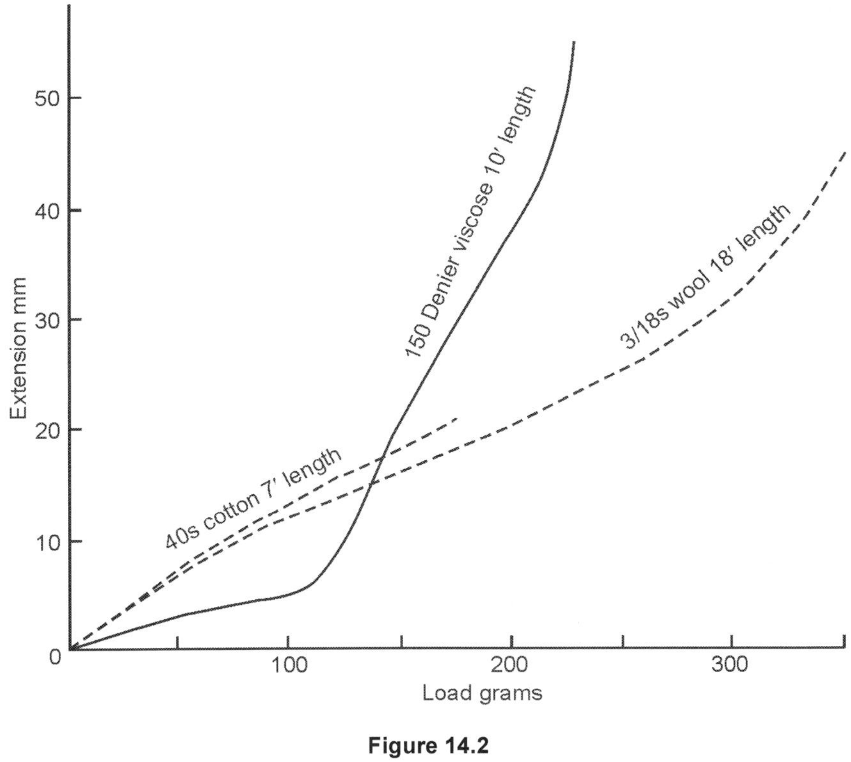

Figure 14.2

14.2.6 Work diagram

The load-extension diagram is also a work diagram in which the ordinates represent the varying resistance and the other the displacement of the resistance. Up to the elastic limit the load or resistance increases with the extension and the work diagram is a triangle, the work being

½ × load at elastic limit × extension at elastic limit

Beyond the elastic limit, the average load or resistance can be found out by Mid-ordinate rule method. The work done up to the breaking point is one of the best measures of the capacity of the material to resist against sudden shocks or loads.

Example 2: The diving belt of a carding engine is 3 in. wide × 0.25 in. thick. During starting up of the machine, the belt transmits 3.6 horse power at a belt speed of 660 ft per minute. If the slipping ratio of the tight side to slack side is 2.5, what is the maximum stress on the belt? What is the approximate stretch at the tight side of the belt, if the original length is 10 ft and $E = 25000$ lb per sq in.?

Solution:

Area of the belt = 3 * 0.25 in.

Work done = 3.6 * 33000/660 = 180 ft lb

Since, the slipping ratio

$$T_t / T_S = 2.5$$
$$T_t = 180$$
$$T_t / T_S = 2.5$$
$$T_S = T_t /2.5$$
$$T_S - T_t/2.5 = 180$$
$$T_t - 0.4\,T_t = 180$$
$$T_t = 180/0.6$$
$$= 300 \text{ lb}$$

Maximum tension in the belt = 300 lb (P)

Maximum stress in the belt = P/A

=300/3 * 0.25 = 400 lb per sq in.

Strain (l) = l/10 ft * 12 = 120 in.

Stress/strain = E

Stretch of the belt at the tight side = stress*l/E

= 400 * 120/25000 = 1.93 in.

Example 3: Each of the drawing-up ropes for pulling the carriage of a spinning mule has to transmit 2 horse power at a carriage speed of 3 ft/s. If the rope has a cross-sectional area of 0.7 sq in. and a breaking stress of 10000 lb per sq in., what is the stress on the rope and the factor of safety?

Solution:

Work done per second = 2 * 33000/3 = 366 ft/s

Working stress = P/A

P = 366 ft/s

Stress = 366/0.7 = 523 lb per sq in.

Factor of safety = Breaking stress/working stress

= 10000/523 = 19.03

Example 4: Two spiral springs each of stiffness 2 lb per in. are attached directly to the underside of a heald shaft to act as a reversing motion. In the bottom position of the shaft, each spring exerts a force of 5 lb on the shaft. How much energy is stored in the springs by lifting the shaft 4 in. from its bottom position?

Solution:

Stiffness of the spring increases uniformly with the stretch.

Force exerted by the two springs (P) = 5 lb * 2 = 10 lb

Stretch of the springs (l) = 4 in. * 2 = 8 in.

Energy stored in the springs = 10 + 8/2 * stretch

= 18/2 * 4 = 36 in. lb per spring

Example 5: The energy given up by a loom shuttle, when stopped at the end of the traverse, was made to store energy by compressing a spiral spring. If the spring is compressed to 3 in., what stiffness of the spring is required to store 30 ft lb?

Solution:

Compression of the spring = 3 in.

Energy stored in the spring = 30 ft lb (30 * 12 = 360 in. lb)

Energy stored for 1 in. spring is 360/3 = 120 in. lb

Energy stored = $Pl/2$

120 = P * 3/2

P = 120 * 2/3 = 80 lb per in.

Example 6: A horizontal lever for lifting the griffe of a Jacquard loom is pivoted at its centre on a 7/8 in. diameter pin, supported at both sides of the lever. If the lever weighs 10 lb, and downward force of 100 lb act at each end of the lever, what is the shearing stress on the fulcrum pin?

Solution:

Downward force acting on both the ends of the lever = 100 lb.

Total load acting on the lever = Weight of the lever + load acting at both ends of the lever = (2 * 100) + 10 = 210 lb.

Average load acting on the lever = 220/2 = 105 lb.

Diameter of the fulcrum pin = 7/8 in.

Area of the fulcrum pin $= \pi * D*D/4$

$= \pi*7/8*7/8/4 = 0.60$ sq in.

Shear stress $= P/A$

$= 105/0.60 = 175$ lb per sq in.

Numerical questions

1. A 3 in. diameter line shaft can transmit safely 150 horse power at 200 rpm. What is the safe horse power if the speed is reduced to 150 rpm? What maximum horse power can a 2 in. diameter shaft of the same material transmit at 150 rpm?

   Ans. 112 *HP*; 33 *HP*

2. A rayon warp has 4000 ends of 100 denier yarn. During weaving, it is subjected to a total tension of 240 lb (108 kg). If the elastic limit load of the yarn is 0.6 grams per denier, and the extension is 1 percent of the original length, how much the warp will be stretched on a length of 50 in. between cloth fell and loom beam, when under the weaving tension?

   Ans. 0.22 in.

3. If the tenacity of the viscose rayon is 1.5 grams per denier, calculate the breaking lengths in kilometers of the yarns of (a) 100 denier and (b) 200 denier.

   Ans. (a) and (b) = 13.5 km

4. If the product count * strength of single yarn for a cotton yarn is 300 oz, what is the breaking length in yards and in kilometers? What is the tenacity in g/denier?

   Ans. Breaking length = 15750 yards; 14.4 km; tenacity = 1.6 grams per denier.

5. The average load up to the breaking point of a dry acetate yarn is 0.89 grams per denier and the extension at break is 35 percent. The same yarn, when wet, has an average load up to the break is 0.41 grams per denier and the extension 45 percent. Compare the work done, up to the breaking point for a 200 denier yarn, 200 mm long, in the wet and dry state.

   Ans. Dry state = 12450; Wet state = 7380 grams.

6. A spiral spring has a stiffness of 3 lb per in. How much work is it capable of doing when stretched to 6 in.?

   Ans. 54 in. lb

7. A loom sley is brought to rest by a fast reed mechanism, by causing it to deflect two flat springs. If the energy of the sley, which has to be transferred to the springs during the stop, is 37 ft lb and the spring deflection is not to exceed 0.25 in. what stiffness of spring is required?

   Ans. 7200 lb

8. A carding engine cylinder weighs 1200 lb and is carried by two bearings each 2.5 in. diameter. What is the shear stress on the shaft at the bearings, due to the weight of the cylinder?

   Ans. Shear stress = 122 lb per sq in.

9. A rayon staple fibre yarn of 30s count on the cotton system has a breaking load of 12 oz. What is the breaking length?

   Ans. Breaking length = 18900 yards

10. A single cotton fibre has a breaking length of 40 km. What is the tenacity in grams per denier?

    Ans. Tenacity = 4.45 grams per denier

11. A 30s count blended yarn has a single thread breaking load of 12 oz. What is the count-strength product, breaking length in yards and in kilometers, and the tenacity in grams/denier?

    Ans. *CSP* = 360; breaking length in yards = 18900; breaking length in kilometers = 17.27; tenacity = 1.92 grams per denier

15

Momentum energy impulsive forces and impact

15.1 Momentum

Momentum can be defined as the total motion possessed by a body. It is measured by the product mass × velocity. Since velocity is a vector quantity, having both magnitude and direction, it follows that momentum is also a vector quantity. This can be represented like a velocity or a force, by the length and direction of a straight line.

Mathematically,

$$\text{Momentum} = \text{Mass} \times \text{Velocity}$$

i.e., m = Mass of the body

u = Initial velocity of the body

v = Final velocity of the body

a = Constant acceleration

t = Time in seconds required to change the velocity from u to v.

Now, Initial momentum $= m\,u$

Final momentum $= m\,v$

Change of momentum $= m\,v - m\,u$

Rate of change of momentum $= m\,v - m\,u/t = m\,(v - u)/t = m\,a$

Since $v - u/t = a$

For example, a loom shuttle moving to and fro across the race board has in the centre of its movement, the same quantity of momentum; if its speed is the same, its momentum when moving in one direction is considered positive, then when travelling in the opposite direction its momentum is negative.

15.2 Law of Conservation of Momentum

Let two bodies A and B collide. According to Newton's third law, the force exerted by A and B must be equal and opposite to that exerted by B on A. Hence the impulse transferred from A to B must be equal and opposite to that transferred from B and A. Since impulse is equal to the change of momentum, it follows that if A loses any momentum, B gains an equal amount, and the total momentum of the system remains unchanged. This is known as law of conservation of momentum, which states that

The total momentum of a system of bodies remains unaltered by mutual action between them.

A familiar example is the firing of a shot from a gun. Since initially, the shot and the gun are at rest, the total momentum of the system is zero. Hence, when the shot is fired, the momentum of the gun must be equal and opposite to that of the shot so that the total sum may remain zero.

It follows that momentum cannot be destroyed, although it may appear to be. If for example, a weight is dropped on the floor. And so stopped, it loses its momentum, which apparently is lost. Actually, it is transferred to the floor and the earth, the mass of which is so great that no apparent change in its motion results.

The loom shuttle moving with positive momentum is stopped at the end of its traverse by the force applied to it during its impact with the swell and picker, and in fact, to the whole loom frame, its momentum is transferred. Thereafter, external force must be applied to the shuttle from the driving through the picking parts to give negative momentum on its return journey.

15.3 Collision of two bodies

Consider the impact between two bodies which move with different velocities along the same straight line. It is assumed that the point of impact lies on the line joining the centers of gravity of the two bodies. The behavior of these colliding bodies during the complete period of impact will depend upon the properties of the materials of which they are made. The material may be 'perfectly elastic or perfectly inelastic'.

In any case, the first effect of impact is approximately the same. The parts of each body adjacent to the point of impact are deformed, and the deformation will continue until the centre of gravity of the two bodies is moving with the same velocity. Assuming that there are no external forces acting on the system, the total momentum must remain constant.

Perfectly elastic bodies

The bodies that rebound after impact are called elastic bodies, e.g. rubber ball.

Perfectly inelastic bodies

The bodies that do not rebound after impact are called perfectly inelastic bodies, e.g. ball of clay.

15.4 Collision of inelastic bodies

When two inelastic bodies *A* and as shown in Fig. 15.1(a) moving with different velocities collide with each other as shown in Fig. 15.1(b), the two bodies will remain together after impact and move together with constant velocity.

(a) Before impact (b) After impact

Figure 15.1(a) and (b)

Let m_1 = Mass of the body *A*

m_2 = Mass of the body *B*

u_1 = Initial velocity of the body *A* before impact

u_2 = Initial velocity of the body *B* before impact

v = Common velocity of the two bodies after impact

A little consideration will show that the impact will take place only if u_1 is greater than u_2. Now according to the principle of momentum,

Momentum before impact = Momentum after impact

$$m_1 u_1 + m_2 u_2 = (m_1 + m_2) v$$

$$v = m_1 u_1 + m_2 u_2 / m_1 + m_2$$

The loss of kinetic energy can be found out by finding out the kinetic energies of the two bodies before and after impact. The difference between the two kinetic energies of the system gives the loss of kinetic energy during impact.

We know that the kinetic energy of the first body before impact.

$$= \tfrac{1}{2} m_1 (u_1)^2$$

and kinetic energy of the second body before impact

$$= \tfrac{1}{2} m_2 (u_2)^2$$

Total kinetic energy of the system before impact is

$$E_1 = \tfrac{1}{2} m_1 (u_1)^2 + \tfrac{1}{2} m_2 (u_2)^2$$

The two bodies move with the same velocity after impact,

Kinetic energy of the system after impact is

$$E_2 = ½(m_1 + m_2) v^2$$

Loss of kinetic energy during impact

$$E_L = E_1 - E_2$$

$$= ½ m_1(u_1)^2 + m_2(u_2)^2 - ½ (m_1 + m_2)v^2$$

15.5 Collision of elastic bodies

When two bodies as shown in Fig. 15.2(a) collide with each other, they suffer a change of form. As soon as the bodies touch each other, the pressure between them is zero. After some time, the bodies continue to approach each other and the pressure exerted by one body over the other body increases (Fig. 15.2b). Thus the two bodies are compressed and deformed at the surface of contact due to their mutual pressures.

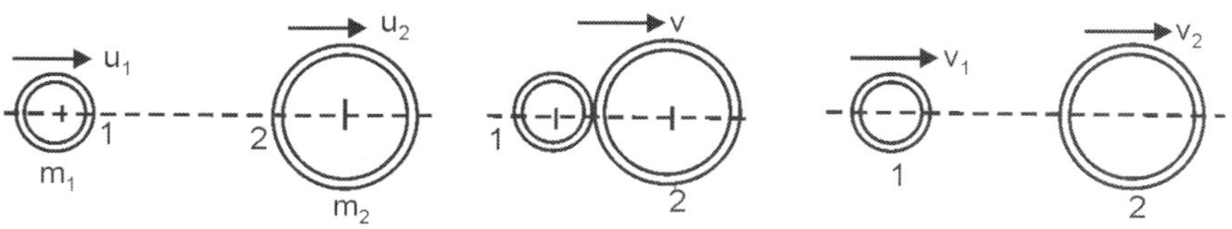

(a) Before impact (b) After compression (c) After impact

Figure 15.2 Collision of elastic bodies

However, if one of the bodies is fixed then the other will momentarily come to rest and then rebound. But, if both the bodies are free to move, then each body will momentarily come to rest relative to the other. At this juncture, the pressure between the two bodies becomes maximum and the deformation is also maximum. At this stage, the two bodies move with a common velocity as shown in Fig. 15.2(c).

The work done in deforming two bodies is stored up as strain energy. Since no energy is absorbed in internal friction; therefore there will be no conversion of kinetic energy in to heat energy. After this collision, the two bodies move with the same velocity, the bodies begin to regain their original shape. This process of regaining their original shape is called *restitution*.

The strain energy thus stored is converted into kinetic energy and the two bodies ultimately get separated as shown in Fig. 15.2. In this case, the change of momentum of each body during the second phase of impact (i.e. when the bodies are separating) is exactly equal to the first phase of impact (i.e. when the bodies are approaching or colliding).

15.6 Coefficient of Restitution(e)

The work done in deforming the two bodies is stored up as strain energy. Since no energy is absorbed in overcoming internal friction, apparently there will be no conversion of kinetic energy into heat energy. Thus immediately after the instant at which the two bodies move with the same velocity, the bodies begin to regain their original shape. This process of regaining their original shape is called 'Coefficient of Restitution', which is denoted by letter "e" (Fig. 15.3).

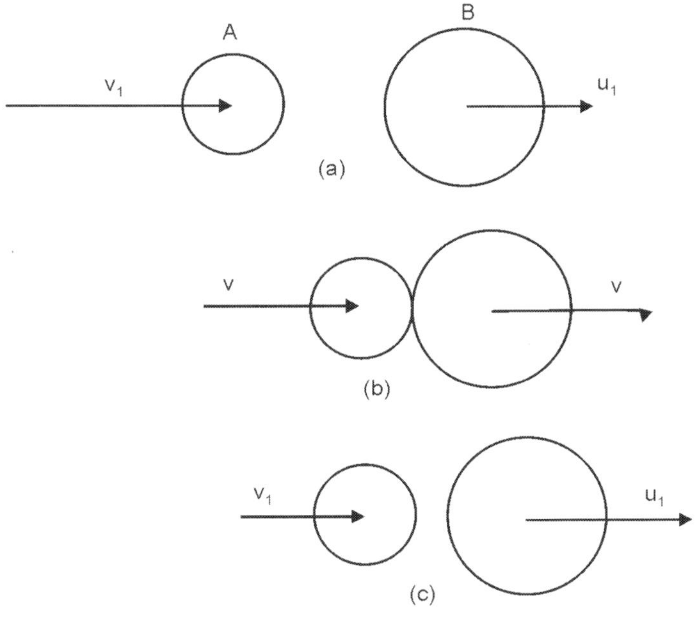

Figure 15.3 Co-efficient of restitution

Let m_1 = mass of the first body

u_1 = velocity of the first body before impact

v_1 = velocity of the first body after impact

m_2, u_2 and v_2 are the corresponding values for the second body, and

v = common velocity of the two bodies after the impact

Change of momentum of the first body during the second phase of impact

$$= m_1(v_1 - v)$$

and change of momentum of the first body during first phase of impact

$$= m_1(v - u_1)$$

$$m_1(v_1 - v) = m_1 (v - u_1) \, pr \, v_1 = 2v - u_1 \qquad \ldots (15.1)$$

Similarly, for the second body, change of momentum of the second body during second phase of impact is

$$= m_2(v_2 - v)$$

And change of momentum of the second body during first phase of impact

$$= m_2(v - u_2)$$

$$m_2(v_2 - v) = (v_2 - u_2) \text{ or } v^2 = 2v - u_2 \qquad \dots (15.2)$$

Subtracting Eq. (15.2) from Eq. (15.1), we get

$$v_1 - v_2 = (u_2 - u_1) = -(u_1 - u_2)$$

Therefore, the relative velocities of the two bodies after impact are equal and opposite to the relative velocities of the two bodies before impact. Due to the fact that physical bodies are not perfectly elastic, the relative velocity of two bodies after impact is always less than the relative velocity before impact. The ratio between the two is called 'Coefficient of Restitution'.

Mathematically (e) = Relative velocity after impact / relative velocity before impact

$$= v_1 - v_2 / -(u_1 - u_2)$$

The value $e = 0$ for the perfectly inelastic bodies and $e = 1$ for perfectly elastic bodies. In case, if the bodies are neither perfectly elastic nor perfectly inelastic, then the value of e lies between 0 and 1, e.g. ball of clay

15.7 Impulse

Impulse may be defined as if a constant force P acts for a time (t) on a body, the product Pt is called the impulse of the force. Impulsive forces and impact are not very common in textile machinery, but they sometimes occur, as example, in picking mechanism of looms, in the fast reed motion of a loom. Moreover, in most cases, other factors as well as impact come in, such as the action of swell in stopping the shuttle. The important point to note is that whenever impact occurs, there is a loss of kinetic energy, but no loss of momentum. It is generally necessary to apply the principle of the conservation of momentum in finding the velocities of the two bodies after impact.

15.8 Energy

Energy may be defined as the capacity of doing a work. The energy exists in many forms like mechanical, electrical, chemical, heat and light, etc. The mechanical energy is equal to the work done on a body in altering either its position or its velocity. The following three types of mechanical energies are important from the subject point of view.

15.9 Potential energy

It is the energy possessed by a body for doing work by virtue of its position. For example, a body raised to some height above the ground level possesses potential energy because it can do some work when falling on earth's surface.

Let W = Weight of the body

m = Mass of the body and

Then potential energy

$$P.E. = W h = m g h$$

If W is the mass in *lb*, and h is the height in *ft*, its potential energy is Wh in *ft lb*.

This is obviously equal to the work done in lifting it from the datum level.

15.10 Kinetic energy

It is the energy possessed by a body for doing work, by virtue of its mass and velocity in motion. If a body of mass m attains a velocity v from rest to time t, under the influence of force F and moves a distances, then

$$\text{Work done} = F s = m a s$$

Kinetic energy of the body

$$K E = m a s = m \times a \times v^2/2a = ½ mv^2$$

If the mass m is W *lb*, v in *ft/s*, and a in *ft/s*^{2}, then F is the *lb* (force) $W/g f$, Kinetic energy $(K E) = Wv^2/2g$

15.11 Strain energy

It is the potential energy stored by an elastic body when deformed. *A* compressed spring possesses this type of energy because it can do some work in recovering its original shape.

If P *lb* is the straining load and x ft the increase or decrease in length due to P, the work done = $P x/2$. Likewise, the strain energy stored in a spring of stiffness S *lb* per unit of extension, when strained by an amount x is $S x /2$. The average straining load is $S x/2$ and it acts through a displacement x, and hence work done = /2 * $x = Sx^2/2$.

15.12 Law of Conservation of Energy

In its most general form the principle of energy states that the total amount of energy in the universe is constant: Energy cannot be destroyed or created although it may be converted into various forms.

A particle of mass m falls from rest from a particular height h above the ground. The sum of potential and kinetic energies remains constant throughout the motion.

Initially the height of the particle above the ground is h and its velocity is zero. Hence its potential energy is m g h and kinetic energy is zero, the sum of the two energies being m g h.

After the particle has fallen through a distance x, its velocity v is given by $v^2 = 2gx$. And its height above the ground is h-x. Hence its potential energy in this position is mg(h-x) and the kinetic energy is $\frac{1}{2}\, mv^2 = \frac{1}{2}\, m.2gx = m\ g\ x$.

When the particle reaches the ground, its velocity is given by $v^2 = 2gh$ and the kinetic energy is $\frac{1}{2}\, m.2gh = m\ g\ h$ while the potential energy is zero. The sum of the two energies is $m\ g\ h$.

Thus it is seen that in all positions the sum of kinetic and potential energies is constant, being equal to the potential energy. In addition to that the initial energy which is wholly potential converted totally into kinetic energy on reaching the ground.

This can be illustrated by an example.

Let a particle of mass m slide along a smooth curve from P to Q. The velocities at P and Q be u and v. Since the reaction everywhere is normal to the direction of motion, no work is done against reaction and the motion is similar to the free motion under gravity.

Sum of Kinetic and Potential energies at P is equal to that at Q. If h be the vertical height at P above Q, and Q be chosen as datum level or zero level of potential energy, we have

Kinetic energy + Potential energy at $P = \frac{1}{2}\, mu^2 + m\ g\ h$

Kinetic energy + Potential energy at $q = \frac{1}{2}\, mv^2 + 0$

$$\frac{1}{2}\, mv^2 = \frac{1}{2}\, mu^2 + m\ g\ h$$

$$V^2 = u^2 + 2gh$$

Examples

1. A shuttle weighing 0.5 lb is given a speed of 45 ft/s in 0.03 s. What is its gain in momentum and what average force has acted upon it? If the maximum acceleration is 2500 ft/s^2, what maximum accelerating force acts on it?

   Solution:

   Mass of the shuttle (W) = 0.5 lb.

   Velocity (v) of the shuttle = 45 ft/s

   Time (t) = 0.03 s

Change in Momentum = $W/g * v$

= 0.5/32 * 45 = 0.703 Engineers units

Force = Change of momentum/time

= 0.703/0.03 = 23.39 *lb*

Maximum accelerating force = Mass* Acceleration

= 0.5/32 * 2500 = 39.01 *lb*.

2. The swell of a loom reduces the speed of the shuttle from 45 to 30 ft/s while the shuttle moves 4 in. The shuttle weighs 0.5 lb. If the retardation is uniform, find the time taken and the value of the checking force applied by the swell.

Solution:

Final velocity (V) = 45 ft/s

Initial velocity (u) = 30 ft/s

Mass of the shuttle (W) = 0.5 lb.

Distance moved by the shuttle (S) = 4 in.

$$
\begin{aligned}
V^2 &= u^2 - 2aS \\
45^2 &= 30^2 = 2 * a * 4/12\ ft \\
a &= 2025 - 900/2*4/12 = 1688\ ft/s^2 \\
V &= u - at \\
45 &= 30 - 1688*t \\
T &= 15/1688 \text{ s or } 0.008 \text{ s} \\
\text{Force} &= \text{Mass * Retardation} \\
&= 0.5/32 * 1688 = 26.4 \text{ lb.}
\end{aligned}
$$

3. A shuttle weighs 1 lb and is given a speed of 40 *ft/s*. How much work is done in accelerating it? If the work is done in 200 times per min, what horse power is expended in it?

Solution:

Mass of the shuttle = 1 *lb*.

Velocity of the shuttle (V) = 40 ft/s

Kinetic energy of the shuttle = $Wv^2/2g$

= $1 * 40^2 / 2*32$ = 25 ft lb

HP = Work done / 33000 = 200/33000 = 0.006

Since the kinetic energy is 25 *ft lb*,

HP = 0.06 * 25 = 0.151 *HP*

4. A shuttle weighing 0.75 lb is stopped from a speed of 40 *ft/s* with a maximum retardation of 200 g. What maximum stopping force is applied? If the retardation is uniform, what time is taken to stop the shuttle and how far does it move during the stoppage?

Solution:

Mass of the shuttle (W) = 0.75 *lb.*

Velocity of the shuttle at stop (v) = 40 *ft/s*

Maximum retardation = 200 g

$$\begin{aligned} \text{Stopping force } (F) &= \text{Mass} * \text{Retardation} \\ &= 0.75 * 200 \text{ g} = 150 \ lb. \\ V &= u + at \\ 40 &= 0 + 200*\text{t} \\ t &= 40/200 * 32 = 0.00625 \text{ s} \\ \text{Distance (S)} &= \text{Average velocity} * \text{Time} \\ &= 40/2 * 0.00625 = 1.5 \text{ in.} \end{aligned}$$

The distance (S) can also be calculated with $V^2 = u^2 - 2aS$

5. The maximum distance available for slowing down and stopping a shuttle is 10 in. The minimum stopping force is that required to retard it at a uniform rate. Find the force if the shuttle weighs 0.75 *lb.* and has a speed of 40 *ft/s*. How many degrees will the loom crankshaft turn whilst the shuttle is stopping, if the crankshaft speed is 200 revs per min?

Solution:

Distance (S) = 10 in.

Velocity of the shuttle (v) = 40 ft/s

Mass of the shuttle (W) = 0.75 lb.

Crankshaft revolutions per minute = 200

$$\begin{aligned} V^2 &= u^2 - 2aS \\ 40^2 &= 0 - 2 * a * 10/12 \\ a &= 960 \text{ ft/s}^2 \\ \text{Force} &= \text{Mass} * \text{Acceleration} \\ &= 0.75/32 * 960 = 22.5 \text{ lb.} \\ \text{Force} &= \text{Mass} * \text{Velocity} / \text{Time} \\ 22.5 &= 0.75/32 * 40/ \text{T} \\ \text{Time } (t) &= 0.041 \text{ s} \end{aligned}$$

Crank shaft revs per min = 200

The crankshaft will stop at

$$\text{Time } (t) = X/360/60/200$$

$$0.041 = 60 * x / 360 * 60 = 49.2 \text{ or } 50 \text{ degrees}$$

6. What average horse power is required to accelerate a mule carriage weighing 6400 *lb* to a speed of 3 *ft/s* in 1.25 s?

Solution:

Mass of the mule carriage (W) = 6400 *lb.*

Velocity of the carriage (V) = 3 *ft/s*

$$\text{Time } (t) = 1.25 \text{ s}$$

$$\text{Force} = \text{Mass} * \text{Velocity/Time}$$

$$= 6400/32 * 3/1.25 = 480 \text{ lb.}$$

$$V = u + at$$

$$3 = 0 + a*1.25$$

$$a = 3/1.25 = 2.4 \text{ ft/s}$$

$$\text{Force} = \text{Mass} * \text{acceleration}$$

$$= 6400$$

7. A mule carriage weighing 6400 lb. is given a speed of 3.75 ft/s in 1 s. The carriage movement during that time is 2 ft. What average force has acted on it during the time of acceleration? Compare this with the average force over the distance moved? Is the acceleration uniform?

Solution:

$$F = \text{Mass} * \text{Velocity/Time}$$

$$= 6400/32 * 3.75/1 = 750 \; lb.$$

$$\text{Kinetic energy } (KE) = Wv^2/2g$$

$$= 6400 * 3.75^2/2*32 = 1397 \; lb.$$

Work done = Kinetic energy stored in the carriage before movement

$$\text{Average force} = \text{Kinetic energy}/2 = 1397/2 = 699 \; lb.$$

$$V^2 = u^2 - 2aS$$

$$3.75^2 = 0 - 2 * a * 2$$

$$a = 3.51 \; ft/s^2$$

$$\text{Force} = \text{Mass} * \text{acceleration}$$

$$= 6400/32 * 3.51 = 703.12 \; lb.$$

Acceleration is not uniform, since the forces are not same.

8. In one method of picking or propelling a shuttle in a loom the shuttle was stopped by causing it to compress a spiral spring. Assuming that the whole energy of the shuttle is utilized in compressing the spring 4 in., what stiffness of the spring is required?

   The weight of the shuttle is 1 *lb.* and speed before stoppage is 45 ft/s

   Solution:

   Mass of the shuttle (W) = 1 *lb.*

   Velocity of the shuttle (V) = 45 *ft/s*

   Compression of the spring = 4 in.

$$V^2 = u^2 - 2aS$$

$$45^2 = 0 - 2\,a*4/12$$

$$a = 21.09\ ft/s^2$$

   Kinetic energy stored in the spring

$$= Wv^2/2g$$

$$= 1*45^2/2*21.09$$

$$= 48\ lb.$$

   Stiffness of the spring = 48 *lb* per inch of spring compression.

9. In the previous Example 8, the energy stored in the compressed spring was used to help to accelerate the shuttle on its next journey. If 50% of the strain energy is available, what speed will it give it to the shuttle?

   Solution:

   Strain energy available is 48 *lb.*

   If only 50% of the strain energy is utilized, then

   Actual strain energy available = 48 lb * 0.5 = 24 *lb.*

   Velocity of the shuttle (v) =

$$KE = W\,v^2/2g$$

$$24 = 1 * v^2/2*21.09$$

$$v = 31.8\ ft/s$$

10. A loom sley with its mass equivalent to 128 lb, and moving at 4 ft/s is stopped by the fast reed motion in 0.5 in. What average force acts on it?

    Solution:

    Mass of the sley (W) = 128 *lb.*

Distance (S) = 0.5 in.

Speed of the sley (V) = 3 *ft/s*

$$V^2 = u^2 - 2aS$$

$$4^2 = 0 - 2 * a * 0.5/12 \text{ ft}$$

$$a = 16/0.083 = 192\ ft/s^2$$

$$\text{Force} = \text{Mass} * \text{Acceleration}$$

$$= 128/32 * 192 = 768\ lb.$$

11. A loom sley driven by a crank and connecting rod has a mass equivalent to 128 lb. concentrated at the sword pin. The sley has to be accelerated from rest at the beating up position to a speed equal to the crank pin speed, while the crank rotates through 90 degrees.

    If the crank pin circle is 5 in. diameter and the crank speed is 180 rpm, how much work is done in accelerating the sley. What average force has acted on it during this time and what average horse power is expended on it? During the acceleration the sley moves 2.75 in., what is the average force over this distance?

Solution:

Mass of the sley (W) = 128 *lb.*

Crank pin circle diameter (D) = 5 in.

Crank shaft speed (N) = 180 rpm.

Crank shaft movement = 90°

Crank shaft speed = π * 5/12 * 180 = 235 ft/min. or 3.92 *ft/s*

$$v = 3.92\ ft/s$$

$$KE = W v^2/2g$$

$$= 128 * 3.92^2/2*32 = 30.54 \text{ lb.}$$

$$\text{Work done} = KE \text{ stored in the sley} = 30.54\ lb.$$

$$\text{Time} = X/360 * 60/180 = 0.0833\ s$$

Average force acting on the sley

$$F = \text{Mass} * \text{Velocity/Time}$$

$$= 128/32 * 3.92/0.0833 = 188 \text{ lb.}$$

Average force during the sley movement of 2.75 in.

$$V^2 = u^2 - 2aS$$

$$= 0 - 2 * a * 2.75/12$$

$$a = 33.53 \text{ ft/s}^2$$

$$F = \text{Mass} * \text{Acceleration}$$

= 128/32 * 33.53
= 134.12 lb.

Horse power = Work done = Force * Distance moved
= 30.54 = 188.5 * D

D = 1.94 in.

Work done = 188.15 * 1.94 = 365.69 ft/s

Work done per min. = 365.69*60 = 21941.4 ft. lb.

$H.P.$ = Work done per min./33000
= 21941.4/33000 = 0.67

12. A carding engine cylinder revolving at 180 rev/min has 25000 ft lb of kinetic energy stored in it. When the belt is thrown off the machine comes to rest in 130 revs. Find the approximate horse power required to drive the cylinder at 180 rpm assuming all the kinetic energy is utilized for all the moving parts of the machine?

Solution:

Speed of the cylinder = 180 rpm

$K.E.$ stored in the cylinder = 25000 *ft. lb.*

V = 180/60 = 3 revs. per sec.

$K.E. = Wv^2/2g$

$25000 = W * 3^2/2*32$ = 5555 lb.

$V = u + at$

$(130/60)*2*\pi = u + 32*t$
= 0.425 s

F = Mass* Velocity/Time
= 5555 * 3/0.425
= 39211.76 *lb.*

Work done = Force * Distance

25000 = 39211.76 * Distance

Distance = 25000/39211.76
= 0.637 *ft.*

Force = Mass * acceleration
= 39211.76 * 0.637 = 24999.9

$H.P.$ = Work done per min/33000
= 33000/24999.9 = 1.32

13. A bale of cotton weighing 100 lb comes down a smooth shoot as shown in Fig. 15.4. Find the potential energy and kinetic energy of the bale at the top and bottom of the shoot and at the floor level. Also find the time to traverse the shoot, the total time of the fall and the horizontal distance *h* (Fig. 15.4).

Solution:

Top of the shoot

Potential energy = $W\,h = 100\;lb * 25ft = 2500\;ft.\;lb.$

Kinetic energy = 0

Bottom of the shoot

Potential energy = $W\,h = 100\;lb * 16 = 1600\;ft.\;lb.$

Kinetic energy = potential energy lost by the bale during traverse of the shoot = 900 *ft. lb.*

$$\text{Kinetic energy} = Wv^2/2g$$

$$900 = 100 * v^2/2g.\; g = 32\;ft/s^2$$

$$v = 24\;ft/s$$

At floor level, Potential energy = 0.

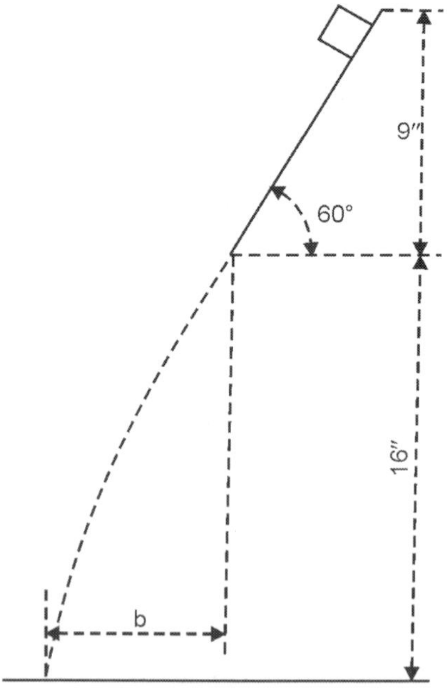

Figure 15.4

$$\text{Kinetic energy} = \text{Potential energy lost} = 2500\ ft\ lb.$$
$$\text{Kinetic energy} = Wv^2/2g$$
$$2500 = 100*v^2/2g$$
$$v = 40\ ft/s$$

This again the speed of the bale would have if it dropped vertically 25 *ft*.

Time for drop

(a) Along shoot, length of shoot = 9/sin 60° = 10.39 *ft*.

Average speed over the shoot = 24 ft/s^2 = 12 *ft/s*

Time taken to traverse the shoot = 10.39/24 = 0.87 *s*

(b) From bottom of shoot to floor.

At the bottom of the shoot, the bale is moving at 40 *ft/s* at an angle of 60° to the horizontal. Therefore, its vertical and horizontal velocities are

Vertical velocity = 24/2 = 12 *ft/s*

Horizontal velocity = 24 * cos 60°

= 20.8 *ft/s*

The time to fall can be calculated by considering the vertical movement under the influence of gravity.

$$u^2 = v^2 - 2as$$
$$= 20.8*20.8 - 2 * 32 * 16$$
$$u = 38.2\ ft/s$$

Also,
$$u = v + at$$
$$38.2 = 20.8 + 32 *t$$
$$t = 0.545\ s$$

$$\text{Total time for drop} = 0.87 + 0.545 = 1.415\ s$$

Horizontal distance (h)

When the bale is leaving the bottom of the shoot, the bale is moving at 12 *ft/s*., which can be treated as constant for the time for the rest of the drop (0.545 *s*)

$$H = 12 * 0.545\ s = 6.54\ ft.$$

14. A spinning mule carriage, weighing 6400 *lb*., is moving at 2.5 *ft/s*. How much work was done in it to give this speed? If the carriage is slowed down and stopped in 1.25 *s* during which time it moves 2 *ft*, find the average force required to stop it during that time and average stopping force over the distance moved.

Solution:

Work done in accelerating the carriage to 2.5 *ft/s* = Kinetic Energy stored in the carriage in 2.5 *ft/s*

$$KE = Wv^2/2g$$
$$W = 6400 \; lb.$$
$$v = 2.5 \; ft/s$$
$$g = 32 \; ft/s^2$$
$$K.E. = 6499 * 2.5^2/2*32 = 625 \; ft. \; lb.$$

Average force during the time of stop

$$F = \text{Mass} \times \text{velocity/time}$$
$$= 6400/32 * 2.5/1.25 = 400 \; lb.$$

Average force over distance moved

Work done = Force × Distance moved = *K.E.* stored before stoppage.

Average force = 625/2 = 312.5 *lb.*

or it can be calculated also by

$$u^2 = v^2 - 2as$$
$$0 = 2.5^2 - 2\,a*2$$
$$a = 1.56 \; ft/s^2$$

Also,
$$u = v - at$$
$$0 = 2.5 - 1.56\,t$$
$$t = 1.6 \; s$$
$$\text{Force} = \text{Change of momentum/time}$$
$$= 6400 * 2.5/32*1.6 = 312.5 \; lb$$

15. A sphere of mass 50 *lb.* moving at 20 *ft/s* overtakes and collides with another of 100 *lb.* moving at 12 ft/s in the same direction. Find the velocities of the masses after impact (a) if they are inelastic, (b) if perfectly elastic, (c) if the coefficient of restitution is 0.5. Find also the loss of kinetic energy of the bodies due to the impact.

Solution:

(a) If v is the common velocity when the two spheres are moving together and R the transfer of momentum.

$$50v = (50*20) - R \quad \ldots (15.3)$$
$$100v = (100 * 12) \quad \ldots (15.4)$$

Adding equations (15.3) and (15.4), we get

$$150\,v = 1000 + 1200$$

$$v = 2200/150 = 14.67\ ft/s$$

Substituting the value of v in equation (15.3)

$$50 * 14.67 = 50*20 - R$$

$$R = 266\ lb\text{-}ft.\ \text{sec units.}$$

The total kinetic energy before impact

$= 50 * 20^2/2*32 + 100 * 12^2/2*32 = 538\ ft.\ lb.$

Kinetic energy after impact

$150 * 14.67^2/2*32 = 500\ ft.\ lb.$

Loss of kinetic energy = Kinetic energy before impact − Kinetic energy after impact

= 539 − 500 = 38 ft. lb.

The common velocity v is 14.67 ft/s and $R = 266\ lb.\ ft.\ s$ units.

(b) If v_2 and u_2 are the final velocities of the two spheres,

$$50\ v_2 = 50 * 20 - 2R \quad \dots (15.5)$$

$$100\ u_2 = 100*12 - 2R \quad \dots (15.6)$$

Substituting the value of R in equation (15.5)

$$50\ v_2 = 1000 - 532$$

$$v_2 = 468/50 = 9.36\ ft/s$$

and from equation (15.6)

$$u_2 = 1732/100 = 17.32\ ft/s$$

Here there is no loss of kinetic energy for, after impact

Total $K.E. = 50 * 9.36^2/64 + 100 * 17.32^2/64 = 538\ ft.\ lb.$

The same as before impact.

(c) Now,

$$50\ v_2 = (50*20) - 1.5\ R \quad \dots (15.7)$$

$$100\ u_2 = (100*12) + 1.5\ R \quad \dots (15.8)$$

$$R = 266$$

$$50\ v_2 = 1000 - 399;\ v_2 = 601/50 = 12.02\ ft/s$$

From equation (15.7)

$$u_2 = 1200 + 399/100 = 15.99\ ft/s$$

Final kinetic energy $= 50 * 12.02^2/2*32 + 100 * 15.99^2/2*32$

$= 512\ ft.\ lb.$

Loss of energy during impact $= 538 - 512 = 26\ ft.\ lb.$

16. A railway truck weighing 20 tons, moving at 10 miles per hour, overtakes and collides with another truck weighing 8 tons, moving at 5 miles per hour. If each of the buffer springs has a stiffness of

1.5 tons per inch, find the speeds of the trucks after the impact and the maximum deflection of each spring, assuming the springs are perfectly elastic.

If the value of $e = 0.6$, what are the final speeds of the trucks and the loss of kinetic energy during impact?

Solution:

When the trucks are moving with the same speeds,

$$\text{Total momentum} = \text{Sum of the momentum before impact.}$$

$$20v = (20*10) - R \qquad \text{... (15.9)}$$

$$8v = (8*5) + R \qquad \text{... (15.10)}$$

Adding equations (15.9) and (15.10)

$$28v = 200 + 40 = 240$$

$$v = 240/28 = 8.57 \text{ miles per hour}$$

R is the transfer of momentum

Substituting the value of v in equation (15.9), the value of R is

$$20 * 8.57 = 200 - R$$

$$R = 28.6 \text{ units}$$

If the springs are perfectly elastic, v_2 and u_2 are the final speeds of the trucks,

$$20\, v_2 = (20 * 10) - 2R \qquad \text{... (15.11)}$$

$$8\, u_2 = (8 * 5) + 2R \qquad \text{... (15.12)}$$

Substituting the value of R in equations (15.11) and (15.12)

$$v_2 = 7.14 \text{ miles per hour}$$

$$u_2 = 12.15 \text{ miles per hour}$$

Before impact, the kinetic energy is

$20 * 14.67^2/2*32 + 8 * 7.34^2/2*32 = 74.35$ *ft ton*

When both are moving at speed $v = 8.57$ miles per hour or 12.55 ft/s,

Kinetic energy $= 28*12.55^2/2*32 = 69$ *ft. tons.*

Loss of kinetic energy $= 74.35 - 69 = 5.35$ *ft tons*

Strain energy stored in the springs $= 5.35$ *ft tons* or 64.2 *in ton.*

If x is the maximum deflection of each spring,

The maximum spring force = 1.5 ton, and work done in compressing one spring $= \frac{1}{2} x * 1.5x = 1.5x^2/2$ in. ton.

Strain energy stored on 4 springs = work done on springs

$= 4 * 1.5x^2/2 = 64.2$ *in. ton.*

From which $x = 4.62$ *in.*

The maximum force acting between each pair of buffers

= *1.5 = 6.93 ton.

If the coefficient of restitution (e) is 0.6, then

$$20\, v_2 = (20 * 10) - 1.6\, R \quad \dots (15.13)$$

$$8\, u_2 = (8 * 5) + 1.5\, R \quad \dots (15.14)$$

Substituting the value of R as 28 in equations (15.13) and (15.14)

$$v_2 = 7.71 \text{ miles per hour}$$

$$u_2 = 10.7 \text{ miles per hour}$$

Kinetic energy after impact =

$20 * 11.3^2/2832 + 8 * 15.7^2/2*32 = 70.8$ *ft tons*

Loss of kinetic energy during the impact =

$74.35 - 70.8 = 3.55$ *ft tons*

Numerical questions

1. A bullet of mass 0.5 *oz* is fired with velocity 2570 *ft/s* in to a block of wood weighing 10 *lb* suspended by a vertical cord. Find the velocity of the bullet and the block after the impact.

   Ans. Velocity = 9.99 *ft/s*

2. A hammer weighing 2 *lb* and moving at 25 *ft/s* strikes a nail weighing 1 *oz.* and drives it in to a piece of wood, find the speed of hammer and nail after impact, the percentage loss of kinetic energy and the resistance offered by the wood, assumed uniform.

   Ans. 24.25 *ft/s*; 2.64 percent; 228 *lb*

3. A ball is thrown horizontally against a vertical wall at a speed of 30 *ft/s*. If the coefficient of restitution is 0.75, with what speed the ball will rebound and what percentage of its original velocity is lost?

   Ans. 22.5 *ft/s*; 43.75 percent

4. A shuttle and picker weighing together 16 *oz.* strike the shuttle box end at 15 *ft/s*. If the coefficient of restitution is 0.1, with what velocity will they rebound? If a retarding force of 2 *lb.* due to the swell, acts on the mass of the shuttle and the picker, how far they will rebound?

   Ans. Velocity = 1.5 *ft/s*; Distance moved = 0.21 *in.*

5. A loom sley together with the parts connected to it, is equivalent to a mass of 256 *lb.* when moving at 4 *ft/s*. It is suddenly stopped by the two fast reed strikers on the sley engaging with the spring

loaded frogs mounted on the loom frame, the drive to the loom being disengaged at the same time.

If each of the springs has a stiffness of 1500 *lb* per inch, find

(a) Distance moved by the sley in coming to rest; (b) if the coefficient of restitution is 0.25, how far the sley will rebound, if a braking force equivalent to 100 lb. is acting on it?

Ans. (a) 0.71 in.; (b) 0.5 in. approximately

6. A shuttle weighing 10 *oz.* moving at 30 *ft/s* strikes a picker weighing 3 *oz.*, which is connected as shown in Fig. 15.5 by a check strap which can be considered inextensible and which weighs 8 oz. to the other end of the sley. Find the speed with which the picker and shuttle strike the end of the shuttle box. How much energy is lost during the impact of shuttle with the picker?

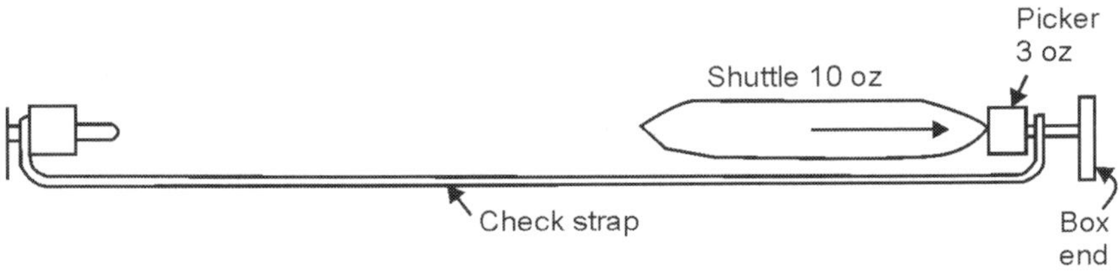

Figure 15.5

Ans. Loss in kinetic energy = 5.14 *ft. lb.*

7. A spinning mule carriage weighing 6400 *lb.* is moving at 1 *ft/s* when it strikes against the stops at the end of the inward run and it is observed to rebound 1 inch.
8. If the coefficient of restitution is 0.5 and assuming constant retardation during the rebound, find the resistance acting on the carriage during rebound.

Ans. Resistance = 300 *lb.*

16

Motion in a circle

16.1 Introduction

We see that whenever a person drives round a curve or circular track on a motorcycle or a bicycle, he has to lean inward in order to maintain a perfect equilibrium. As a matter of fact, the angle at which the man leans with the vertical is more, if he is running at a faster speed than that when he is running at a lower speed. This type of motion is called motion along a circular path.

16.2 Centripetal force

A body moving in a circle or along a circular path, with a constant velocity, suffers a continuous change in direction at every point of its motion though the magnitude of the speed remains same. Since the velocity involves both magnitude and direction, the velocity of the body is continuously changing due to change in direction. According to Newton's First Law of Motion, an external force must act continuously upon the body to produce a change in the direction of the moving body.

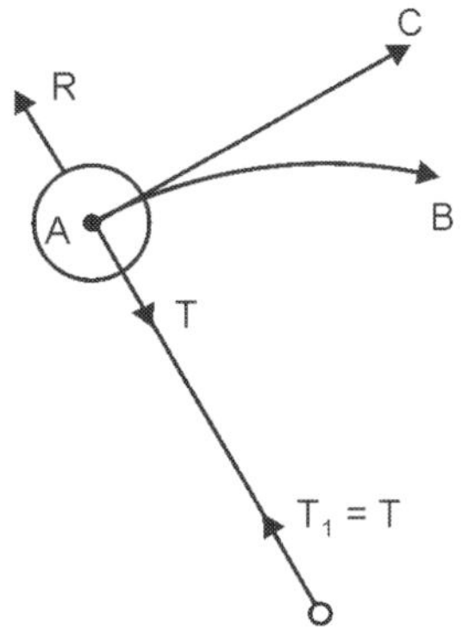

Figure 16.1

Consider a stone swing in a circular path *AB* in Fig. 16.1, at constant speed by means of an inextensible string. If it is free from constraint of the string, it would move from *A* to *C*, along the tangent to the circular path at *A*. The string is under tension (*T*) and is applying a radially inward pull to change the direction of stone's velocity.

The force (T_O), applied in this way, is usually called centripetal force. The other end of the string is under the same tension $T_1 = T$ and is applying the force radially outwards to the pivot at the centre of the circular path. This outward force is called centrifugal force.

It is the centripetal force acting on the stone and not centrifugal force.

There is no externally applied outward force acting on the stone. It offers an inertia resistance *R* radially outwards and equal to tension *T*, the resistance of the stone to change its velocity. In the same way, a body having its speed changed, when moving in a straight line path, offers resistance, proportional to its mass and to the acceleration, the resistance being equal and opposite to the accelerating force.

The acceleration of the particle = Change in velocity/Time

Accelerating force required, i.e. the centripetal force

$$= \text{mass} \times \text{acceleration}$$

$$= W/g * v/r$$

or putting $v = 2 * 3.14 * r * N/60$

where N = revolution per min.

and $g = 32\ ft/s^2$

$$\text{Centripetal Force} = W/\ 32\ r\ (2 * 3.14 * r * N/60)^2$$

$= 0.00034\ WrN^2$, i.e. centripetal or centrifugal force is directly proportional to the mass, radius and square of the angular speed. The radius and velocity must be taken at the centre of gravity of the mass.

16.3 Centrifugal force

According to Newton's Third Law of Motion, the force which acts opposite to the centripetal force is centrifugal force. It may be noted that centrifugal force always acts away from the centre of the path, or in other words, the centrifugal force always tends to throw the body away from the centre of the circular path.

16.4 Conical pendulum

The simplest form of a centrifugal governor is a Watt governor as shown in Fig. 16.2. It is basically a conical pendulum with links attached to a sleeve of negligible mass. The arms of the pendulum are connected to the spindle. This governor mechanism can be seen in most of the steam engines, the simplest form is the original Watt governor.

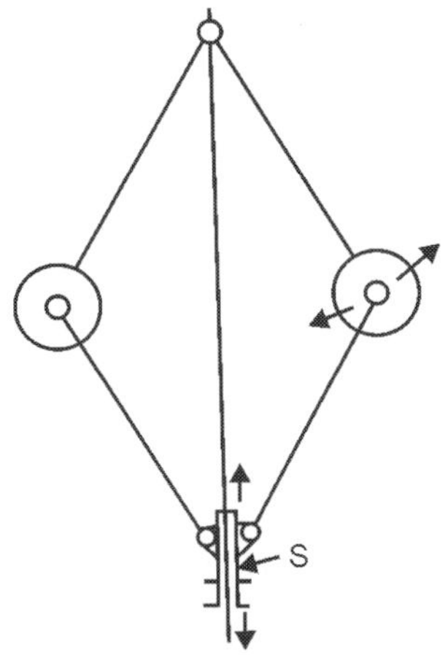

Figure 16.2 Conical Pendulum

A vertical shaft, driven from the engine, has two arms pivoted on it, each carrying a heavy ball. The latter are connected by arms to a sleeve, *S*, which is connected to the steam or gas supply. If the speed increases, the ball moves further from the shaft, lifting the sleeve and thus reducing the amount of steam or gas supply to the engine.

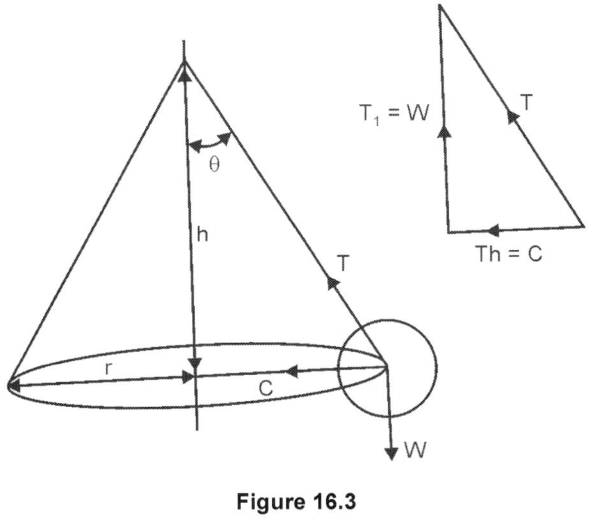

Figure 16.3

In Fig. 16.3, when the ball rotates, the required centripetal force, C, is provided by the horizontal component of the tension T, in the string or arm. The vertical component of T balances the force of gravity. If the speed of rotation is increased, the inward force C is no longer sufficient to keep the ball in circular path and the ball moves out by its own inertia, until the increased force, C, due to greater inclination of T to the vertical is sufficient to keep the ball moving at the new radius.

From the triangle of forces $C/W = r/h$, h is called the height of the cone of revolution.

$$h = 12rW/C = 12rW/0.00034WrN^2 = 35300/N^2.$$

Where h is in inches, N = revs per min.

From this, it will be seen that the height of the cone of revolution depends only on the speed of rotation. It is not dependant on the weight of the balls nor on the length of the arms.

16.5 Tension in rotating masses

Consider a short length x (Fig. 16.4) of a belt or thread subtending a small angle $d°$ at the centre, moving at speed v in a circular path of radius, r. In order to keep it in a circular path, a radially inward force, C, is required. This can be applied only by tensions T, at either end of the length x. These tensions have a resultant, R, which provides the necessary inward force, C.

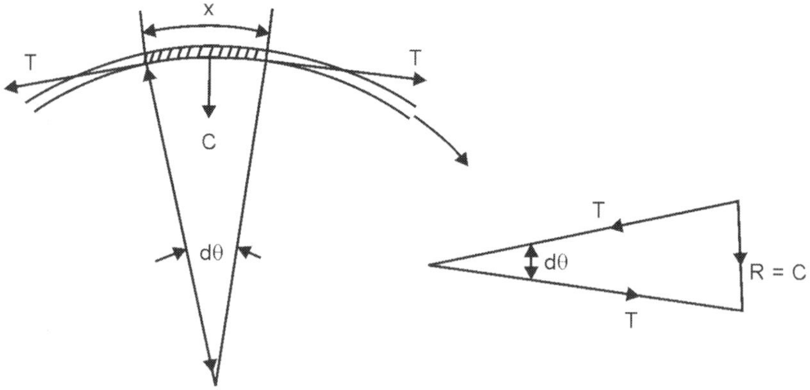

Figure 16.4

If $d\theta$ is small, then $C/T = x/r$

If w is the mass per unit length, the mass of length x is $w\,x$, then

$$C = wxv^2/g = Tx/r$$

Hence $$T = wv^2/g$$

Examples

1. A centrifugal drier consists of a perforated shell of 2 *ft* 6 in. diameter, mounted on a vertical spindle and rotates at 300 rpm. With what force 1 *lb*. of yarn mass against inside of the drier?

   Solution:

   Centripetal force, $C = 0.00034\ wr\ N^2$

   $$W = 1 \text{ lb}; N = 300 \text{ rpm}$$

   The whole mass is at the inner radius, $d/2 = 1.25$ *ft*; the centripetal force is provided by the reaction from the inside of the drier is

   $$C = 0.00034 * 1 * 1.25 * 300^2$$

   = 38.5 lb, which is the pressure between the yarn mass and inside of the shell.

2. A centrifugal clutch coupling shown diagrammatically in Fig. 16.5 has 4 blocks each weighing 0.75 *lb*. The inner diameter of the clutch face is 6 *in*. in radius and the radius to the centre of gravity of each block, when the clutch faces are in engagement is 2.5 *in*; μ between the clutch faces is 0.15; what maximum horse power the clutch can transmit at a speed of 960 rpm?

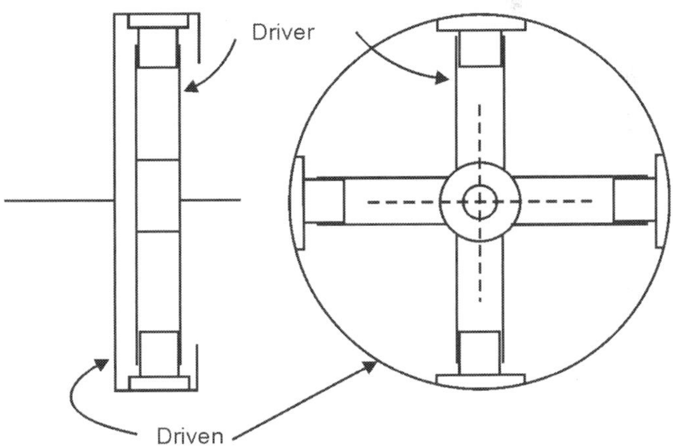

Figure 16.5 Centrifugal clutch

Solution:

At a speed of 960 rpm, the centripetal force, C, of one block =

$= 0.00034 * 0.75 * 2.5/12 * 960^2 = 49$ lb

Total driving force available by friction = 4*49 *0.15 = 29 *lb*.

Maximum horse power that can be transmitted =

29 * π * 6/12*960/33000 = 1.34

3. A ring frame traveler shown in Fig. 16.6 mounted on the ring flange weighs 1 grain, and its centre of gravity moves in a circular path of 0.8 in. radius at 8000 rpm. The traveler is acted upon by a vertical pull of 100 grains and a horizontal radial inward pull of 260 grains. What total radial inward force is required to change the direction of its rotation? Also what is the direction and magnitude of reaction from the ring to the traveler? Ignore the force of gravity 1 grain.

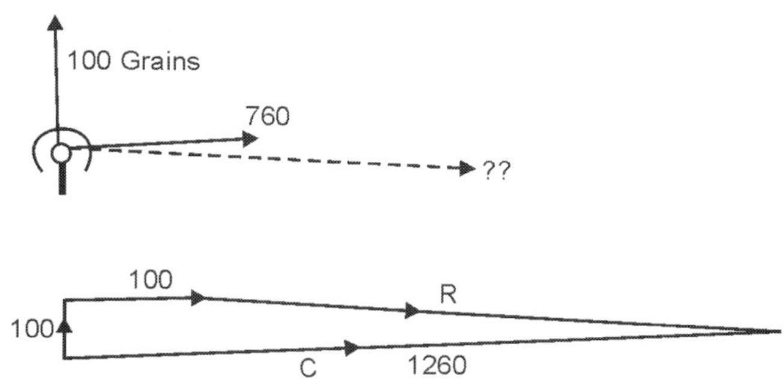

Figure 16.6

Solution:

Total inward force required = centripetal force, *C*

$$\begin{aligned} W &= 1 \text{ grain} \\ \text{Radius} &= 0.8 \text{ in.} \\ N &= 8000 \text{ rpm} \\ C &= 0.00034 * 1 * 0.8/12 * 8000^2 \\ &= 1450 \text{ grains} \end{aligned}$$

The two yarns pulls together with the reaction, *R*, must have a resultant *C* = 1450 grains

From the law of polygon of forces, *R* = 1002 + 11902 = 1194 grains.

μ between the ring and the traveler is 0.12, the frictional resistance of the traveler is

$$\begin{aligned} F &= \mu R \\ &= 0.12 * 1194 \\ &= 143.28 \text{ grains} \end{aligned}$$

Inside ring diameter is 1.5 in.

$$H.P = \pi * 1.5/12*8000 *143.28/7000*33000$$
$$= 0.0019 \text{ per traveler}$$

If the machine has 400 spindles, *HP* required is

$$0.0019 *400 = 0.77$$

4. A carding engine cylinder has a mass of 1120 *lb*. Its centre of mass is midway between the bearings, but it is 1/20 *in*. from the centre of rotation. If it were left unbalanced, what would be the maximum and minimum loads on the two bearings, if the cylinder speed is 180 rpm? To balance the cylinder, balance masses are fixed to the inside of the rim, one at each side of the at 22 *in*. radius. What weight is required in each balance mass?

Solution:

When the centre of the mass is vertically below the centre of rotation, the upward reaction from the bearings to the cylinder shaft must be sufficient to balance the force of gravity and to apply the force needed to change the direction of motion of the cylinder mass, acting at the centre of the mass.

Hence, the upward reaction = $W + C$

$= 1120\ lb + (0.00034 *1120 * 1/12*20*180^2)$

$= 1120 + 51.5\ lb.$

When the centre of mass is vertically above the centre of rotation, the necessary downward force to change the direction of motion is 51.5 *lb*, which is supplied by the force of gravity.

Hence $1120 - 51.5 = 1068.5\ lb.$ is to be carried by the bearings.

The balance masses must be kept to the cylinder at points diagrammatically opposite to the centre of the mass of the cylinder, so that, when rotating the force needed to accelerate the balance masses is equal and opposite to the force required for the cylinder.

Let W be the mass of the two balance masses, then

$0.00034*2W*22/12*180^2 = C$

$0.00034*2W*22/12*180^2 = 0.00034*1120*1/12*20*180^2$

$$2W *22/12 = 1120*1/240$$
$$W = 1.27\ lb.$$

5. One end, A, of a horizontal cord, 10 *ft* long, is fixed and the other end carries a mass of 10 *lb*. The mass is allowed to fall from this position and when the cord is vertical, it strikes against a peg 6 *ft*. below A. Prove that the tension in the cord, just after it strikes the peg, is twice, what it was before striking?

   Solution:

   When the cord falls from horizontal to vertical position, the mass acquires a velocity, v, the potential energy being converted in to kinetic energy.

   Hence, from the equations of motion, $v^2 = 2gh$, where h is the vertical fall of the mass.

   $$v^2 = 2*32*10 = 640$$

   The centripetal force, C, just before the cord strikes the peg, is

   $$C = Wv^2 / g\,r$$
   $$= 10 * 640/32*10 = 20\ lb.$$

   This force must be supplied by the upward pull of the cord, and in addition, the cord must support the mass against gravity.

   Therefore, before striking the cord tension is 20 *lb* + 10 *lb* = 30 *lb*.

   Before striking the peg, the mass has the same velocity, v, as before, but the radius is 10 − 6 = 4 *ft*.

   Centripetal force $C = Wv^2/ g\,r$

   = 10 *640/32*4 = 50 l*b*.

   The cord tension is now 50 *lb* +10 *lb* = 60 *lb*.

   Hence, it is proved that the cord tension after striking the peg is twice that before it strikes the peg.

6. In a simple Watt governor, the lower arms are equal in length to the upper arms. Ignoring friction, how much will the sleeve rise if the speed changes from 50 to 52 rpm?

   Solution:

   When the speed changes from 50 rpm to 52 rpm, h changes from (35300/50*50 − 35300/52*52) = 14.1 − 13.05 = 1.05 in.

7. What is the tension, due to its motion in a circle, in a 20s cotton yarn on the surface of the revolving cheese which has a surface speed of 1200 yd/min.

Solution:

$$\begin{aligned}\text{20s yarn} &= 20 * 849 \text{ yd. in } 1\ lb.\\ \text{Weight per } ft &= 1/60*840 = 0.0000198\ lb.\\ \text{Surface speed} &= 1200\ yd/\text{min.}\\ V &= 1200*0.9144/0.3048 = 3600\ ft/\text{min.}\end{aligned}$$

or 60 ft/s

$$\begin{aligned}\text{Tension in yarn} &= Wv^2/g\\ &= 0.0000198*60^2 / 32\\ &= 0.0022\ lb \text{ or } 0.035\ oz.\end{aligned}$$

8. The doffer comb of a carding engine has a mass of 0.6 *lb*, and its centre of mass is 4 *in.* from the rocking shaft centre. If its maximum angular velocity is 56 radians per sec. what is the maximum tension in the arms which connect it to the rocking shaft?

Solution:

Mass of the doffer comb (w) – 0.6 *lb*

Radius from the centre of the rocking shaft = 4 *in.*

Angular velocity (v) – 56 radians per second –

$$\begin{aligned}\text{Tension } T &= Wv^2/g\\ &= 0.6 * 56^2/32 = 58.8\ lb.\end{aligned}$$

Since its centre of gravity acts at 4 in. radius, the maximum tension in the arms is

$T * r$ –

$$=58.8 * 4/12 = 19.6 \text{ lb.}$$

9. A loom crankshaft has two arms, each of which together with the part of the connecting arm is equivalent to a mass of 10 *lb*, at a distance of 2.2 *in.* from the shaft centre. What is the difference in the load on the shaft bearings when the cranks are in their up and down position respectively due to leaving of the cranks unbalanced? Crank shaft revolution is 200 revs per min.

Solution:

Mass of the cranks (W) = 10 *lb.*

Since there are two cranks, therefore total mass = $2W$

Radius or distance from the shaft centre (r) = 2.2 *in.*

Centripetal force, $C = 0.000342WrN^2$

$= 0.00034*2*10*2.2/12*200^2$
$= 49.86$ lb.

10. What is the centripetal force required to keep a ring traveler, weighing 1.5 grains, moving in a circular path of 1 *in.* radius at 7000 rpm?

    Solution:

    Mass of the traveler (W) = 1.5 grains

    Radius of the ring (r) = 1 in.

    Speed of the traveler (N) = 7000 rpm.

    Centripetal force, $C = 0.00034*1.5*1/12*7000^2$
    $= 2080.5$ grains

Numerical questions

1. A mass of 100 *lb.* is attached by a horizontal chain 16 *ft.* long to a fixed hook. If the mass is allowed to fall, find its speed vertically below the hook, and the tension in the chain in that position.
2. A cord supporting a mass of 4 *lb.* rotates as a conical pendulum. If the centre of mass describes a circle, 1 *ft.* diameter and the cord is 1 *ft.* long, what is the speed in revs per min. and what is the tension in the cord?

   Ans. 54.23 revs per min; 2 *lb*
3. A fly frame bobbin has a speed of 1300 rpm when the diameter is 1.1875 *in.* and a speed of 1192 rpm when the diameter is 4 *in.* If the count of the material being wound is 4s cotton material, what is the tension on the material at two bobbin diameters, due to centripetal force?

   Ans. 0.0026 *oz.* at 1.1875 *in.*; Diameter: 0.0217 *oz.* at 4 in. diameter.
4. A hydro extractor for removing moisture from yarn consists of a vertical cylindrical casing with an inside perforated container, which is rotated at high speed. If the container has a diameter of 30 *in.* and height of 15 *in.* and a speed of 1000 rpm, contains a charge of 50 *lb.* of wet yarn, what is the pressure per sq. in. between the yarn and the container? Assume the yarn is evenly distributed over the perforated container and that the radius to the centre of the yarn mass is 14 *in.* If the metal container weighs 0.3 *lb.* per cubic inch, what is the stress due to rotation?

   Ans. 14.1 lb per sq in.; 1930 *lb* per sq in.

5. The presser for a flyer frame bobbin is mounted on one of the flyer arms as shown in Fig. 16.7. The presser is kept in contact with the bobbin by the force, P, between bobbin and presser supplying the necessary centripetal force to the balance mass, M. The total mass of the presser, including M is 0.25 *lb.* and its centre of mass is 3 in. and 2.85 in. from the bobbin centre.

   The bobbin diameters are 1.5 in. and 4 in., respectively. If the speed of the flyer is 800 revs per min, find the centripetal force for the presser at the two bobbin diameters. If the perpendicular distances, from the fulcrum F to the lines of action of the centripetal force, are 0.15 in. and 0.35 in. for the bobbin diameters 1.5 in. and 4 in., find out the pressure in each case.

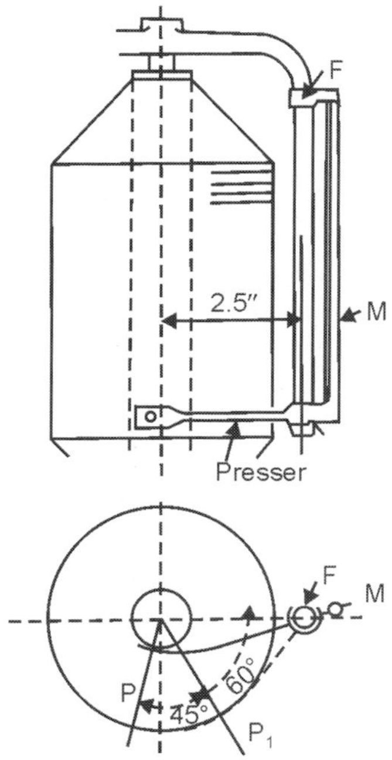

Figure 16.7

Ans. Centripetal force = 17.2 and 16.3 *lb*; Pressure P = 1.07 and 2.64 *lb*.

6. A ring frame traveler (Fig. 16.8) is pulled round the ring by the tangential component of the horizontal yarn tension P of 450 grains between traveller and bobbin. The traveler resistance to motion is due to friction between traveler and ring (μ = 0.12), and to a horizontal backward pull of 56 grains. In addition to that a vertical yarn tension of 225 grains acts on the traveller.

Calculate the reaction force between the traveller and the ring and the mass of the traveller, if its speed is 6000 revs per min and its centre of mass moves in a circle of 0.9 in. radius.

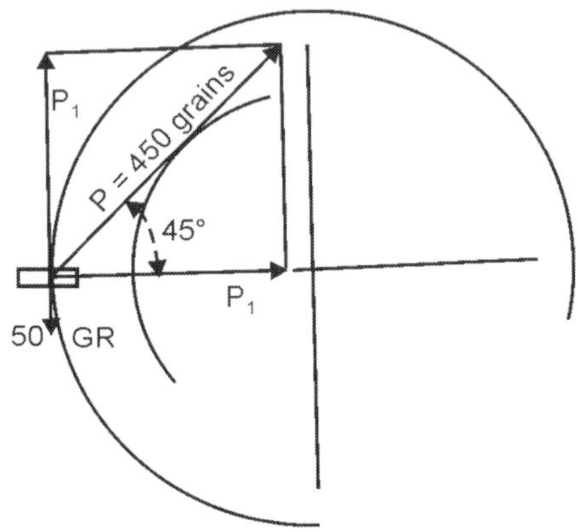

Figure 16.8

Ans. 2190 grains; traveler weight = 2.74 grains

17

Simple harmonic motion

17.1 Introduction

Consider a particle moving round the circumference of a circle in an anti-clockwise direction with a constant angular velocity. As shown in Fig. 17.1. Let P be the position of the body at any instant and N be the projection of P on the diameter XX' of the circle.

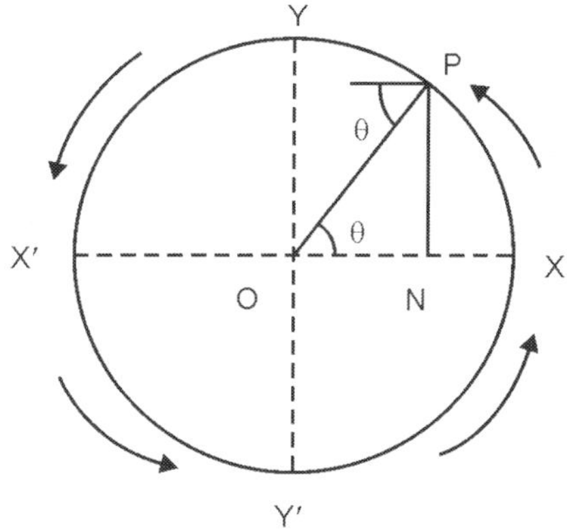

Figure 17.1 Simple harmonic motion

It will be noticed that when the point P moves round the circumference of the circle from X to Y. N moves from A to O. When P moves from Y to X', N moves from O to X'. similarly when P moves from X' to Y', N moves from O to X. Hence as P completes one revolution, the point N completes one vibration about the point O. This to and for motion of N is known as Simple Harmonic Motion (simply written as *S.H.M.*)

17.2 Definitions of few terms

1. Amplitude

The maximum displacement of the point N (which is having simple harmonic motion) from its mean position is called the amplitude. In Fig. 17.1, O is the mean position of the point *N* and maximum displacement of *N* from the mean position is *OX* or *OX'*.

Hence amplitude = *OX* = Radius of the circle

2. Oscillation

The motion of *N* from *X* to *X'* and back from *X'* to *X* is called oscillation.

3. Period of S.H.M.

The period of S.H.M. is the time taken by the point *N* for one complete oscillation i.e. time taken by *N* in moving from *X* to *X'* and then from *X'* to *X*. The time taken by the point *N* for one complete oscillation is the same as the time taken by the body *P* moving along the circumference for one complete revolution.

Let *T* = Period of S.H.M.

ω = Constant angular velocity of the body P

Now by definition,

T = Time taken by the body (*P*) moving along the circumference for one complete revolution

= Angular displacement for one revolution / Constant angular velocity

= 2*3.14/ω

Periodic time *t* = 2*π × √ Displacement/ Acceleration

4. Frequency

Frequency is defined as the number of cycles per second and mathematically it is given as

$$i/T = 1/2*\pi/\omega$$
$$= \omega/2 *\pi$$

Examples

1. A thread guide bar on a reeling machine is reciprocated by the crank and slot mechanism as shown in Fig. 17.2. Its mass is 8 *lb*. and the crank circle radius is 2.5 in. diameter. Crank speed is 140 rpm. What is the maximum speed of the bar and its speed when the crank arm makes 45 degrees with the horizontal? If frictional resistance of the bar is 2 lb, what force is required to keep it in motion: (a) at the start of its movement from an end position; (b) when the crank is at 30 degrees with the horizontal? How much work is done per minute in reciprocating the bar?

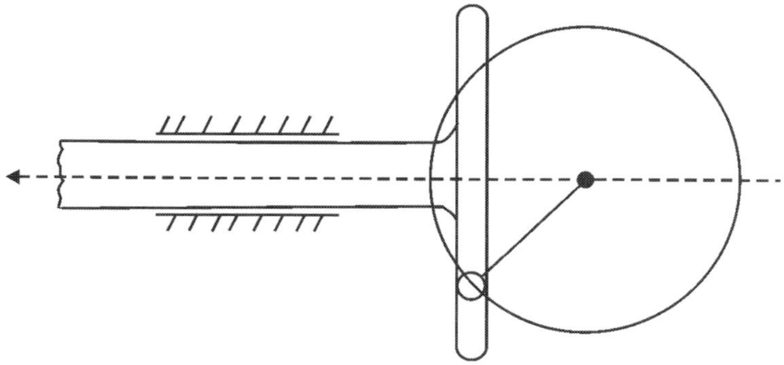

Figure 17.2 Reciprocating guide bar in reeling machine

Solution:

The mechanism obviously shows the simple harmonic motion.

The maximum speed of the bar is when the crank is vertical and is then equal to crankpin speed.

$$v = \pi * \text{crank pin radius} * \text{crank speed}/60$$
$$= \pi * 2.5 * 140/60*12$$
$$= 1.53 \text{ ft/s}$$

Crank speed at 45 degrees is =

$$* \sin 45° = 1.53 * 0.707$$
$$= 1.08 \text{ ft/s}$$

(a) At the start of the movement, the acceleration of the bar = V^2/r

$$\text{Acceleration} = 1.53^2 * 12/1.25$$
$$= 22.5 \text{ ft/s}^2$$

Where V is the crank pin speed and r = crank radius

Accelerating force required (F) = ma/g

Mass of the bar (m) = 8 *lb*

$$a = 22.5 \text{ ft/s}^2$$
$$F = 8 * 22.5/32$$
$$= 5.6\ lb.$$

$$\text{Total force required} = \text{Accelerating force} + \text{Frictional force}$$
$$= 5.6 + 2$$
$$= 7.6\ lb.$$

When the crank makes 30° with the horizontal,

$$\text{acceleration} = V^2/r * \cos 30°$$
$$= 22.5 * 0.866$$
$$= 19.5 \text{ ft/s}^2$$

Total force required = (8 /32 + 19.5) + 2 = 6.9 *lb.*

Work done per minute = friction force per min. × displacement per minute.

$$= 2 \times 5/12 \times 140$$
$$= 117 \textit{ ft. lb.}$$

2. The griffe of a single lift jacquard loom has a mass of 96 *lb*. It is moved 4 *in.* vertically up and down by means of a crank and connecting rod, 120 times per minute. Friction at the griffe guides amounts to 20 *lb* (Fig. 17.3).
   Draw curves (a) showing the speed at any position. (b) The force, *T*, which the connecting rod must apply to keep the griffe moving.
   What is the force at the start of the upward and downward movement and when it has been lifted 1 in. and 3 in. from its lowest position?
   How much work is done on it?

Solution:

Mass of the griffe = 96 lb.

Crank diameter = 4 in.

Crank speed = 120 rpm.

Friction force at the griffe = 20 lb.

The crank and the connecting rod give the griffe a simple harmonic motion. Hence, its maximum speed is equal to that of a point moving in a circle of 4 in. diameter and speed of 120 times per min.

Maximum speed of griffe = π * crank diameter * crank speed / 60 * 12

$$= \pi * 4 * 120/60 * 12$$
$$= 2.099 \text{ ft/s}$$

Maximum acceleration is at the start and end of the movement, when acceleration

$a = V^2/r$, where $V = 2.099$ ft/s and r = crank radius = 2/12 ft.

$$a = 2.1^2 * 12/2 = 26.43 \text{ ft/s}^2$$

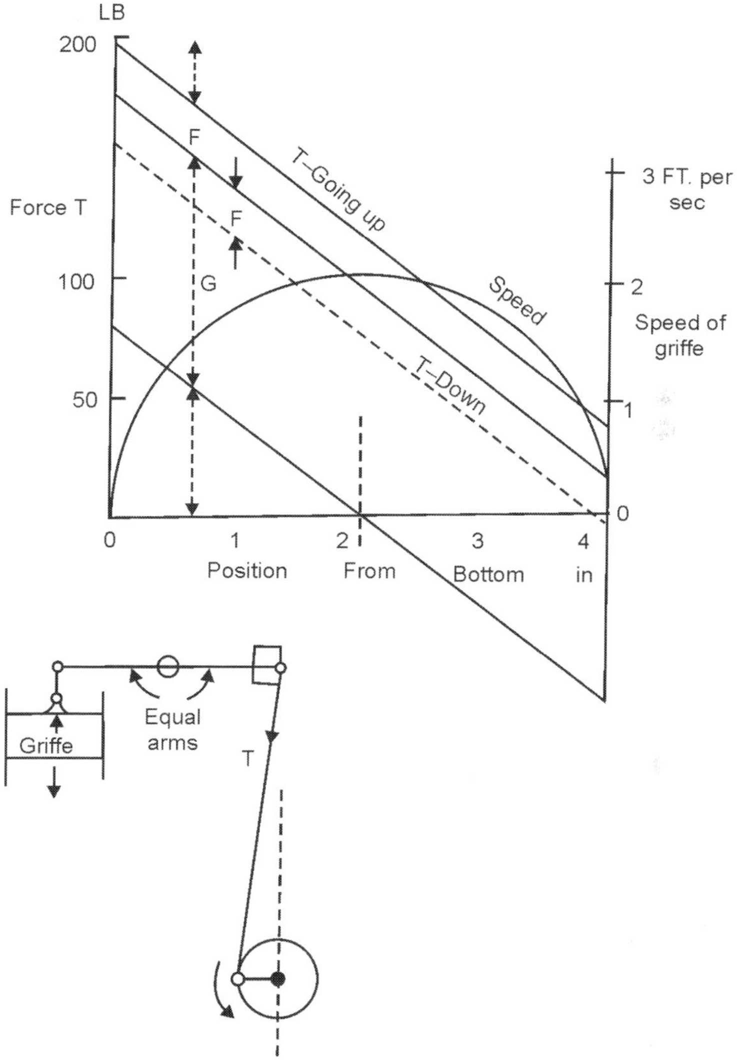

Figure 17.3

Maximum accelerating force (F) = ma

$$= 96/32 * 26.43$$

$$= 79.3 \text{ lb.}$$

The forces required at different positions are shown in Table 17.1.

Downward forces are taken positive and the upward forces are taken as negative. All the forces are in lb wt as shown in Table 17.1.

Table 17.1 Movement of griffe

Motion of the griffe	Gravity (G)	Friction (F)	Inertia force (I)	Net force (T)
Start of up movement	+96	+20	+79.3	+195 5
Start of down movement	+96	−20	−79.3	-5.5
When 1 in. up	+96	+20	+39.65	+155.75
When 3 in. up	+96	+20	−39.65	+76.25

The curves of forces are all straight line curves and are shown in Fig. 17.3.

The speed curve is a semi-circle and the maximum speed is at the centre of the lift.

It has been assumed that the loom speed is constant throughout. Actually it is not the case, since when the griffe is lifted, the loom speed will fall and the speed will rise during the downward movement of griffe.

3. In the above problem, suppose if balancing masses of 80 lb. are added, the object is for steadying the loom at the opposite end of the griffe connecting link, calculate the forces as before.

Solution:

The addition of balance masses reduces the forces required to overcome the gravity.

Hence

Mass of the griffe − Balance masses

$$= 96 - 80 = 16 \text{ lb.}$$

But the balance masses also have to be accelerated and the force for acceleration would be increased in the proportion

$$= 96 + 80/96$$

The maximum accelerating force will be

$$79.3 * 176/96 = 146 \text{ lb.}$$

At the beginning of the upward movement, the connecting rod T must apply a force of

$$16 + 20 + 146 = 182 \text{ lb}$$

This force is slightly lesser than the previous one of 195.5 lb.

Again when the griffe is 1 in. up, the force T would be

$$16 + 20 + 73 = 109 \text{ lb, instead of } 155.5 \text{ lb.}$$

17.3 Simple pendulum

A simple pendulum in its simplest form consists of a heavy bob suspended at the end of a light inextensible and flexible string. The other end of the string is fixed at *O* as shown in Fig. 17.4.

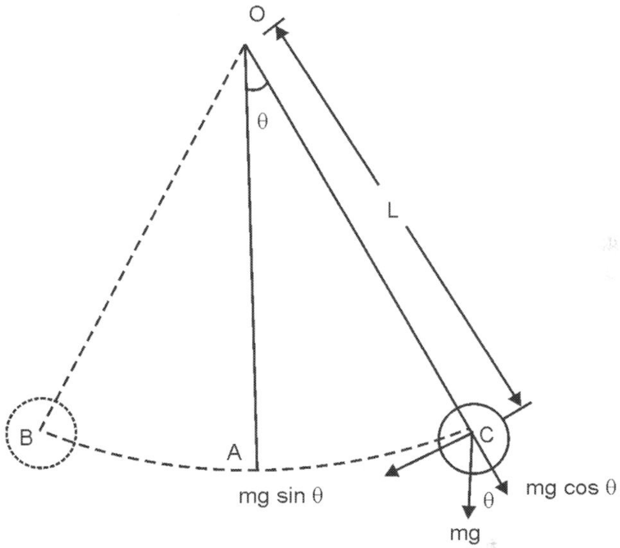

Figure 17.4 Simple pendulum

Let L = Length of the string

m = Mass of the bob in kg

W = Weight of the bob Newton's

= $m\,.g$ and

θ = Angle through which the string is displaced

When the bob is at *A*, the pendulum is said to be in equilibrium. If the bob is brought to *B* or *C* and released, it will start oscillating between the two positions *B* and *C* with *A* as the mean position. If the angle (θ) is very small, say (less than 4 degrees), then the bob will have simple harmonic motion. Now the couple tending to restore the bob in equilibrium position or restoring torque,

$$T = m.\ g.\ \sin\theta \times L$$

Since angle is very small, therefore $\sin\theta = \theta$ radians.

$$T = m.\ g.\ L\,\theta$$

We know that the mass moment of inertia of the bob about an axis through the point of suspension

$$I = \text{mass} \times (\text{length}^2) = mL^2$$

Angular acceleration of the string

$$\alpha = T/I$$
$$= m.\ g.\ L\ \theta/m.L^2$$
$$= g.\ \theta/L$$
$$= \theta/\alpha = L/g$$

i.e., Angular displacement / Angular acceleration

$$= L/g$$

Periodic time,

$$t = 2 * \pi \times \sqrt{}\ \text{Displacement/ Acceleration}$$
$$= 2 * \pi \times \sqrt{}\ L/g$$

and frequency of oscillation

$$n = 1/t = \tfrac{1}{2}*\pi \times \sqrt{}\ g/L$$

From the above, it may be noted that the periodic time and the frequency of oscillation of a simple pendulum depends only upon its length and acceleration due to gravity and the mass of the bob has no effect on it.

17.4 Helical springs

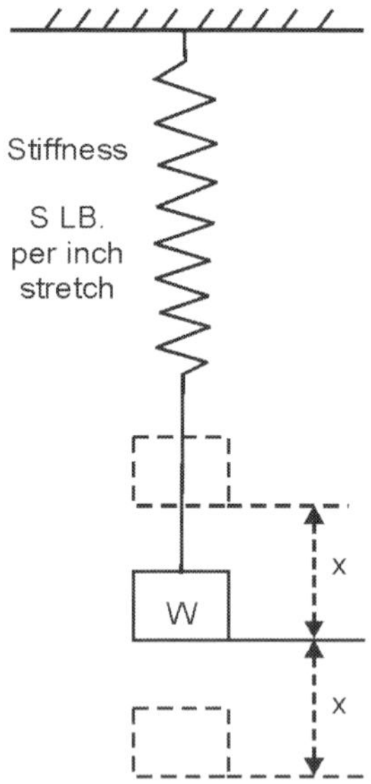

Figure 17.5 Helical springs

Consider in a helical spring, if a mass W is suspended by a spring as shown in Fig. 17.5. It comes to rest in an equilibrium position such that the upward force applied by the spring is equal to the force of gravity on W.

If we consider S the stiffness of the spring per inch of stretch and D the stretch in the equilibrium position, then

$$W = SD$$

Where W is the force of gravity on the mass.

If the mass is displaced downwards from its equilibrium position it oscillates up and down when it is released and moves with simple harmonic motion.

At any displacement, x, below the equilibrium position, the upward pull of the spring is greater than W by S_x, which is the force available to accelerate W upwards. Similarly, at a displacement, x, above the equilibrium position, there is a downward force of S_x available for acceleration.

The upward pull of the spring is = $(W - S_X)$

Downward force on the mass is the force of gravity = W

The force available for acceleration = $W - (W - S_X) = S_X$

Thus the accelerating force acting on the mass is directly proportional to the displacement from the equilibrium position and is directed to that position with the result that the mass moves with simple harmonic motion.

When the mass is displaced downwards by a distance, x, and released, it rises to the same displacement x, above the equilibrium position and keeps on oscillating with simple harmonic motion.

As the mass rises and falls, energy is being transferred between it and the spring, but the total energy at any instant is constant. In the bottom position, all the energy stored in the spring is strain energy and this position is referred as datum level. In the top position, the strain energy in the spring is less than the bottom position, but the mass has gained potential energy.

When the mass passes the equilibrium position, the mass has potential energy and also kinetic energy, but the spring has only strain energy.

In the bottom position the total energy is the strain energy in the spring, which is equal to the work done in stretching it $D + x$.

Total energy is then $\frac{1}{2} S (D + x)^2$

If the mass rises to the same height, x, above the equilibrium position, then

Potential energy in $W = W \times 2x = SD \times 2x$.

Strain energy in spring = $\frac{1}{2} S (D - x)^2 = \frac{1}{2} SD^2 + \frac{1}{2} Sx^2 - SD_X$

The total energy of the system =

$2SD_X + \frac{1}{2} SD^2 + \frac{1}{2} Sx^2 - SD_X$

$= \frac{1}{2} SD^2 = \frac{1}{2} Sx^2 + SD_X$, which is same as in bottom position.

When passing the equilibrium position,

Potential energy in $W = W_X = SD_X$

Energy stored in the spring = ½ SD^2

Kinetic energy in W must be ½Sx^2, since total energy is

$$\tfrac{1}{2}SD^2 + \tfrac{1}{2}Sx^2 = SD_X$$

Also, kinetic energy = $Wv^2/2g$, where v is the velocity of the mass passing the equilibrium position.

$$Wv^2/2g = \tfrac{1}{2}\,Sx^2$$

and $$v^2 = gSx2/W$$

The periodic time

$$t = 2 * \pi \times \sqrt{\ } \text{displacement/acceleration}$$

At displacement, x, the accelerating force is $S_X = W/g \times$ acceleration

$$\text{And acceleration} = g\,S\,x/W,$$

Hence $$t = 2 * \pi * \sqrt{x}\ W/g\,S\,x$$

$$= 2 * \pi \times \sqrt{\ } W/S\,g$$

It is to be noted in springs, the periodic time or the number of oscillations per minute is independent of x, but depends only on the stiffness of the spring and on the size of the mass.

4. A mass of 10 lb is suspended from a spring, extends it 10 in. If the mass is pulled down another 1 in. and released, find the periodic time of the motion and the highest speed of the mass. What is the spring tension when the spring is in its highest position?

Solution:

The periodic time $t = 2*3.14 *\sqrt{\ } W/S\,g$, where S is the stiffness of the spring and is given by 1 lb. per inch or 12 lb. per ft. $g = 32$ ft/s

Then, $t = 2 * \pi *\sqrt{\ } 10/12 * 32 = 1.01$ s

The maximum speed of the mass occurs when the mass is passing through the equilibrium position.

$v = x * \sqrt{\ } g\ S/W$, where x is the maximum displacement from the equilibrium position,

1 in. or 1/12 ft

$$v = 1/12 * \sqrt{\ } 32 \times 12/10 = 0.516 \text{ ft/s}$$

Therefore the mass will rise 1 in. above the equilibrium position. The stretch of the spring will then be 9 in. and the spring tension is 9 lb. wt.

Numerical questions

1. A mule carriage in a conventional spinning machine rail makes an inward run with a movement similar to simple harmonic motion. It moves 5 ft. in 3 seconds. What is its maximum speed?

   If the carriage has a mass of 2 tons, what is the maximum accelerating force acting on it and the accelerating forces when it is (a) 1 ft. and (b) 3 ft. from its outer position?

   Ans. Maximum speed (v) = 2.62 ft/s

   Acceleration = 2.745 ft/s^2

   Maximum accelerating force = 384 lb wt

   Accelerating force at 1 ft = 230 lb. wt

   Accelerating force at 3 ft = -77 lb wt. (at midway position, the force becomes negative).

2. A loom heald shaft is lifted 4.5 in. vertically, in the time the crankshaft takes to turn 180° at 120 revs per min. It is moved with motion similar to simple harmonic motion. A second shaft is lowered simultaneously with the same kind of movement.

   At what speed relatively with each other, in miles per hour, do the warp threads controlled by the heald shafts pass each other at the centre of the shed? (Fig. 17.6)

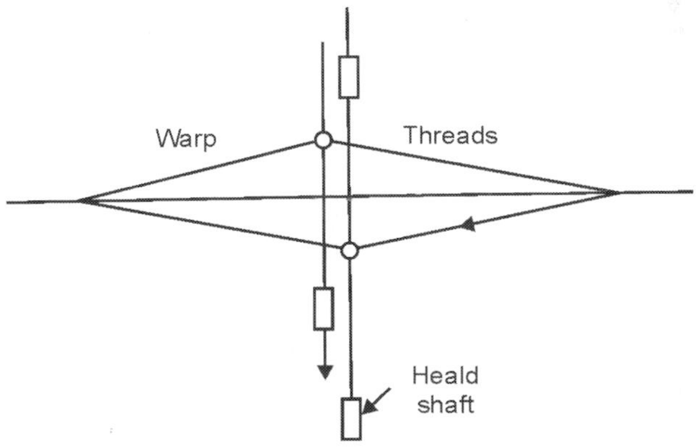

Figure 17.6

   Ans. 3.22 miles per hour

3. The two card cylinders in a double cylinder jacquard loom are rigidly connected together and are moved horizontally with S.H.M.

The mass moved weighs 42 lb. and the maximum displacement is 3 in. from the end position. The mass gets its motion from a crank which revolves at 100 rpm. If the friction at the slides, which guides the mass is 10 lb., draw the curve showing the force required to move the mass, at any position, what is the maximum force and when it is required?

At what speed is the cylinder moving when the card on the cylinder face strikes the needle ends 0.25 in. from the end of the movement?

Ans. 28 lb. at the start of the movement; Speed = 8.65 in. per sec.

4. A pirn winding machine has a spindle speed of 3600 rpm. The spindle and pirn are reciprocated to traverse the yarn, the speed of traverse remains constant except at the reversal of the movement. The distance moved per double traverse is 3 in. and the spindle makes 14 revolutions per double traverse. What is the average traverse speed?

   During reversal from full speed in one direction to full speed in opposite direction takes place during 0.5 rev. of the spindle. Calculate the average acceleration during the reversal and the average force required, if the reciprocating parts weigh 1 lb.

   Ans. 1.07 ft/s; 8.05 lb; 1.41 lb for SHM

5. A drop box in a multiple shuttle loom is moved positively down a distance of 6 in. During the time the crank shaft turns 150 degrees and the loom speed is 150 rpm. The movement given to the box is similar to SHM. The box is controlled by a vertical control rod which weighs 10 lb. in addition, a frictional force of 4 lb. acts on it. How far is the box moving when halfway down? What are the magnitudes of the forces required at the beginning and at the end of the movement?

   Ans. 4.71 ft/s; At the start 21.75 lb downward and 33.75 lb. upward at the end.

6. A heald shaft in a dobby loom is lifted 4 in. to form the shed, in 2/3rds of the time for one revolution of the crankshaft, which revolves at 160 rpm. The mass of the shaft is 2.5 lb, and the movement is S.H.M. What maximum accelerating force is required to move it?

   A spring reversing motion applies a constant downward pull of 10 lb. on the shaft throughout the lift. The tensions in the warp threads apply a maximum upward and downward force of 6 lb to the shaft and this force is in direct proportion to the distance of the warp threads from the centre of the shed.

What are the maximum and minimum forces? Calculate the work done in lifting the shaft.

Ans. 2.95 lb; 16.45 lb; maximum force = 8.55 lb; minimum force = 50 in. lb.

7. A light helical spring stretches 1 in. for a load of 6 lb. When the spring is suspended from affixed point and not stretched a load of 9 lb. is suddenly hooked on it. How far will the weight descend? What is the spring tension at the lowest point? Find the periodic time of oscillation. What is the length of the equivalent simple pendulum which has the same periodic time?

   Ans. 3 in.; 18 lb. wt; 0.3925 s; 1.5 in

18

Moments of inertia and kinetic energy of rotating bodies

18.1 Mass moment of inertia

A body is composed of small elements. If the mass of every element of the body is multiplied by the square of its distance from its axis, the sum of the product of the mass of the element and square of the distance is known as mass moment of inertia of the body about that axis. It is denoted by letter *I*.

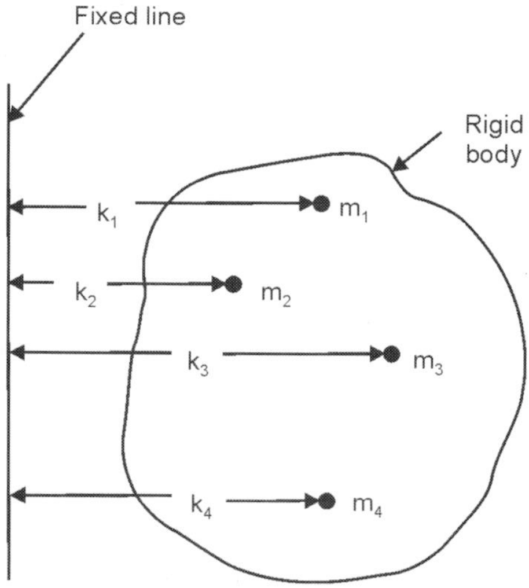

Figure 18.1 Mass moment of inertia

Figure 18.1 shows a body of small elements.

Let m_1 = Mass of the first element of the body

k_1 = Perpendicular distance of the first element from the axis

Similarly m_2, m_3, m_4 ... = Mass of other elements

k_2, k_3, k_4 ... = Perpendicular distances of respective elements from the axis

Now the mass moment of inertia of the body is given by

$I = m_1 k_1^2 + , m_2 k_2^2 + , m_3 k_3^2 + \ldots$

Let m is the total mass of the body. The total mass m of the body is assumed to be concentrated at a point known as centre of mass or centre of gravity. Let k is the radius of such length that mk^2 is the mass moment of inertia, then

$I = mk^2 = m_1 k_1^2 + m_2 k_2^2 + m_3 k_3^2 + \ldots$

The distance k is known as radius of gyration of the body. Hence the radius of gyration of the body is defined as the distance from the axis where the total mass of the body is assumed to be concentrated to give the same value of I.

Therefore $I = mk^2$

18.2 Fundamental equation for a rotating body

Consider a particle or small body of weight W or mass $M = W/g$, rotating about an axis O, at radius K, and being accelerated from linear velocity v_1 to v_2, in time t. It will be recollected that the relationship between linear velocity, v, and angular velocity, ω, is $v = \omega r$ or $\omega = v/r$ where r is the radius of the circle in which the particle or body is moving.

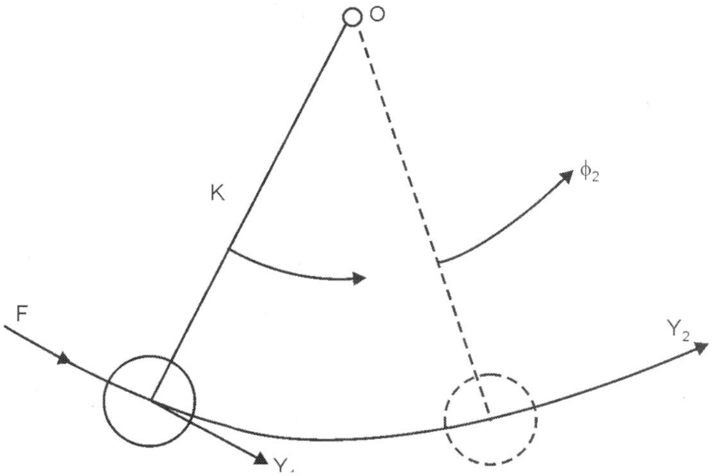

Figure 18.2

Hence the linear acceleration of the particle or body is given by

$v^2 - v_1/t = \omega^2 K - \omega_1 K/t = (\omega^2 - \omega_1)\ K/t =$ angular acceleration $\times K$

The force F required for the linear acceleration = mass × linear acceleration $= W/g\ K \times$ angular acceleration. This is equal to the inertia resistance offered by the body to have its velocity changed.

The torque or moment required to overcome this resistance = FK, hence torque required for acceleration = FK.

= $W/g\ K \times$ angular acceleration x K

= $W/g\ K^2 \times$ angular acceleration

The product $W/g\ K^2$ or MK^2 is the moment of inertia of the particle or body, and is generally denoted by I.

This equation Torque = Moment of Inertia × Angular Acceleration is a fundamental equation of a rotating body. This equation is corresponding to Force = Mass × Acceleration for a body moving with straight line motion.

18.3 Kinetic energy of a rotating body

The kinetic energy of a mass M or weight W,

$$KE = ½Mv^2 \text{ or } Wv^2/g$$

If the mass is a rotating one, then if the mass is small and so far from the axis that all the particles are moving t nearly at the same speed, e.g. in simple pendulum, there will be no any great error in taking speed (v) of the centre of gravity of the mass.

Example 1: A mass of 10 lb. is swinging as simple pendulum. It is suspended by affine wire 10 ft. long from the point of suspension to the centre of the mass. If it swings 30° on either side of the vertical, what is the speed and kinetic energy in its lowest position?

Solution:

Mass = 10 lb; length of the wire = 10 ft long; angle of swing = 30°

As the mass swings from highest to its lowest position, the potential energy gained is lost and the mass gains an equal amount of kinetic energy.

Vertical fall of the mass from highest to lowest position is =

$$10 - 10 \cos 30° = 10 - 8.66$$
$$= 1.34 \text{ ft}$$

Kinetic energy at its lowest position = Loss of potential energy

$$= 10 \times 1.34$$
$$= 13.4 \text{ ft. lb.}$$

If v is the speed of the centre of the mass,

Then,
$$KE = Wv^2/2g$$
$$13.4 = 10 \times v^2/\ 2 * 32$$
$$v = 9.26 \text{ ft/s}$$

Example 2: Two masses, each of 1 lb, are mounted on a light lever. The lever is fulcrumed at one end and one mass is 1 ft from the other is 2 ft from the axis. If the lever rotates at 1 radian per second, what is the radius of gyration of the weighted lever?

Solution:

Weight of the mass (W_1) = 1 lb.

Weight of the mass (W_2) = 1 lb.

Distance of the W_1 from the axis = 1 ft.

Distance of the W_2 from the axis = 2 ft.

Angular velocity of the lever = 1 radian per second

Kinetic energy (KE) = $W\ \omega^2\ r^2/2g$, where ω = angular velocity, and r = radius

Total kinetic energy =

$(W_1 + W_2) \times 1^2 \times 2\ K^2/2*32$

$$= (1 + 1) \times 1 \times 2\ K^2/64$$

$$2K^2 = \sqrt{}5$$

$$K = 1.58 \text{ ft.}$$

Example 3: In the above Example 2, if the lever is allowed to fall from a stationary horizontal position, what is its speed and kinetic energy when passing the vertical position?

Solution: The centre of gravity of the system will be in the midway between the centre of gravity of the masses. i. e 1.5 ft from the axis.

Gain in kinetic energy when the lever falls from the horizontal to vertical position − loss of potential energy

= Weight × fall of centre of gravity

= 2 lb × 1.5 ft = 3 ft. lb.

Then $KE = \text{W}\ \omega^2\ K^2/2 *g$

$$3 = 2 \times \omega^2 \times 1.58^2/64$$

$$\omega^2 = 3 \times 64/2 \times 1.58\ 81.58$$

$$\omega = 6.2 \text{ radians per second, or } 6.2 \times 60/2*3.14$$

$$= 59.2 \text{ revs per min.}$$

Example 4: The rope flywheel on a mill engine is 25 ft. diameter to the centre of the ropes and weighs 7 tons. How much kinetic energy does it possess at 60 rpm. assuming the radius of gyration to be the radius to the centre of the ropes? If the energy of the flywheel rises and falls 10000 ft. lb. from its value at 60 revs. per min., what are the highest and lowest speeds?

Solution: First, we have to calculate the kinetic energy at 60 revs. per min.

$$= W\omega^2 K^2/2*g$$
$$= 7 * (2*3.14)^2 \times 12.5^2/64$$
$$= 675 \text{ ft. tons or } 1{,}510{,}000 \text{ ft. lb.}$$

The kinetic energy of any body is equal to a constant × square of the speed. Assuming the constant as M, $KE = MN^2$, where N = revs per min.

$$M = 1{,}510{,}000/60^2$$

If the highest speed is N_1 and lowest speed is N_2,

Then $$MN_1^2 - MN_2^2 = 20000 \text{ ft. lb}$$
$$M(N_1 - N_2)(N_1 + N_2) = 20000$$

Since $N_1 + N_2$ is twice the average speed of 60 revs. per min.

$$120\,M(N_1 - N_2) = 20000$$

and $$N_1 + N_2 = 20000 \times 3600/120 \times 1510000$$
$$= 0.398$$

$$N_1 + N_2 = 120 \qquad \ldots (18.1)$$
$$N_1 - N_2 = 0.398 \qquad \ldots (18.2)$$

Adding equations (18.1) and (18.2), we get

$$2N_1 = 120.398$$
$$N_1 = 60.2 \text{ and } N_2 = 59.8 \text{ revs per min.}$$

Example 5: A carding engine cylinder has a mass of 1120 lb. and its radius of gyration is 2 ft. Find the kinetic energy if its speed is 180 rpm.

Solution:

$$W = 1120 \text{ lb}$$
$$\omega = 180 \text{ rpm}$$
$$= 180/60$$
$$= 3 \text{ revs per second}$$
$$= 3 * 2 * \pi \text{ radians per second}$$
$$K = 2 \text{ ft.}$$
$$KE = W\omega^2 K^2/2*g$$
$$= 1120 * 6^2 * \pi^2 * 2*2/64$$
$$= 24900 \text{ ft. lb.}$$

Example 6: In Example 5, if the moment of inertia of the cylinder is 90 percent of the moment of inertia of the whole machine, what torque and maximum horse-power are required to accelerate from rest to a speed of 180 rpm in 30 sec?

Solution:

The angular acceleration during the start up

= Change in angular velocity/ time

= $6 * \pi / 30$

= 0.628 radians/s^2

As we know that

Torque = Moment of Inertia × Angular acceleration

= $1120 \times 2^2 * 0.628/ 32 *0.90$

= 97.68 lb ft.

Horse power =

Rate of doing work per sec./550

= torque × angular displacement per sec./550

= $97.68 \times 6 * \pi / 550$

= 3.36 HP

Example 7: The rim shaft of a spinning machine together with its pulleys fixed to it weighs 128 lb. and has a radius of gyration of 6 inches. During backing-off it has to be brought to rest from a speed of 960 rpm in 1.5 sec. What average retarding torque must be applied to it?

Solution:

Weight of the rim shaft and pulleys (W) = 128 lb.

Speed N = 960 rpm

Angular velocity (v) = 960/60 = 16 2* π radians/s^2

Time (t) = 1.5 second

Radius of gyration (K) = 6 inches or 0.5 ft.

Angular retardation = Angular velocity/time

= $16 * 2* \pi / 1.5$

= 67 radians/s^2

Torque required = Moment of Inertia × Angular retardation

= $128 * 0.5^2 * 67/32$

= 67 lb ft.

18.4 Compound pendulum

When a rigid body is suspended vertically, and it oscillates with a small amplitude under the action of force of gravity, the body is known as compound pendulum.

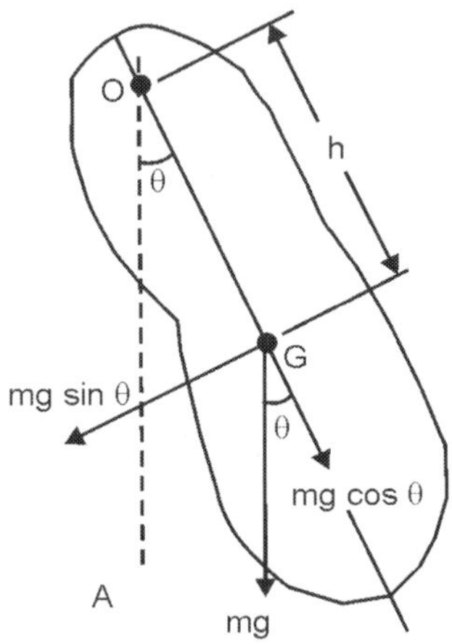

Figure 18.3 Compound pendulum

Let m = Mass of the pendulum in *lb.*

W = Weight of the pendulum in *kgf. mg.*

K_G = Radius of gyration about an axis through the centre of gravity G and perpendicular to the plane of motion, and

h = Distance of point of suspension O from the centre of gravity G of the body.

If the pendulum is given a small displacement (θ), then the couple tending to restore the pendulum to the equilibrium position OA.

$$T = mg \sin\theta \times h$$
$$= mgh \sin\theta$$

Since θ is very small, substituting θ = θ radians, we get

$$T = mgh\theta$$

Now, the mass moment of inertia about the axis of suspension o,

$$I = I_G + mh^2$$
$$= m(KG^2 + h^2)$$

Angular acceleration of the pendulum

$$\alpha = T/I$$
$$= mgh\theta/m(KG^2 + h^2)$$
$$= gh\theta / KG^2 + h^2 = \text{constant} \times \theta$$

We see that the angular acceleration is directly proportional to the angular displacement, therefore the pendulum behaves in simple harmonic motion (S.H.M)

$$\theta/\alpha = KG^2 + h^2/g.h$$

We know that the periodic time (t) is

$$t = 2 * \pi * \sqrt{\text{Displacement/ Acceleration}}$$
$$= 2 * \pi \sqrt{\theta/\alpha}$$
$$= 2 * \pi * \sqrt{KG^2 + h^2/g.h}$$

and frequency of oscillation is

$$t = ½ * \pi \times \sqrt{g.h / KG^2 + h^2}$$

18.5 Centre of percussion

The centre of percussion is defined as the point at which a blow may struck on a suspended body so that the reaction at the support is zero.

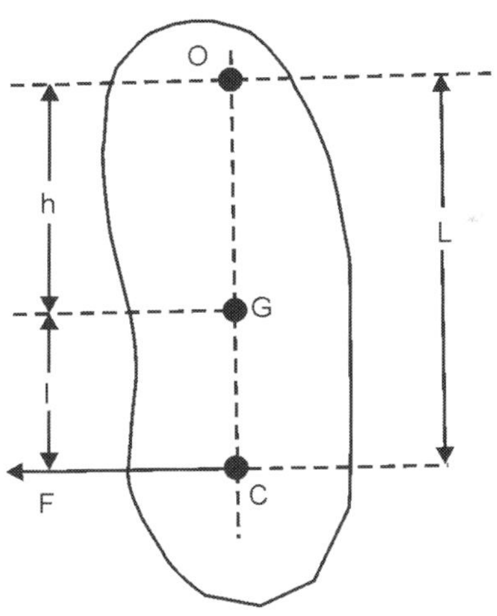

Figure 18.4 Centre of percussion

Consider the case of a compound pendulum suspended as shown in Fig. 18.4. Suppose the pendulum is at rest in the vertical position, and a blow is struck at a distance L from the centre of suspension. Let the magnitude of the blow is *F kgf.* The blow will have two effects on the body.

1. A force (F) acting at C will produce a linear motion with an acceleration (a), such that $F = m.a.$ … (18.3)

   where m is the mass of the body.

2. A couple with moment equal to ω ($F \times l$) which will tend to produce a motion of rotation in the clockwise direction about the centre of gravity G. Let this turning moment ($F \times l$) produce an angular acceleration (α) such that

$$F \times l = I \times \alpha \quad \ldots (18.4)$$

Where I is the moment of inertia of the body about an axis passing through G and parallel to the axis of rotation.

From Equation (18.3) $a = F/m$ ... (18.5)

And from Equation (18.4) $\alpha = F\,l\,/I$

Now, the linear acceleration of O

$$= aO = \alpha.\ h = F.\ l.\ h/I = F.\ l.\ h/mK^2 \quad \ldots (18.6)$$

Where K is the radius of gyration of the body about the centre of gravity G.

Since there is no reaction at the support when the body is struck at the centre of percussion, therefore a should be equal to aO.

Equating equations (18.5) and (18.6)

$$F/m = F\,.l\,.h/mK^2$$

or

$$K^2 = I.\ h \text{ and } l$$
$$= K^2/h$$

We know that the equivalent length of simple pendulum

$$L = K^2 + h^2/h = K^2/h = h = l = h$$

From the above, there are three points of importance in compound pendulum be considered.

(i) The potential energy of the compound pendulum as it changes its position being its weight x centre of gravity above the datum level.

(ii) The radius of gyration, where the whole mass may be considered concentrated, when considering its moment of inertia or when finding its kinetic energy.

(iii) The centre of percussion

Example 8: The Ballistic type pendulum of a cloth testing machine is shown in Fig. 18.5. has a mass of 12 lb. and its centre of gravity is 11.25 in. from the axis. When allowed to swing freely it makes 46 oscillations in one minute. What is its radius of gyration about its axis?

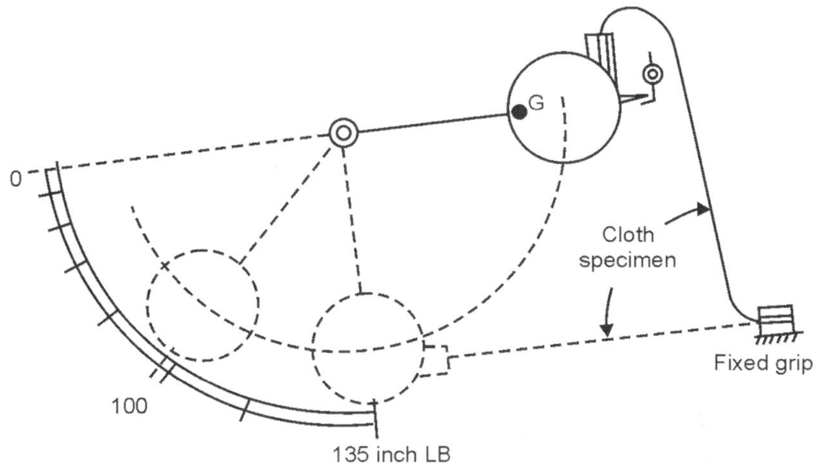

Figure 18.5 Ballistic testing machine

Solution:

Mass of the pendulum (W) = 12 lb.

Centre of gravity (h) = 11.25 in. (0.94 ft)

Number of oscillations = 46

Periodic time (t) = 60/46 = 1.3 sec.

$l = 2 * \pi \times \sqrt{K^2/gh}$, where K is the radius of gyration about the axis and h is the distance of the centre of gravity from the axis.

$$K^2 = ght^2/4 * \pi^2$$
$$= 32 \times 0.94 \times 1.3^2 / 4 * \pi^2$$
$$K = 1.13 \text{ ft}$$

Example 9: In the above Example 8, if the pendulum swings 45° past the vertical after breaking the specimen, how much work is done in breaking it? Also calculate the angular velocity of the pendulum when it reaches the vertical position. At what radius the grip on the pendulum be placed?

Solution:

Potential energy stored in the pendulum in the starting position =

12 lb × 11.25 in = 135 in. lb. or 11.25 ft. lb.

This 11.25 ft. lb. potential energy is considered taking the vertical position of the centre of gravity as datum level. Hence, when the pendulum reaches the vertical position, the pendulum has 135 in. lb or 11.25 ft. lb of kinetic energy.

(a) Work done on the specimen =

Kinetic energy − Potential energy in the pendulum at the end of its swing

135 − 12 (11.25 − 11.25 cos 45°)

= 135 − 39.6 = 95.4 in. lb. or

Work done = 12 lb × 11.25 *cos 45°

= 95.4 in. lb.

(b) Angular velocity of the pendulum is

Kinetic Energy = $W \omega^2 K^2/2g$

ω^2 = $2g\ KE/WK^2$

= 2 * 32 * 11.25/12 * 1.288

ω = 6.82 radians per sec or 65 revs per min.

(c) The centre of percussion is at the length, l, of the equivalent simple pendulum. i.e. one which has a periodic time of 1.3 sec.

For the simple pendulum, $t = 2 * \pi * \sqrt{l/g}$

l = $gt2/4 * \pi^2$

= $32 * 1.3^2 / 4 * \pi^2$

= 1.37 ft. or 16.45 in.

Example 10: A fly wheel on the driving shaft of a warp knitting machine weighs 160 lb. and has a radius of gyration of 7 inches. How much work is done in accelerating to a speed of 250 rpm. If the speed is reached from rest in 3 seconds, what is the average acceleration and what torque is required at the driving shaft to accelerate the flywheel?

Solution:

Weight of the flywheel (W) = 160 lb.

Radius of gyration (K) = 7 in. (7/12) =

Speed of the knitting machine (N) = 250 rpm

Time to accelerate from rest (t) = 3 sec.

Angular velocity (ω) = 250/60 = 4.166 × 2 π radians per sec.

KE = $W \omega^2 K^2/2g$

= $160 \times (4.166 \times 2 * 3.14)^2 \times (0.583)^2/ 2*32$

= 581.72 ft. lb

Angular acceleration = Change of angular velocity/ time

= v/t

$= 4.166 \times 2 * \pi / 3$

$= 8.729$ radians per sec.

Torque = Moment of Inertia × Angular Acceleration

$= 160/32 \times 0.583^2 \times 8.729$

= 14.82 ft. lb.

Example 11: The moment of inertia of a carding machine is equivalent to that of a mass of 1200 lb, with a radius of gyration 2 ft. from the driving shaft. What is its kinetic energy at a machine speed of 165 rpm? if the machine requires 0.8 horse power to drive it at 165 rpm, in approximate how many revolutions, and in what time the machine will come to stop when the belt is disengaged from the drive, assume uniform retardation.

Solution:

Weight of the carding cylinder (W) = 1200 lb.

Radius of gyration (K) = 2 ft.

Speed of the machine (N) = 165 rpm.

Horse power = 0.8

Angular velocity (ω) = 165/60

$= 2.75 \times 2 * \pi$

= 17.27 radians per sec.

$K.E. = W\omega^2 K^2/2\,g$

$= 1200 \times 17.27^2 \times 2*2/2*32$

= 22368.96 ft. lb.

Horse power = Work done per sec/550

Work done per sec = Horse power × 550

$= 0.8 \times 550 = 440$

Work done per sec = torque × angular displacement per sec

Torque = Work done per sec/ Angular displacement per sec.

$= 440 / 17.27 = 25.40$ ft. lb.

We know that,

Torque = Moment of inertia × angular acceleration or retardation

$25.40 = 1200/32 * 2*2 \times$ Angular acceleration

Angular acceleration = 0.169 radians/s^2

Thus time (t) = Angular velocity/Angular acceleration
= 17.27/0.169
= 102 seconds or 1.7 minutes

Example 12: A cylindrical yarn package (cheese) on a paper tube centre has a diameter of 5 in. and weighs 1.65 lb. Its radius of gyration is radius/square root of 2.

During warping process, the yarn has to be withdrawn from the surface of the cheese which is free to rotate on the creel peg. How much work is done in accelerating 400 such cheeses from rest to a warping speed of 70 yd. per min. if the time taken to accelerate is 3 seconds, what yarn tension is required to accelerate each cheese?

Solution:

Diameter of the cheese = 5 in.
Warping speed = 70 yard per min.
Angular velocity (ω) = $\pi \times 5/12 \times N$
= 70
N = 161.53 rpm
= 161.53 × 2*3.14/60
ω = 16.90 radians per sec.
No. of cheeses = 400
Weight of the cheese (W) = 1.65 lb.
Radius of gyration (K) = Radius of the cheese/√ 2
= 2.5 in/1.414 *12
= 0.1473 ft.
$K.E.$ = $W\omega^2 K^2/2g$
= $1.65 * 16.90^2 * 0.1473^2/2*32$
= 0.159 lb ft per cheese
For 400 cheeses = 0.159 × 400
= 63.96 ft. lb

Tension required to accelerate each cheese is
Time for acceleration = 3 seconds
Velocity (v) = 70 × 3/60 = 3.5 ft/s

From the equations of motions,

$$v = u + \omega t = 0 + \omega \times 3$$

$$= 7.326 \text{ radians per sec.}$$

Tension or kinetic energy ($K.E.$) = $W\,\omega^2\,K^2/2g$

$$= 1.65 \times 7.326^2 * 0.1473^2/2*32$$

$$= 0.03 \text{ lb/cheese or } 0.48 \text{ oz per cheese}$$

Numerical questions

1. A carding machine has a moment of inertia of 150 engineer's units (lb/g ft^2) and requires 0.9 horse power to drive it at 170 rpm. If the tension in the driving belt is such that the maximum tension on the tight side does not exceed 300 lb, and the belt slips when the slipping ratio is 2.5. What maximum torque is required at the driving pulley of 18 in. diameter? In approximately, what time will the machine reach its maximum speed of 170 rpm, from rest when the belt is shifted from fast to loose pulley?

   Ans. Torque = 135 lb ft; Time = 25 seconds

2. A cone with its cone holder weighs 2.5 lb. and has a radius of gyration of 1 in. The cone is dropped in to contact with the driving roller of the winding machine, the surface speed of the roller is 750 yd. per min. The frictional driving force applied by the driving roller to the cone surface is 1.5 lb.

   The driving force is applied at a point where the cone surface is 4 in. diameter, and that this point on the cone is accelerated to the roller surface speed. Calculate the time taken to get the cone to full speed, the revolutions made by it during the acceleration and the relative movement during acceleration, between the driving roller surface and the cone surface where the driving force is applied.

   Ans. Time = 0.49 sec.; no. of revolutions = 8.8; relative movement = 9.2 ft.

3. The picking stick with its picking shoe, on an under pick loom weighs 5 lb. and has a radius of gyration of 10 in. about its fulcrum. The picker weighing 2 oz. fits loosely on the top of the picking stick. During picking the stick and the picker are accelerated to a maximum picker speed of 30 ft/s. How much work is done per pick in accelerating the stick and the picker respectively?

   Ans. 9ft. lb; 1.75 ft. lb.

4. The upright picking stick with its cone and picking stick in a cone over pick loom weighs 14 lb, and its radius of gyration is 2.75 in. during picking, it is accelerated to a maximum angular velocity of 13.2 radians per sec. How much work is done on accelerating it? During acceleration, a spring is attached to the rim of a collar on the shaft is stretched to 0.5 in. and its tension increased from 1 lb. to 4 lb. After the shaft has been stopped at the end of its acceleration, it is returned to the starting point by the spring. Find the angular velocity of the picking shaft when it reaches the starting position.

   Ans. Work done for acceleration = 2 ft. lb; angular velocity of the picking shaft = 3.3 radians per sec.

5. The lever of a ballistic testing machine weighs 23 lb. and its centre of gravity is 13.8 in. from the axis. When allowed to oscillate about its axis, it makes 50 oscillations in 62 seconds. Calculate radius of gyration, K, about its axis.

   The lever is released from the horizontal position and when it reaches the vertical it begins to apply load to a specimen of cloth. This breaks the specimen when the lever is 30° past the vertical and the lever swings further 30° before beginning to swing back. Find the work done in breaking the specimen and the angular velocity of the lever when it reaches vertical position and also when the specimen breaks.

   Ans. Radius of gyration $K = 17.21$ in.; Work done = 158.5 lb; Angular velocity of the lever = 7.15 and 4.31 radians/s

6. The sley of an under pick loom is pivoted above the fell of the cloth, the centre of gravity of the sley being 4 ft. vertically below the axis at beat-up. The beating-up point on the reed is 4.25 ft. from the axis.

   The centre of gravity of the sley is displaced 10 in. horizontally from the beat-up position, when the sley is then released and allowed to swing freely, under the action of gravity to beat up the weft. If the beat-up point on the reed is at the centre of percussion of the sley, calculate the radius of gyration, K, about the axis. If the mass of the sley is 2 tons, how much energy is available for beating-up the weft? What is the speed of the reed at the beat-up?

   If the time for sley movement from release to beat-up must not exceed the time for one-third of a revolution of the crankshaft, find the maximum possible speed of the shaft.

   Ans. $K = 4.12$ ft.; Kinetic energy = 2 in. tons; $v = 2.38$ ft/s; speed of the shaft = 35 rpm.

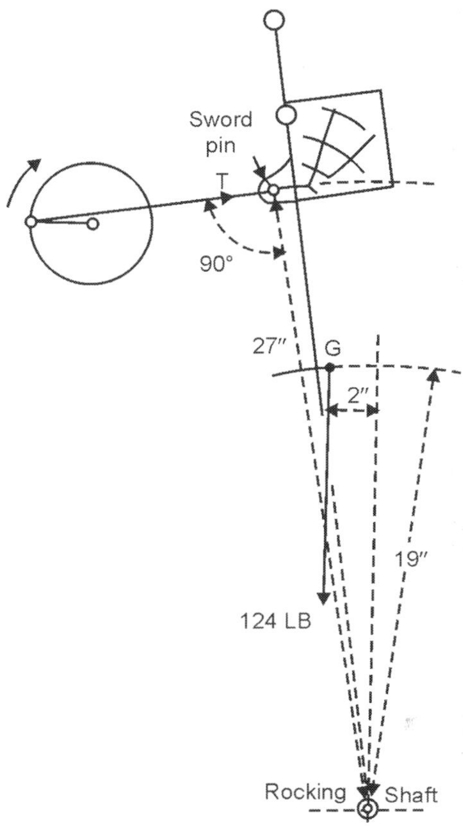

Figure 18.6

7. A loom sley has a mass of 124 lb. and its centre of gravity, *G*, is 19 inches from the rocking shat or axis. When removed from the loom and kept on knife edge, it oscillates 36 times per minute. In the position shown in Fig. 18.6 where the sley is starting its forward movement, its angular acceleration is 26.5 radians/s^2. Calculate the thrust, *T*, required from the connecting arms to keep the sley moving. **Ans.** Torque $T = 173$ lb wt.

19

Practical methods of finding moment of inertia

19.1 Introduction

In the previous chapters, we have seen the methods of finding out the Moment of Inertia in simple and compound pendulum. The pendulum is allowed to oscillate about its axis, under the force of gravity, and the periodic time t is found out. If the Moment of Inertia about the centre of gravity is required, it can be obtained from the relationship $Kg^2 = K^2 - h^2$, where k and Kg are the radii of gyration about the axis and the centre of gravity, respectively.

19.2 Flywheel

It is important to know about the importance of flywheel before we discuss in the next section about the Moment of Inertia of the flywheels. A flywheel is used to control the speed variations caused by the fluctuations of energy during each cycle of operation. In general, it acts as a reservoir of energy which stores energy during the period when the supply of energy is more than the requirement and releases the energy during the period when the supply of energy is less than the requirement. When the flywheel absorbs the energy, the speed of the flywheel increases; whereas when the flywheel releases the energy, the speed of the flywheel decreases. Flywheels are found in speed frames and in looms.

19.3 Methods of finding Moment of Inertia of fly wheels

The flywheel, or other mass, is accelerated from rest by falling weight acting through a rope or cord coiled round the shaft or a pulley on it, so arranged that the cord disengages itself when the weight reaches the floor.

From the time taken by weight to reach the floor, speeds of the falling weight and the flywheel, at the instant when the weight reaches the floor, can be calculated and hence the kinetic energy. The difference between the kinetic energy and the potential energy of the weight before it started is the amount of energy transferred from weight to wheel. If the total number of revolutions from start to stop is observed, the average amount of energy expended per revolution can be found and the energy spent due to friction in the bearings can also be obtained. Then the kinetic energy in the wheel at the instant the weight reaches the floor, when its speed is "n" revolutions per second, its potential energy in weight at start – kinetic energy in weight at floor level – energy spent in overcoming bearing friction during fall.

Therefore, kinetic energy on wheel =

$\frac{1}{2} I \omega^2 = \frac{1}{2} I (2^* \pi^* n)^2$, from which the Moment of Inertia can be found out.

If the mass M of the wheel is known, the radius of gyration is got by $I = MK^2$

Example 1: The kinetic energy of a carding engine cylinder is 26100 ft. lb at 170 rpm. If the mass of the cylinder is 1300 lb, calculate the radius of gyration.

Solution:

KE = 26100 ft. lb.

Mass of the cylinder (M) = 1300 lb

Speed of the cylinder = 170 rpm

Angular velocity of the cylinder (ω)

= $170 \times 2^*3.14/60$

= 17.8 radians per sec.

Kinetic energy = $\frac{1}{2} I\omega^2$

= $\frac{1}{2} I \times 17.8 \times 17.8$

= 26100 ft lb.

I = $2 \times 26100/17.8^2$

= 165 engineer's units or (lb/g ft^2 units)

I = MK^2

K^2 = $165 \times 32/1300$

K = 2.01 ft

Example 2: A carding engine cylinder is disconnected from its drive and from the other parts of the machine so that it can rotate freely in its bearings. A rope is coiled round the pulley of the cylinder shaft, passed round a frictionless pulley near the roof and attached to a weight of 56 lb. The weight is allowed to come down to a floor level from a height of 8 ft accelerating the cylinder, the rope disengaging itself when the weight reaches the floor level. From the following test results, find the kinetic energy in the weight when it reaches the floor and the kinetic energy at that instant. Also find the kinetic energy in the cylinder rotating at 170 rpm.

Time taken for the weight to reach the floor = 9 s
Revs made by the cylinder during fall of weight = 1.5
Total revs of the cylinder from start to stop = 8

Solution:

Velocity, v, of the falling weight can be got by using equations of motion

$$S = ut = \tfrac{1}{2}\,\text{f}t^2$$
$$8 = 0 + \tfrac{1}{2}\,\text{f} \times 9^2$$
$$f = 0.1975 \text{ ft/s}^2$$

Also,
$$v = u + ft$$
$$v = 0 + 0.1975 \times 9$$
$$v = 1.775 \text{ ft/s}$$

Kinetic energy in weight when it reaches the floor level,

$$K.E. = Wv^2/2g$$
$$= 56 \times 1.775^2/2*32$$
$$= 2.77 \text{ ft. lb.}$$

Energy transferred from weight to cylinder =

Actual potential energy in weight − kinetic energy of the weight in floor level

$$= (W\,h) - 2.77$$
$$= (8 \times 56) - 2.77 = 445.23 \text{ ft. lb.}$$

During the fall of the weight, some part of the energy is absorbed in friction at the bearings. The average energy absorbed per revs. of the cylinder = Total energy given to cylinder − Revolutions made by the cylinder from start to stop

$$= 445.23/8 = 55.7 \text{ ft. lb.}$$

Energy absorbed in friction during falling of weight =

$$55.7 \times 1.5 = 83.5 \text{ ft. lb.}$$

Therefore, the kinetic energy in the cylinder at that instant when the weight reaches the floor =

$$445.23 - 83.5 = 361.7 \text{ ft. lb.}$$

Average cylinder speed during fall of weight = 1.5/9 revs. per sec.

Speed when weight reaches the floor =

$$2 \times 1.5/9 = 0.33 \text{ revs. per sec or 20 revs. per min.}$$

Kinetic energy at the cylinder speed of 170 rpm = kinetic energy varies at square of the speed.

K.E. at 170 rpm = $361.7 \times 170^2/20^2 = 26100$ ft. lb.

19.4 Bifilar suspension

The Moment of Inertia of a body may be determined experimentally by an apparatus called bifilar suspension. For a small rotating mass such as spindle of a ring frame or a war per's bobbin, this method is quite easy to find out the Moment of Inertia. The body whose Moment of Inertia is to be determined is suspended by two parallel cords or threads as shown in Fig. 19.1.

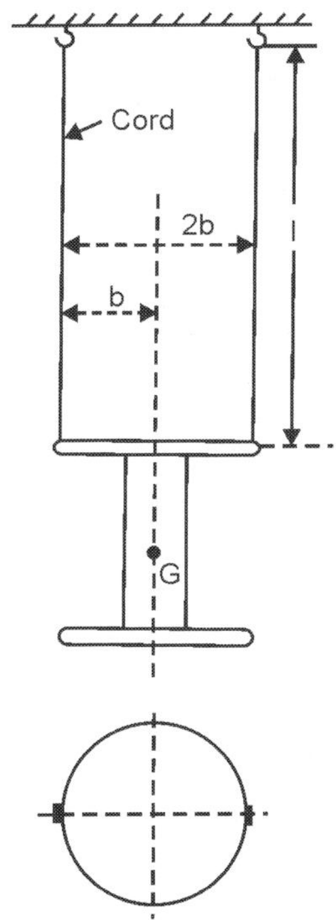

Figure 19.1 Bifilar suspension

The method can also be used for a mass such as a bar or a compound pendulum, the mass being hung so that the centre of gravity is midway between the points of suspension.

It can be proved that if the mass is allowed to oscillate through a small angle, the periodic time of oscillation,

$$t = 2*3.14 \times \sqrt{K^2 l/gb^2}$$

where K = Radius of gyration about the vertical axis through the centre of gravity,

$2 =$ Distance between the threads

$l =$ Length of each thread

K, b, and l should be in ft. if g is in ft/s^2, but may be in inches and g must be 32 × 1 in. per sec

Example 3: A partly filled war per's bobbin weighs 0.5 lb. It is suspended vertically by two parallel threads, as shown in Fig. 19.1. The distance $2b = 3.5$ in. and the length (l) = 30.5 in. when given a small displacement, it oscillates 60 times in 61 seconds. What is its radius of gyration? When the bobbin is mounted in the warping frame, the yarn is withdrawn from the bobbin by the rotation of the warping beam. During startup, the bobbin has to be accelerated to full speed in one second and a yarn tension of 1 oz. is required to overcome the friction at the creel peg. Find the yarn tension during the acceleration. The bobbin radius to the yarn surface is 1.2 in. and the warping speed is 75 yd. per min.

Solution.

Mass of the war per's bobbin (M) = 0.5 lb

$2b = 3.5$ in.; $b = 3.5/2 = 1.75$ in. or 0.1458 ft.

$l = 30.5$ in. (30.5/12) = 2.54 ft

Oscillations = 60 times

$t = 1$ s

Warping speed = 75 yd. per min.

(i) Periodic time $t = 61/60 = 1.02$ s

and also $t = 2*3.14 \times \sqrt{K^2 l/gb^2}$

$t = 2 * 3.14 \times \sqrt{K^2 \times 2.54/32 \times 0.1458^2}$

$K = 1.01$ in.

(ii) The tension in the yarn has to overcome (a) friction and (b) bobbin inertia

Bobbin surface speed at 75 yard per min. = 75 × 3/60 = 3.75 ft/s

Linear acceleration of bobbin surface (assumed uniform) =

3.75/1 = 3.75 ft/s^2

Angular acceleration of bobbin = linear acceleration/radius

= 3.75 × 12/1.2 = 37.5 radians per sec

$$\text{Torque required} = I \times \text{Angular acceleration}$$
$$= MK^2 \times a$$
$$= 0.5/32 \times (1.01/12)^2 \times 37.5 \text{ lb. ft}$$
$$= 0.00415 \text{ ft. lb. ft.}$$

Yarn tension for acceleration = Torque/ Bobbin radius

= 0.00415/0.1 = 0.0415 lb. or 0.0415 × 16 = 0.664 oz.

Total yarn tension during acceleration = 1 oz + 0.664 oz = 1.664 oz

$$= MK^2 \times a$$
$$= 0.5/32 \times (1.01/12)^2 \times 37.5 \text{ lb. ft}$$
$$= 0.00415 \text{ ft. lb. ft.}$$

Yarn tension for acceleration = Torque/ Bobbin radius

= 0.00415/0.1 = 0.0415 lb. or 0.0415 × 16 = 0.664 oz.

Total yarn tension during acceleration = 1 oz + 0.664 oz = 1.664 oz.

Example 4: When winding yarn from the side of a ring frame bobbin to a war per's bobbin in an upright spindle winding frame, the supply bobbin is mounted on a revolving spindle, fitted with a band brake as shown in Fig. 19.2.

Te weight of the ring bobbin and the spindle is 0.4 lb. Find the Moment of Inertia of the bobbin and the spindle as they are suspended with the spindle vertical by parallel cords with the following particulars.

Distance between cords = 1.5 in.

Length of cord = 24 in.

When given a small displacement, they oscillate 86 times per min.

Calculate the radius of gyration.

When the winding speed is 240 yd per min, the acceleration of bobbin and spindle, as the unwinding point approaches the nose of the chase, where the diameter is 0.8 in., reaches a maximum acceleration of 900 radians per sec per sec. If a braking force of 1 oz. is acting on the 1 in. diameter brake wharve, what is the approximate yarn tension when unwinding near the nose?

Solution:

Periodic time of oscillation $t = 60/86 = 0.697$

Distance $2\,b = 1.5$ in.; $b = 1.5/2 = 0.75$ in. or 0.0625 ft.

Length of cord = 24 in. or 2 ft.

Radius of gyration $K = ?$

$$t = 2 * \pi \times \sqrt{K^2 l / g b^2}$$
$$0.697 = 2 * 3.14 \times \sqrt{K^2 \times 2/32 \times 0.0625^2}$$

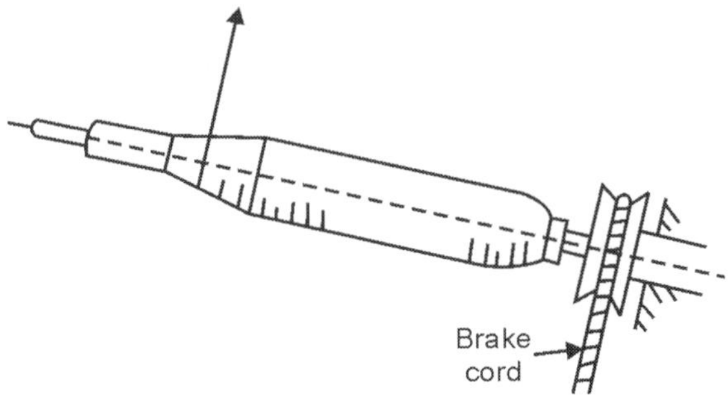

Figure 19.2

Solution:

Squaring on both sides, we get

$$K^2 = 0.697^2 \times 32 \times 0.0625^2 \times / 4 * 3.14^2 * 2$$

$$K = 0.027 \text{ ft. or } 0.333 \text{ in.}$$

Winding speed = 240 yard per min.

Linear velocity = 240 × 3/60 = 12 ft/s

Diameter at the nose of the chase = 0.8 in.

Acceleration = 900 radians per sec. per sec.

Torque = I × Angular acceleration

Angular acceleration = Linear velocity / bobbin radius

= 12/ 0.4/12 = 363 radians per sec. per sec.

Average acceleration = 900 + 363 = 1263/2

= 631.818 radians per sec. per sec.

Hence, Torque = I × Angular acceleration

= 0.4/32 × (0.333/12)2 × 631.81

= 0.00606 lb. ft.

Yarn tension at the nose of the chase = torque / bobbin radius

= 0.00606/0.033 = 0.1838 lb. ft.

Frictional braking force acting on the wharve = 1 oz. (0.0625 lb)

Diameter of the spindle wharve = 1 in.

Torque = 0.0625 × 1/12 = 0.0051 lb.

Yarn tension at the unwinding point of nose diameter 0.8 in. = 0.0051/radius of the nose = 0.0051/0.033 = 0.1571 lb.

Total yarn tension = Torque at the spindle wharve + yarn tension at the nose

= 0.1838 + 0.1571 = 0.340 lb. or 5.45 oz.

Example 5: A loom sley has a mass of 144 lb. and its centre of gravity is 20 in. from the rocking shaft about which it oscillates. When suspended from the rocking shaft and allowed to oscillate, it is found to have the same time of oscillation as a simple pendulum of length 24 in. What is the radius of gyration of the sley about the rocking shaft? What mass, assumed to be concentrated on the sword pin, 28 in. from the rocking shaft, is dynamically equivalent to the sley? (Fig. 19.3)

If when leaving the beat-up position, the linear acceleration of the sword pin is 100 ft/s^2, what torque must be applied to the sley to accelerate it in that position?

Solution:

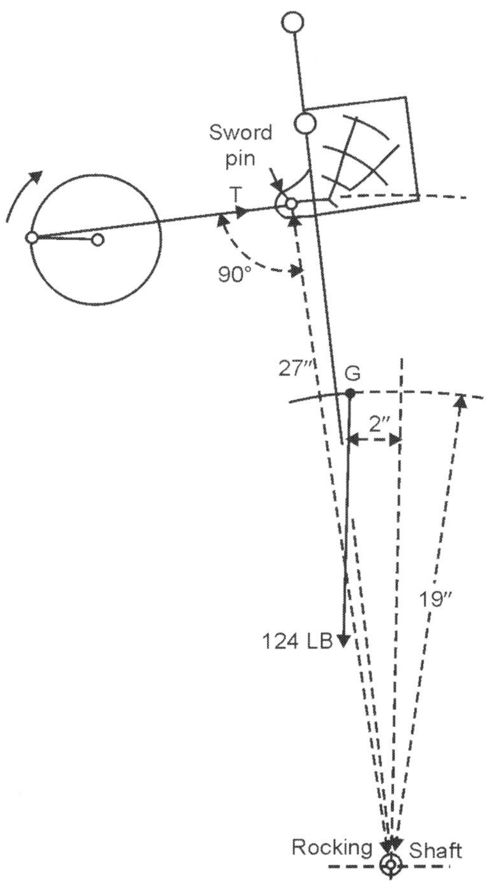

Figure 19.3

Mass of the sley = 144 lb.

Centre of gravity = 20 in. (h)

Period of oscillation = 24 in. (24/12 = 2 ft).

Linear acceleration of the sword pin = 100 ft/s2

Periodic time of oscillation $t = 2*3.14 \times \sqrt{l/g}$

$= 2*3.14 * \sqrt{2/32}$

$= 1.57$ s

Now we have to find radius of gyration K.

$t = 2*3.14 * K^2/g\,h$

$= 2*3.14 \times K^2/32 \times 20/12$

From which K = 1.825 ft.

Mass of the sley = 144 lb.

Since the radius of gyration is 1.825 ft, the mass of the sley with the centre of gravity

= Mass/K

= 144/1.825 = 78.81 lb.

When the mass is concentrated at the sword pin, the dynamic mass equivalent to sley = 144/32 × 28/12 = 10.5 ft. lb.

Total mass of the sley = 78.81 + 10.5

= 89 lb.

Linear acceleration of the sley = 100 radians per sec. per sec.

Torque = $I \times$ Angular acceleration

= $MK^2 \times$ Angular acceleration

Angular acceleration = linear acceleration/radius

= 100 × 12/28 in. = 42.91 rad. per sec. per sec.

Torque T = $144/32 \times 1.827^2 \times 42.91 = 644.57$ lb. ft.

Example 6: A warping machine draws yarn from cheeses at a warping speed of 480 yard per min. Full cheeses are 8 in. diameter and each cheese weighs 2.25 lb. The Moment of Inertia of a full cheese was found by suspending two parallel cords. Distance between cords is 8 in. and length of cord is 40 in. the cheese was found to make 50.5 oscillations per min when given a small displacement. Calculate the radius of gyration.

When the machine is stopped, each cheese is retarded by a spring loaded brake to prevent the yarn from over running. If not more than 2 yard is to run on to the beam during a stoppage, what minimum braking force is required at each cheese?

Solution:

Warping speed = 480 yard per min.

Diameter of full cheese = 8 in.

Weight of the cheese = 2.25 lb.

Distance between the cords = 8 in.; b = 4 in. or 0.33 ft.

Length of the cord = 40 in. (40/12) = 3.33 ft.

Number of oscillation per min = 50.5

Periodic time of oscillation t = 60/50.5 = 1.18 s

Radius of gyration, K

$$t = 2^* \pi \times \sqrt{K^2\, l/2\, b^2}$$

$$= 2^* \pi \times \sqrt{K^2 \times 3.33/2 \times 0.33^2}$$

Squaring on both sides, we get

$$K^2 = 1.18^2 \times 2 \times 0.33/2^*3.14^2{}^*3.33$$

$$K = 2.33 \text{ in.}$$

Warping speed = 480 yard per min.

= 480 × 3/60

= 24 ft/s

Angular velocity of the beam = linear velocity / radius

= 24/4/12

= 72 radians per s

During stoppage of the beam, length of the yarn wound on to the beam should not be more than 2 yard.

Hence, Average speed = 72/0.5

= 144 radians per sec per sec

Time to stop = 6/12

= 0.5 s

Angular retardation of the beam = Angular velocity / time

= 24 / 0.5

= 48 rad per sec. per sec.

Torque = I × Angular acceleration

$$= 2.25/32 \times (2.33/12)^2 \times 144$$

= 0.385 lb.

Braking force required at each cheese =

Torque / radius of the cheese in which the braking force is applied

$$= 0.385/4/12 = 1.166 \text{ lb}$$

19.5 Referring Moment of Inertia

It is often convenient to consider the inertia of the machine parts as though they were fixed on the driving or other shaft and rotating at that shaft speed. This is called as *referring Moment of Inertia* to another shaft. The referring Moment of Inertia offers the same resistance to acceleration as that offered by the actual part. The kinetic energy of the referred Moment of Inertia at that speed of the shaft, to which it is referred must be the same as that of the actual Moment of Inertia, rotating at its own speed.

Therefore, Kinetic energy = ½ $I\omega^2$, where I is the actual Moment of Inertia, ω its speed.

I_r = Moment of Inertia referred to another shaft, rotating at ωr its speed,

Then $$I_r \omega r^2 = I\omega^2$$

And $$I_r = \omega^2/\omega r^2$$

Example 7: The bottom shaft of a loom is geared to the driving or crank shaft and rotates at half its speed. The crank shaft weighs 36 lb. and has a radius of gyration of 2 in. the bottom shaft weighs 70 lb. and it radius of gyration is 4.5 in. What is the Moment of Inertia of the shafts, referred to the crankshaft, and how much work is done in accelerating them from rest to a crank shaft speed of 180 rpm?

Solution:

Moment of inertia (I) of the crankshaft =

$$36 \times (0.166)^2 = 1 \text{ lb. ft}^2$$

Actual Moment of Inertia (I) of the bottom shaft =

$$70 \times 0.375^2 = 9.84 \text{ lb. ft}^2$$

I of bottom shaft referred to crankshaft =

$$1.84 \times 1*1/2*2 = 2.47 \text{ lb. ft}^2 \text{ units}$$

Total I referred to the crank shaft = I of the bottom shaft + I of the crank shaft

$$= 2.47 + 1 = 3.47 \text{ lb. ft}^2 \text{ units}$$

Work done in accelerating to 3 revs. per min = Kinetic energy at 3 revs. per min.

$= \frac{1}{2} I \omega^2$

$= 1 \times 3.47 \times 36 * 3.14^2/2 * 32$

$= 19.25$ ft. lb.

Example 8: A high speed warping machine has a warping speed of 400 yd/min. The warping beam is frictionally driven by surface contact between the yarn and the motor driven driving roller. When the beam is full, it has a mass of 640 lb, diameter 30 in, and radius of gyration 10 in. The roller has a mass of 500 lb., diameter 24 in., radius of gyration 11.5 in. the machine requires 3 H.P. for normal running.

Calculate the following:

(a) What torque is required at the driving roller for normal running?

(b) If the machine has to stop with a beam surface movement of not more than 36 in., what retarding torque the brake must apply to the driving roller, assuming the total inertia of the machine to be that of the beam and the roller.

(c) If the machine has to accelerate from rest to full speed in 5 s, what maximum horse power must the motor apply, assuming uniform acceleration.

(d) If μ between the yarn surface of the beam and the driving roller is 0.25, what minimum pressure is required between beam and the roller to prevent slip during starting and stopping.

Solution:

The surface speed of the beam and the roller is equal to the warping speed.

Warping speed = 400 yard per min. = $400 \times 3/60 = 20$ ft/s

Angular velocity of driving roller = linear velocity/radius

= 20/1 = 20 radians per sec. (Diameter of the roller = 24 in.; Radius = 24/2 = 12 in. or 1 ft.)

Angular velocity of the beam = 20/1.25 = 16 radians per sec. (Diameter of the beam = 30 in.; Radius = 30/2 = 15 in or 1.25 ft)

(a) Horse power = Torque × Angular displacement in radians per sec / 550

Torque = Horse power × 550 / Angular displacement in radians per sec.

Torque $= 3 \times 550/20$

$= 82.5$ lb. ft.

This torque T_f is required to overcome the frictional resistance throughout the machine.

(b) During stop, the beam surface moves 3 ft. at an average speed of 10 ft/s

Hence, time to stop = Distance moved/Average speed

t = 3/10

= 0.3 s

The angular retardation of the driving roller during stop

= Warping speed/time

= 20/0.3

= 67 radians per sec. per sec.

The Moment of Inertia of the roller = MK^2

Mass M = 500 lb.; Radius of gyration (K) = 11.5 in. or 0.96 ft.

Moment of inertia of the roller =

$500 \times 0.960^2/32$ = 14.4 engineer's units (lb/g ft^2 units)

Moment of inertia of the beam =

Mass of the beam = 640 lb.; Radius of gyration (K)

= 10 in. or 0.833 ft.

Moment of inertia of the beam =

$640 \times 0.833^2/32$ = 13.9 engineer's units

Referred to the roller, the beam's Moment of Inertia =

$13.9 \times 16*16/20^2$ = 8.9 units

Total Moment of Inertia (I) referred to the roller =

14.4 = 8.9 = 23.3 engineer's units

Torque required to retard the roller = $I \times$ Angular retardation

= 23.3×67 = 1560 lb .ft.

Frictional torque (T_f) = 82.5 ft. lb.

Torque to be supplied by the brake = Retarding torque – frictional torque

= 1560 – 82.5 = 1478 lb. ft.

(c) Acceleration of the roller during startup =

20/5 = 4 radians per sec. per sec.

Torque required for acceleration =

23.3×4 = 93.2 lb. ft.

In addition, the frictional torque is 82.5 lb. ft.

Therefore, total torque required during startup =

93.2 + 82.5 = 175.7 lb. ft.

The maximum horse-power is required at the end of the acceleration period when the roller speed is reaching its maximum angular velocity of 20 radians per sec.

Maximum horse-power = 175.7 × 20/550 = 6.4 horse power

(d) Slip between the beam and roller is likely to occur during stop.

Beam retardation during stop = Angular velocity of the beam/time

= 16/0.3 = 53.5 radians per sec. per sec.

Retarding torque required by the beam = Beam Moment of Inertia × retardation

= 13.9 × 53.5

= 745 lb. ft.

Frictional force required at beam surface to produce this torque =

Retarding torque at the beam / radius of the beam in ft.

= 745/1.25

= 595 lb.

Pressure required between beam and roller = 595/μ

595/0.25 = 2380 lb.

19.6 Moment of inertia referred to its driving shaft

Moment of inertia can easily be found on steadily running machines like carding, drawing, spinning machines. It is more difficult to do so, accurately on looms where the parts are constantly stopped and started. When the machine is steadily running, the resistance it offers is entirely due to the frictional resistances at the bearings and other rubbing parts throughout the machine. The resistance due to any material being processed is relatively small and in any case can be considered as additional frictional resistance.

When the machine is running normally and suddenly the belt is disengaged from the fast pulley or by switching off the motor, the frictional resistance continues to act and immediately begins to slow down the machine. The kinetic energy in the moving parts of the machine keeps it running at a gradually reducing speed, until all the energy is absorbed. The rate at which the machine slows down depends on the magnitude of the frictional resistance (T_f), and on the Moment of Inertia of the machine. If the time taken by the machine to stop is known, the average value of the retardation will be got, but it will not be accurate. It is because that the frictional resistance values will be changing every time when the machine slows down.

An example in this case is the carding cylinder in a carding machine. If the machine slows down very rapidly, there would be no material change in the resistance for the first few seconds. In carding machine the machine does not stop for several minutes, the resistance can be considered constant, for, say 20 to 30 seconds. In such cases, it would be sufficiently accurate to count preferably by revolution counter combined with a stopwatch, the revolutions made by the machine shaft in successive 5 to 10 seconds after the disengagement of the drive.

From these speed curves, during the first part of the stoppage time can be drawn. From the curve, the retardation due to T_f can be got with accuracy.

Since, T_f = Moment of inertia × Angular retardation, the machine Moment of Inertia referred to the driving shaft is

T_f/angular retardation

Example 9: A carding engine cylinder requires 1 horse-power to keep it running at 170 rpm. When the belt is shifted from fast pulley to loose pulley, the driving shaft is found to make 27,24,20,17 and 14 revs. in 5 successive intervals of 10 sec. each. What is the Moment of Inertia referred to the driving shaft?

The average speeds during each of the 5 time intervals are shown in Table 1.

Table 19.1 Average speeds

Time interval no.	Revs. per 10 s	Average speed revs. per sec
1	27	2.7
2	24	2.4
3	20	2.0
4	17	1.7
5	14	1.4

These average speeds have been plotted in the centers of the time intervals, so as to get a falling speed curve. As shown in Fig. 19.4., the starting point of the curve, at zero time, is 2.84 revs. per sec = 170 rpm. It has been observed that all the points lie almost on a straight line and hence the retardation throughout the period of observation is assumed to be uniform.

From the curve, the speed falls from 2.84 to 1.88 revs. per sec i.e. 0.96 revs. per sec, in the first 30 s. The average retardation is therefore 0.96 rev. per sec or 6 radians per sec divided by 30 s = 0.2 radians per sec. per sec.

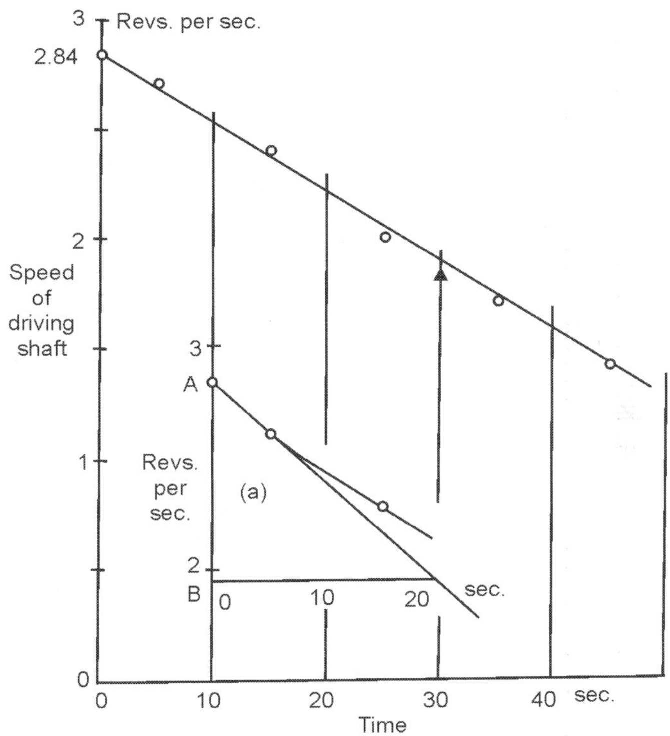

Figure 19.4

When the machine is running at full speed of 170 revs per min, the angular velocity is 17.8 radians per sec.

Horse power required = 1 H.P.

Frictional retarding torque =

550/angular velocity = 550/17.8

= 30.9 lb. ft.

Machine Moment of Inertia referred to the driving shaft =

Frictional torque (T_f)/Angular retardation

= 30.9/0.2

= 154.5 engineer's units (lb/g ft^2)

Example 10: A conventional flyer frame requires 2 horse-power to drive it at 320 rpm. When the diving belt is thrown off, the movement of the driving shaft is found to decrease by 0.1 radian in 0.1 sec successive intervals of time, after the first (Fig. 19.5).

Calculate the Moment of Inertia of the machine referred to the driving shaft and the kinetic energy stored in the machine at full speed. Also find out the effective pull must the belt apply to 12 in. driving pulley and what maximum horse-power is required to start up the machine from rest to full speed, in the same time, that the machine takes to stop when the belt is thrown off?

Solution:

Angular velocity of the driving shaft at full speed =

$$320 \times 2^* \pi/60 = 33.5 \text{ radians per sec.}$$

Displacement in 0.1 sec =

Angular velocity × time = 33.5 × 0.1

= 3.35 radians.

When the belt is thrown off, displacement in first 0.1 sec = radians and average speed is 33 radians per sec.

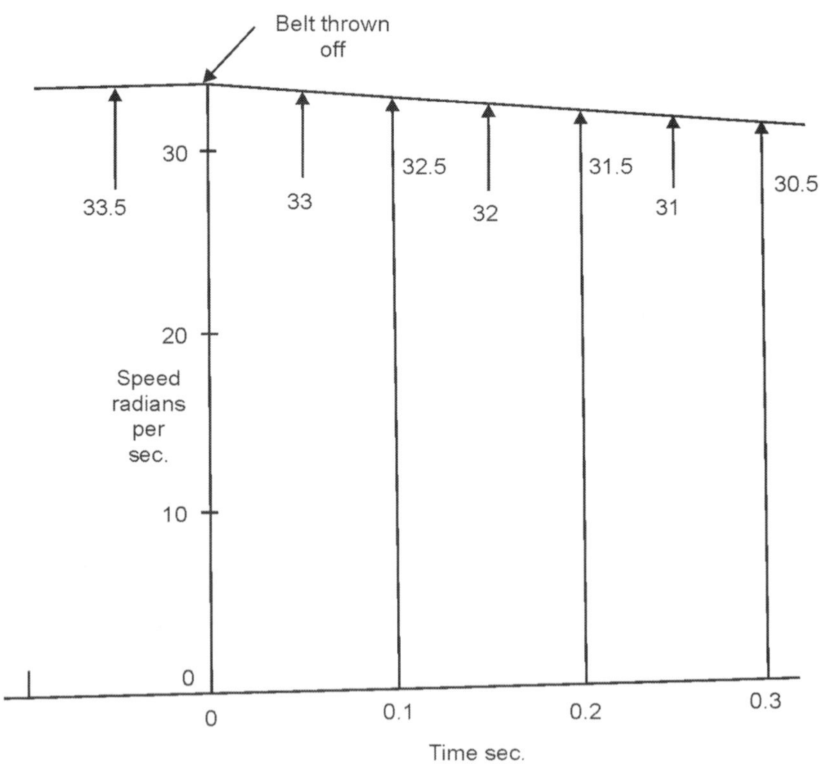

Figure 19.5

Displacement in 2nd s = 3.2 radians; average speed
= 32 radians per sec.

Displacement in 3rd s = 3.1 radians; average speed
= 31 radians per sec. and so on.

From the above, it may be noted that angular velocity is then changing by 1 radian per sec in each 0.1 s

Angular retardation = Change in angular velocity /time
= 1/ 0.1
= 10 radians per sec. per sec.

Torque required during normal running (T_f) =
2 × 33000/320 × 2 *π = 32.8 lb. ft.

Moment of inertia of machine (I)
= Torque / Retardation
= 32.8/ 10
= 3.28 lb/g ft^2 units

Kinetic energy at 320 revs. per min.
= ½ $I\,\omega^2$
= ½ 3.28 * 33.5^0
= 1830 ft. lb.

The torque required for acceleration = I × acceleration
= 3.28 × 10 = 32.8 lb. ft., which is same as frictional torque (T_f)
Hence, total torque required for acceleration = 2 × 32.8 = 65.6 lb. ft.
Effective pull on the belt = Total torque/radius of the pulley
= 65.6/0.5
= 131.2 lb.

Maximum horse-power required =
Torque × Angular displacement per sec/550
= 65.6 × 33.5/550
= 4 horse-power

19.7 Winding mechanics

In studying the winding of yarn on bobbins, pirns, etc., it is often desirable to show graphically variations in winding and bobbin speeds. In many modern winding machines, the traverse rate is relatively fast and the angularity of the coils must be allowed for solving the problems considerably difficult.

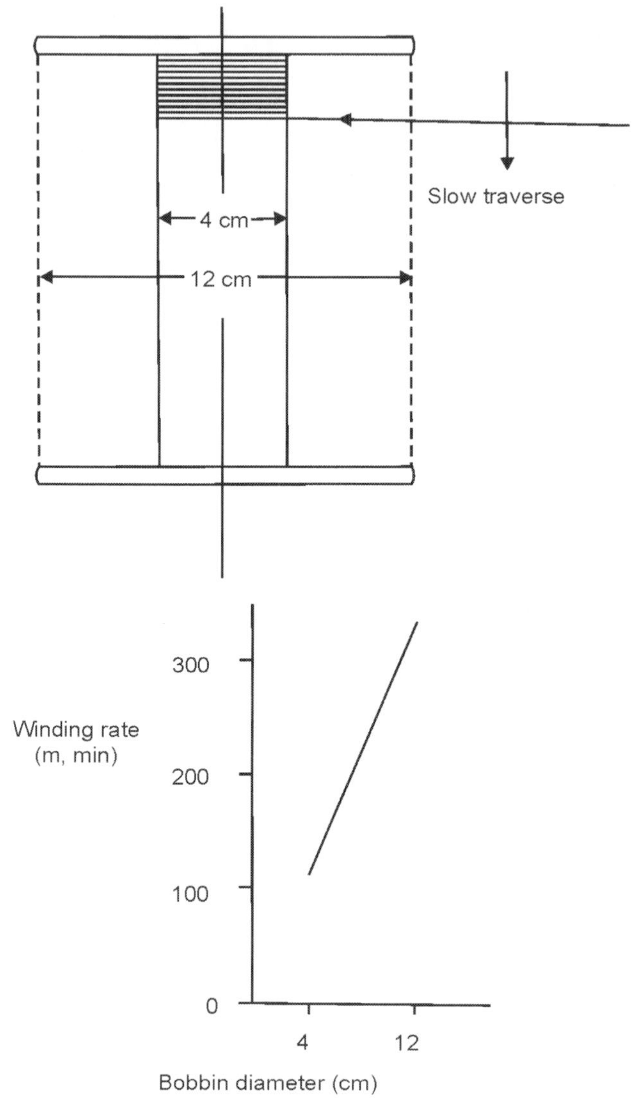

Figure 19.6 Winding mechanics

Example 11: A war per's bobbin is 1.25 in. diameter when bare and 4 in. diameter when full. It is wound by mounting it on a precision winder running at 700 rpm. with a constant rate of traverse. What is the average winding speed and the time required to fill the bobbin with a yarn count of Ne 30s and the mass is 0.75 lb. (Fig. 19.7)?

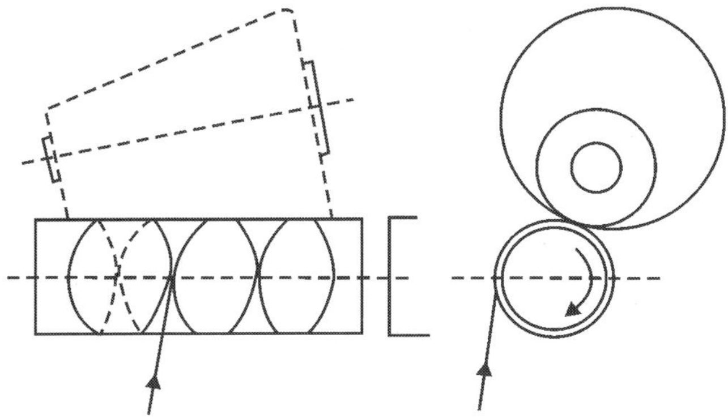

Figure 19.7

Solution:

The average winding speed will be the speed at the mean diameter.

Hence the average diameter is 1.25 + 4 / 2 = 2.625 in.

Winding rate at 2.625 in. diameter = $\pi\, D \times N/36$

$= 3.14 \times 2.625 \times 900 / 36$

$= 160$ yard per min.

Time taken to fill the bobbin of 0.75 lb. mass =

$30 \times 840 \times 0.75$ / winding speed in yard per min.

$= 30 \times 840 \times 0.75 / 160$

$= 118$ min.

It may be noted in the above Example 11, since the time per traverse is constant, the diameter increases uniformly with time and the base line of the diagram representing bobbin diameter will also represent time. If for example, the traverse rate is such that successive yarn coils touch each other, the bobbin diameter will increase by twice the yarn diameter at each traverse.

Example 12: A warper's bobbin of 1.25 in. bare and 4 in. full diameter. It is wound on spindle at 900 rpm until its diameter is 2.625 in. Then the bobbin is removed from the spindle and transferred to a spindle running at 450 rpm for completion. Traverse rate is same for both the spindles. Calculate the average winding rate and running time to fill the bobbin of mass 0.75 lb. of cotton count 30s.

Solution:

Mean diameter of the bobbin is 1.25 + 2.625 / 2 = 1.9375 in.

At the start, the bobbin is wound on a spindle of 900 rpm,

Therefore, winding rate = $\pi \times D \times N / 36$

= 3.14 × 2.625 × 900 / 36 = 152.09 yard per min.

Up to the diameter 2.625 in., the yarn is wound on the spindle which runs at 900 rpm.

After that, the partly filled bobbin is transferred to second another spindle which runs at 450 rpm.

Mean diameter = 2.625 = 4 / 2 = 3.3125 in.

Winding rate = $\pi \times D \times N / 36$

= 3.14 × 3.3125 × 450 / 36

= 148.27 yard per min

Average winding speed = 152.09 + 148.27 / 2

= 148.09 yard per min

Time to fill the bobbin of mass 0.75 lb

Diameter of the bobbin at the completion of winding on first spindle which runs at 900 rpm =

− 1.25 = 1.375 in.

Winding rate = 3.14 × 1.375 × 900 = 107.09 yard per min.

Diameter in which the winding starts in the second spindle up to completion is = 2.625 in.

Winding rate at 2.625 in. diameter = 1.14 × 2.625 × 450

= 114.47 yard per min.

Total winding rate = 107.09 + 114.47

= 222.07 yard per min.

Time to fill the bobbin = 30 × 840 × 0.75 / Total winding speed in yard per min.

= 30 × 840 × 0.75 / 222.07

= 84.97 or 85 minutes

Example 13: The "chase" of a pirn varies from 0.5 to 1.125 in. diameter. During winding of the pirn, the spindle on which it is mounted is varied in speed to keep the winding rate constant. If the winding rate is 200 yards per min., what are the spindle speeds at the nose, the shoulder and at the centre?

Solution:

(a) When the pirn diameter is 0.5 in., the spindle speed will be

Winding rate × 36 / 3.14 × D

Winding rate = 200 yards per min.

200 × 36 / 3.14 × 0.5 = 4585 rpm

(b) When the pirn diameter is 1.125 in., the spindle speed will be
Winding rate × 36 / 3.14 × 1.125
200 × 36 / 3.14 × 1.125 = 2038.21 rpm

(c) When the diameter is 0.8125 in. i.e. (0.5 + 1.125 /2 = 0.8125 in.)
Winding speed × 36 / 3.14 × 0.8125 in.
200 × 36 / 3.14 × 0.8125 = 2822 rpm

Numerical questions

1. A partly filled war per's bobbin weighing 0.5 lb. is mounted on a horizontal spindle and is accelerated by weight suspended by a thread wrapped round the barrel of the bobbin, where the circumference is 8.4 in. a weight of 0.04 lb. is just sufficient to overcome the friction at the spindle. With a weight of 0.1 lb., the bobbin accelerates and the weight is found to fall 30 in. to floor level in 1 second.

   When the weight reaches the floor level, find the
   (a) Kinetic energy in the weight
   (b) Energy required to overcome friction at the spindle, and
   (c) Kinetic energy of the bobbin. Also calculate the radius of gyration.

   Ans. (a) 0.79 in. lb.; (b) 1.56 in. lb.; (c) 1.55 in. lb.; K = 0.84 in.

2. The tin roller of a spinning machine weighs 500 lb. and has a radius of gyration of 2.1 in. and the speed is 1300 rpm. It is driven by bands, 1300 spindles each weighing with partly filled cops of 0.4 lb., with radius of gyration 0.28 in. and spindle speed of 9000 rpm. What is the total Moment of Inertia of tin roller and spindles, referred to tin roller shaft, assuming no slip on the bands? How much work is done and what is the average horse-power expended to accelerate these parts from rest to full speed in 2 seconds, at the beginning of each lift?

   Ans. 28.8 lb. ft^2 units; 8100 ft. lb; 7.55 HP

3. A slubbing frame requires 5.75 H.P. to drive it at 780 revs. per min of the main shaft. A test showed that when disconnected from its drive, the speed of the shaft dropped 135 revs. per min. in 1 s.

   Find the Moment of Inertia referred to the main shaft and the kinetic energy stored in the machine at 780 revs. per min. If the machine has to be accelerated from rest to full speed in 4 seconds, what maximum horse-power is required in starting the machine?

   Ans. 2.75 lb/ft^2 units; Kinetic energy = 9175 ft. lb; H.P. = 14.1

4. A loom running at 180 rpm requires 0.4 H.P. When the starting handle was knocked-off, the loom was found by means of Fletcher trolley apparatus to stop in 1.65 s. Find the approximate value of Moment of Inertia referred to the driving shaft. If in order to stop before beat-up, the loom must stop in not more than 0.6 s. What additional torque must be applied to the rim of the 12 in. diameter brake pulley?

   The loom is started from rest to full speed in 0.5 sec, what maximum horse-power is required?

   Ans. 1.02 lb/g ft^2 units; Braking torque = 40.7 lb; H.P. = 1.72.

5. A winding frame for war per's bobbin has one row of spindles running at 900 rpm. and the other at 450 rpm. Bobbins are 1.25 in. bare and 4.25 in. full in diameter. The traverse rate is uniform and same for both the spindles. When a full bobbin is doffed from a slow spindle, it is replaced with a partly filled bobbin from fast spindle, and an empty bobbin is started on fast spindle.

   Calculate by how much does the diameter of the bobbin increase in each spindle.

   Also calculate the average, highest and lowest winding rates in which the bobbin is wound.

   Ans. From 1.25 in. to 3.25 in. in fast spindle

   From 3.25 in. to 4.25 in. in slow spindle.

   Winding rates:

   Average = 160 yd per min.

   Highest = 255.2 yd per min.

   Lowest = 98.12 yd per min.

6. In a conventional flyer frame with bobbin lead, the roller delivery is 390 in. per min. and the spindle or flyer speed is 306.4 rpm. The bobbin is 1.5 in. bare and 5 in. full in diameter. Calculate the following.

   (a) Winding speed, revs per min. of the bobbin at 1.5 in. and 5 in. diameter

   (b) Total speed of the bobbin

   Ans. (a) For 1.5 in. diameter, the bobbin speed is 82.6 rpm; for 5 in. diameter, the bobbin speed is 24.8 rpm.

   (b) Total speed at 1.5 in. diameter is 389 rpm; total speed at 5 in.; diameter is 331.2 rpm

7. A cone is wound by driving it through surface contact with a revolving roller, grooved to traverse the yarn. The roller is 3 in. diameter and its speed is 3000 rpm. If 54 in. of yarn is wound per double traverse,

and there are 6 spiral grooves on the roller, find the average winding speed and compare with the roller surface speed.

The cone is 1 in. to 3 in. diameter at the start and 4 in. to 6 in. at the end. What are the highest and lowest winding rates at the start and at full cone?

Ans. Average winding speed = 750 yd. per min.

Roller surface speed = 785 yd. per min.

Winding rate at 1 in. diameter of cone = 375 yd per min

Winding rate at 3 in. diameter of cone = 1120 yd per min.

Winding rate at 4 in. diameter of cone = 600 yd per min.

Winding at 6 in. diameter of cone = 900 yd per min.

8. A double flanged bobbin is wound on a vertical spindle at 900 rpm. The bobbin diameters are 4 cm at the centre and maximum bobbin diameter is 12 cm. The yarn is wound with a slow traverse. How does the winding speed vary from an empty to full bobbin?

   Ans. At 4 cm, diameter of the winding rate = 113 m/min.

   At 12 cm, diameter of the winding rate = 340 m/ min.

9. A cylindrical package is wound on a centre of 5 cm diameter. The spindle speed is constant at 3200 rpm. If the traverse velocity is 205 m/min, determine (a) the net winding rate at the start of winding, (b) net winding rate at the package diameter of 16 cm, (c) angles of wind at the start and at a diameter of 18 cm.

   Ans. (a) Net winding rate at 5 cm diameter = 553 m/ min.

   (b) Net winding rate at 16 cm diameter = 1610 m/min.

   Angle of wind at 16 cm diameter = 7° 16 min.

   Surface velocity (Vs) = 503 m/min.

   Angle of wind at the start = 21° 46 min.

10. The full diameter of a pirn wound from cotton yarn is 32 mm, and the bare pirn diameter at the nose of the chase is 14 mm. Determine the chase angle when the traverse is 34 mm.

    Ans. tan α = 14° 49 min

20
Cams

20.1 Introduction

A *cam* is a rotating machine element that gives reciprocating or oscillating motion to another element known as follower. The cam and the follower have a line contact and constitute a higher pair. It frequently happens that some part or parts of a mechanism require to be moved in a wee-defined manner that differs from the traverse produced by a crank motion. In some machines, a very irregular traverse motion may be desired or an intermittent motion. Such movements are reversible and variable motions may be required such as a quick forward uniform motion and a slow return movement. Cams are the mechanical appliances used for producing movement of this kind. Cams are widely used in operating inlet and exhaust valves of internal combustion engines, cutting machines, etc. In textile engineering, cams are used in spinning preparatory, spinning machines, weaving machines and in knitting machineries.

20.2 Classification of cams

There are many classifications of cams used in many ways, but the following are the two types of cams important from the subject point of view.

1. Radial cam or disc cam
2. Cylindrical cam

They are discussed in detail in the next section.

1. Radial cam or disc cam

In radial cams, the follower reciprocates or oscillates in a direction perpendicular to the cam axis. The cams shown in Fig. 20.2 are all radial cams.

2. Cylindrical cams

In cylindrical cams, the follower reciprocates or oscillates in a direction parallel to the cam axis. The follower rides in a groove at its cylindrical surface. A cylindrical grooved cam with a reciprocating and an oscillating follower is shown in Fig. 20.1(a) and (b).

In actual practice, radial cams are widely used.

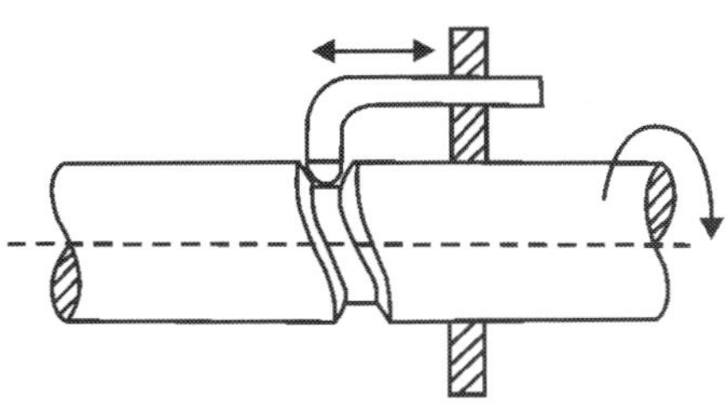

(a) Cylindrical cam with reciprocating follower.

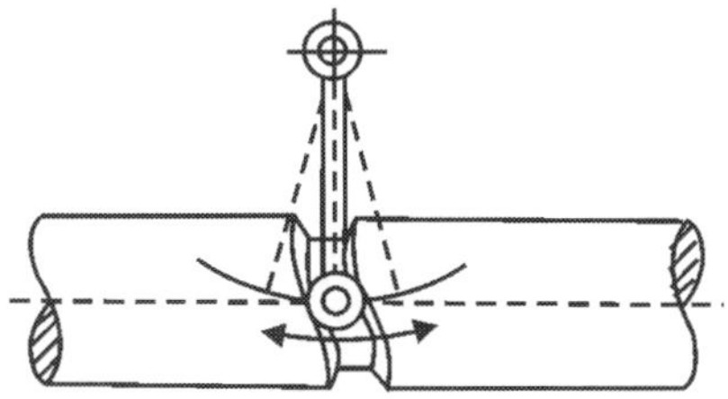

(a) Cylindrical cam with oscillating follower.

Figure 20.1 (a) and (b)

20.3 Classification of followers

The followers may be classified as discussed below.

(a) Knife edge follower

When the contacting end of the follower has a sharp knife edge, it is called knife edge follower as shown in Fig. 20.2(a). The sliding motion takes place between the knife edge and the cam surfaces. This type of follower is a rare case because the small area of contacting surfaces results in excessive wear.

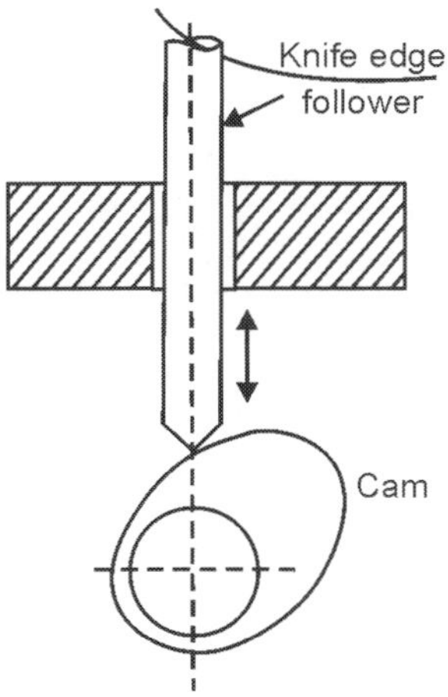

(a) Cam with knife edge follower

Figure 20.2 (a)

(b) Spherical faced follower
When the connecting end of the follower is of spherical shape, it is called a spherical faced follower as shown in Fig. 20.2(a). It may be noted that high stresses develop in automobile engines with flat faced follower is reduced by the use of spherical faced follower (Fig. 20.2b).

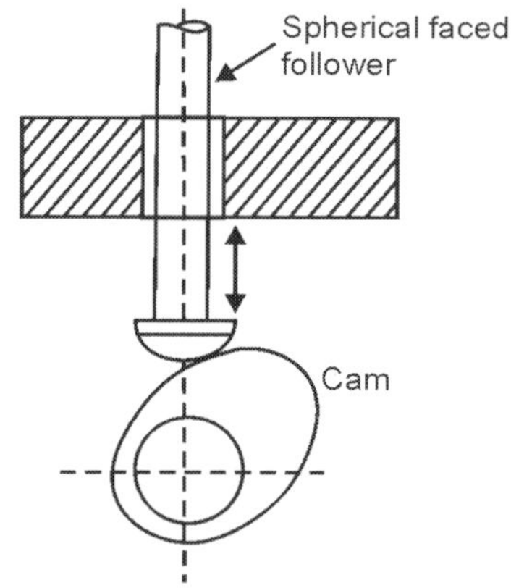

(b) Cam with spherical faced follower

Figure 20.2 (b) Spherical faced follower

When the contacting end of the follower is a perfectly flat faced, it is called flat-faced follower as shown in Fig. 20.2(c).

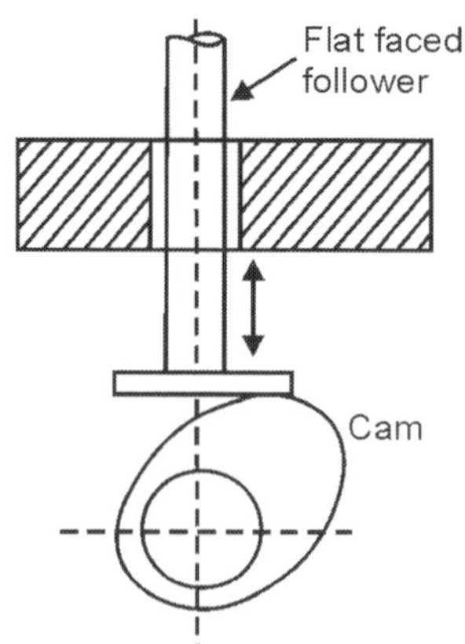

(b) Cam with flat

Figure 20.2 (c) Flat faced follower

In this type of follower, the side thrust between the follower and the guide is much reduced. The only side thrust is due to friction between the contact surfaces of the follower and the cam. This flat faced follower is generally used where space is limited such as in valves of automobile engines.

20.4 Cams in textile machines

Models are readily obtainable or easily made to illustrate the use of cams in textile machines.

Figure 20.3 represents a flat piece of metal *B* cut out in the form of a heart and fixed on a shaft *A*. *A* and *C* is kept in contact with the periphery of the cam, so that as the cam revolves the pin *C* or follower will rise and fall between the smallest and largest diameter of the cam. It will be seen that the shape of the cam can be made to give to *C* a straight line to-and-fro uniform motion and of equal extent in both directions. Such type of cams are used in ring frame machines to traverse the ring rail up and down for winding the yarn on the cop or bobbin.

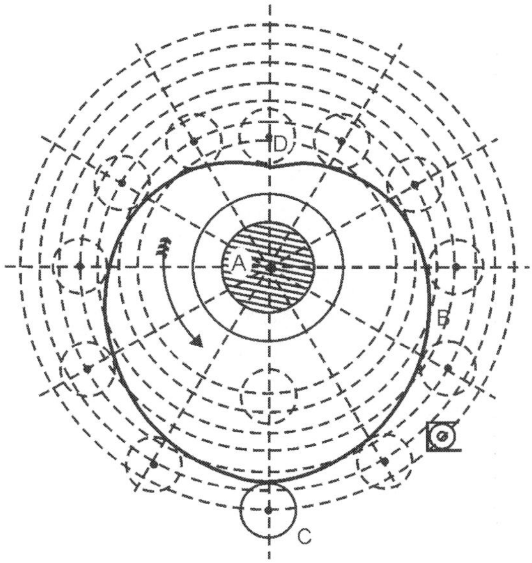

Figure 20.3

In Fig. 20.4, the pin or follower *C* will have the same uniform traverse as in Fig. 20.4 and will make the to-and-fro motion in the same time, but the cam will make three-quarters of a revolution in moving *C* its full distance and only one-quarter of a revolution for *C* to return to its original position.

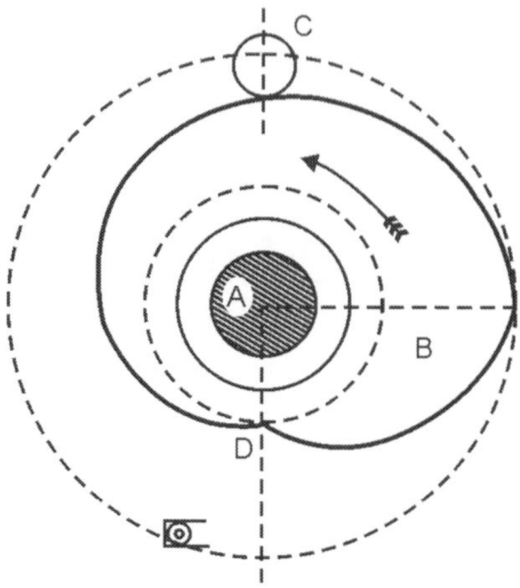

Figure 20.4

Figure 20.5 gives uniform motion to the pin *C* or follower, but it requires one complete revolution to do this, and immediately the highest point *E* of the cam moves from under the pin *C* and the pin will fall at once to *D* and commence again its upward motion.

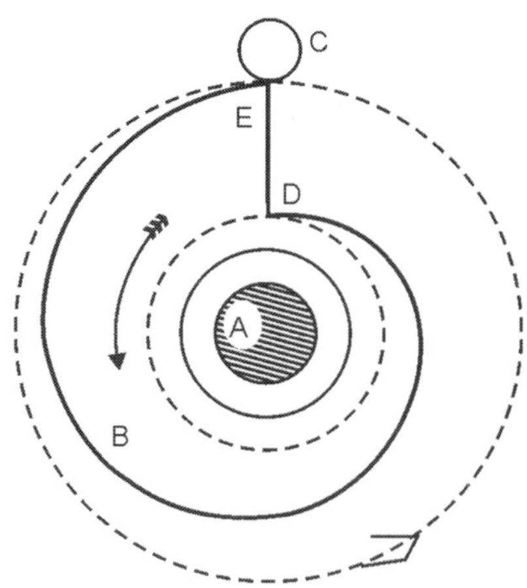

Figure 20.5

In Fig. 20.5, the cam *B* is divided into three parts: the position from *D* to *E* raises the pin *E* from the lowest to the highest point: then comes a portion of the cam *E* to *F*, which is circular, so that during this portion of the revolution of the cam the pin *C* will not move. When this portion has passed, the remainder of the cam from *F* to *D* will permit the pin *C* to return to its starting point.

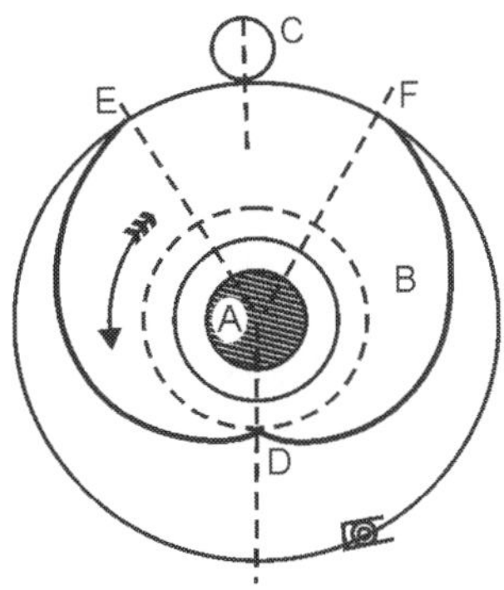

Figure 20.6

Whilst in some cases the pin is kept in contact with the surface of the cam by pressure or weight, which ensures the pin rising and falling, it is sometimes necessary to make the cam in the form of a groove in to which the pin has a sliding fit as shown in Fig. 20.6.

Figure 20.7 illustrates a cam of this type used for detaching rollers in comber machine. The inside cam is the essential portion: the bowl or pin *X* moves along an arc drawn from the centre stud *B* and its motion produces a corresponding effect on a portion of a wheel centered at *B*. *B* is termed as quadrant which gears in to a small wheel *P* on the roller *D*. This wheel *D* thus receives a forward notion then rest due to the circular portion of the cam and then a reverse or backward motion.

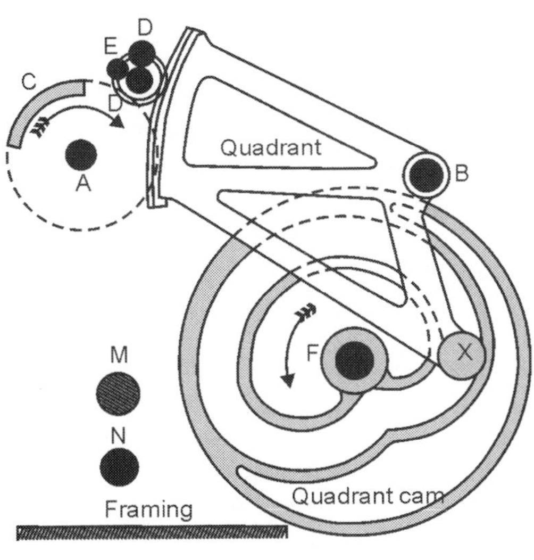

Figure 20.7

Figure 20.8 shows another form of cam used in winding machine. It consists essentially of a cylinder with a groove cut in it in to which fits a pin or bowl. This pin fixed to a slide gives a uniform traverse to the bar and is a frequent form for winding yarn in cheeses etc. on quick winding frames (Fig. 20.8).

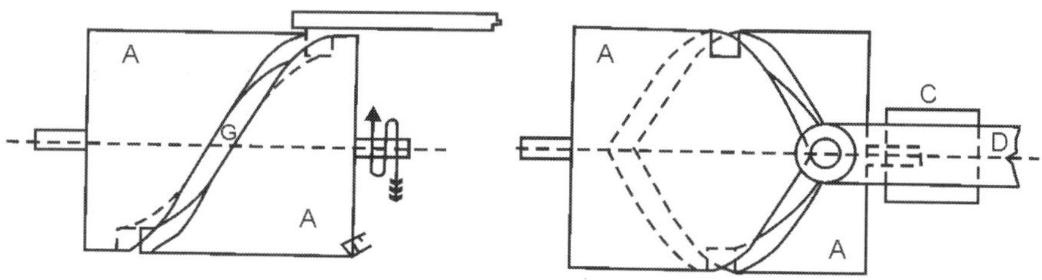

Figure 20.8

Figure 20.9(a) and (b) shows cams used in knitting machines. Revolving cams may be further classified in to positive or negative ones. A positive cam controls the movement of the part that it activates at all times of the cycle. In negative cams, the movement is controlled in only one direction whereas the return movement is dependent upon gravity, or a device such as a spring, to return to its original position for the completion of the cycle. Typical examples of both types are shown in Fig. 20.9(a) and (b).

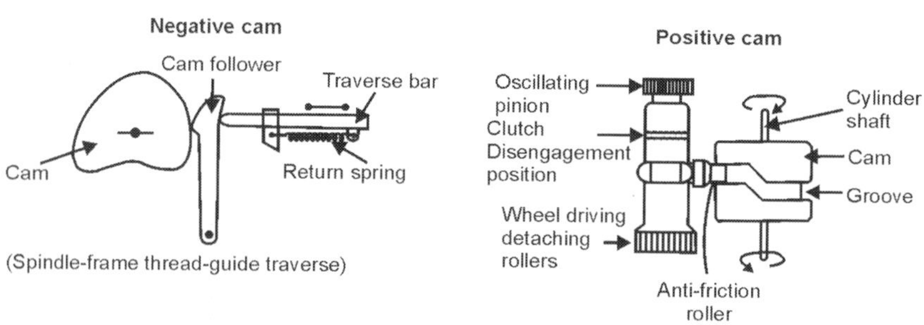

Figure 20.9 (a) Negative cams (b) Positive cams

Another example of a direct action cam is shown in Fig. 20.10. This type of cams are used for needle operations in knitting machines. In one type, the cam is a positive one, controls the needle movement by means of a groove in which the needle butt moves. In the other one, the needles are simply lifted by the cam and must rely on gravitational force to come back to its original position when the cam action ceases. However, the whole topic of cam mechanism in knitting cam dynamics is very complicated.

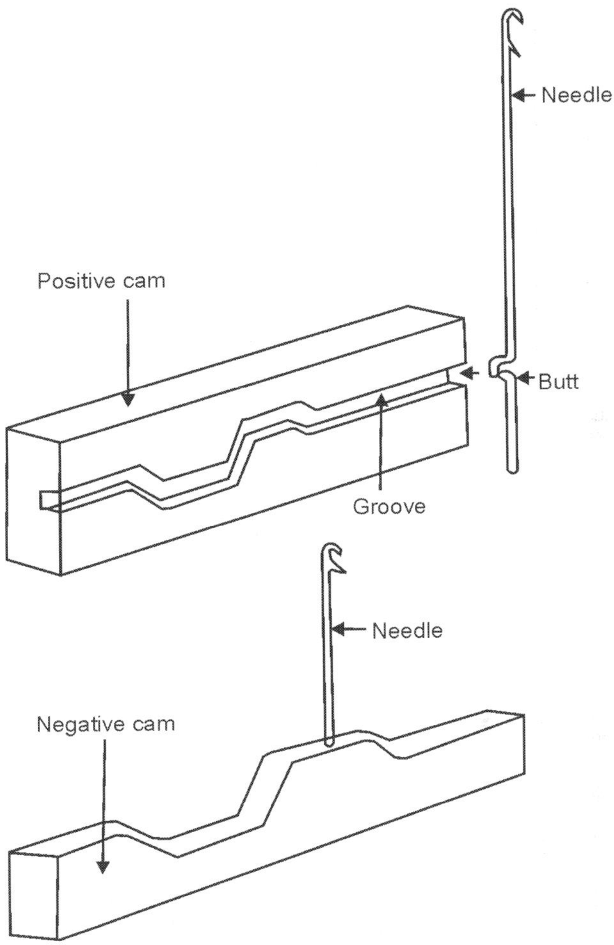

Figure 20.10 Positive and negative cams

In many instances, the nature of the movement given by the cam must be known in order to achieve successful design, or working of the knitting machine.

Plate cam and face cam

A rubbing of the cam shape is taken either by wrapping the paper around the cam (in the case of face cam) or by pressing the paper against the cam against the cam to obtain a trace of the edge or groove (in the case of plate cam). A face cam is one in which the cam follower is lifted by the contour of the face of the surface, whereas in the case of plate cam, the edge of the cam actuates the follower: the two types of cam are illustrated in Fig. 20.11.

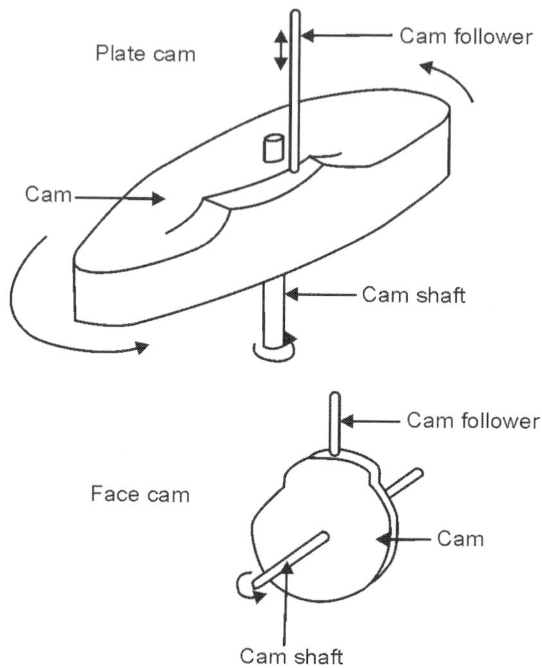

Figure 20.11 Plate and Face cams

Cams in drum winding machine

Cam motion in Fig. 20.12 shows the arrangement of reciprocating the thread guide bar on a drum winding machine.

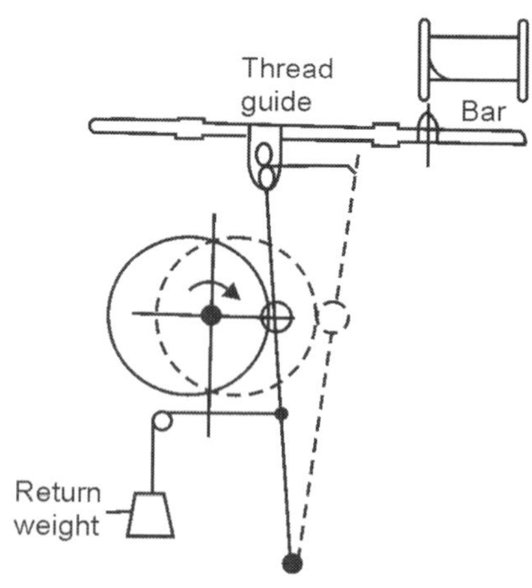

Figure 20.12 Cam motion for reciprocating thread guide

As the cam shaft rotates, the lever controlling the sideways movement of the thread guide bar is pushed laterally until the cam reaches its maximum radius position. As further rotation occurs, reversal of the bar's motion takes place, as a result of the effect of weight, until the cam's position of minimum radius is achieved, when the motion starts once again.

20.5 Construction of the cam design

Construction of the cam design is shown in Fig. 20.13. Contact between the cam and the lever takes place in an anti-friction bowl, and the movement of the centre of the bowl that is to be determined. However, the movement takes place in a circular arc, and it is assumed that the locus of bowl centre movement is a straight line. This line is first constructed by connecting the centre of the cam shaft, as located on the rubbing at *A*, with the centre of the anti-friction roller *B*, on the assumption that *A* and *B* are at the nearest point of approach at the starting time.

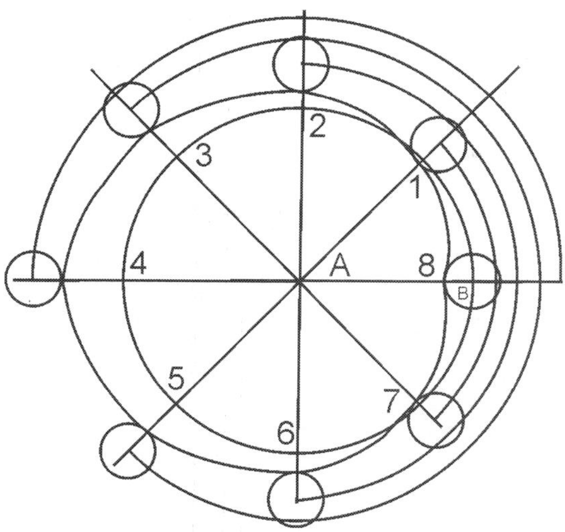

Figure 20.13 Displacement diagram from cam shape

With the centre A and radius *AB*, a circle is now drawn around the cam; and this circle is divided in to any number of equal parts, a large number of parts will give accurate results. *A* radius from A to each of the circle dividing points is then drawn as shown in Fig. 20.13.

As the crank shaft revolves at uniform speed, these radii are brought to the horizontal position, occupied originally by *AB*, at regular intervals. The centre point of the roller, at each position in turn, is found by constructing the circle of diameter equal to that of the roller with the centre on a radial line, which generally touches the cam outline. For each centre so constructed, an arc of a circle, with centre *A*, and radius equal to the distance between *A* and the centre of the particular circle, is drawn to cut the extended portion of the original line *AB*. The distance, *BC*, between the nearest and farthest approaches of the roller centre to A is known as the "Lift of the cam" and the cam displacement can be plotted as a function of cam position by measuring off the proportion of the lift that takes place as each radial position reaches the base line.

In Fig. 20.13(a), the line is drawn as abscissa for a graph, for which the ordinate shows the bowl displacement from the inner most position. All displacements for one complete revolution of the cam are plotted, and the movement of the thread guide bar at any position of the cam can easily be deuced by taking in to fact the relating movement of the bar to roller centre movement.

Displacement of bowl centre from *A* (Y-axis)

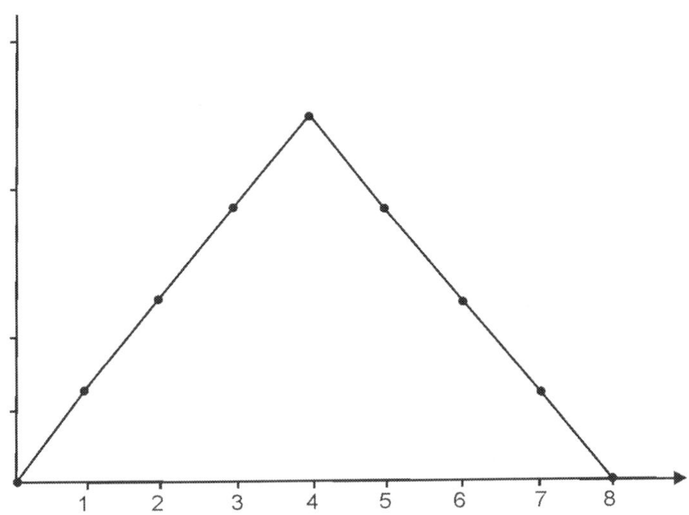

Figure 20.13 (a) Cam position in X-axis

20.6 Terms in Cam profile

Figure 20.14 shows a radial cam with reciprocating roller follower. The following terms are important in order to draw a cam profile.

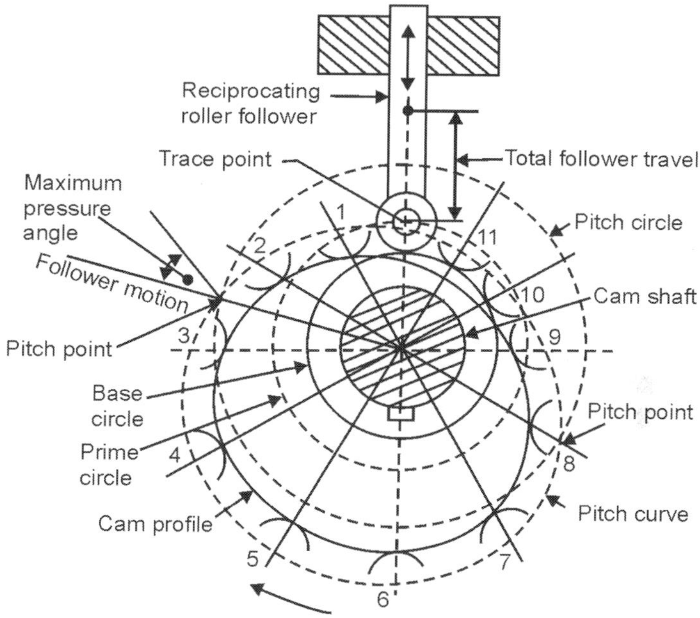

Terms used in radial cams

Figure 20.14

1. *Base circle:* It is the smallest circle that can be drawn to the cam profile.
2. *Trace point:* It is the reference point on the follower and is used to generate a pitch curve. In roller follower, the centre of the roller represents the trace point.
3. *Pressure angle:* It is the angle between the direction of the follower motion and a normal to the pitch curve. The angle is very important in designing a cam profile. If the pressure angle is too large, the reciprocating follower will jam in its bearings.
4. *Pitch point:* It is the point on the pitch curve having the maximum pressure angle.
6. *Pitch circle: It* is a circle drawn from the centre of the cam through the pitch point.
7. *Pitch curve:* It is the curve generated by the trace point as the follower moves relative to the cam. In a roller follower, they are separated by the radius of the roller.
8. *Prime circle:* It is the smallest circle that can be drawn from the centre of the cam and tangent to the pitch curve. For a roller follower, the prime circle is larger than the base circle by the radius of the roller.

20.7 Scutcher feed regulating mechanism (Fig. 20.15)

Control of regularity at the scutcher begins with the method of feeding and provided that the stock as it reaches the scutcher has been adequately opened and is uniformly delivered by the feeding hopper or chute, the feed system will be able to present a uniform fringe of cotton to the action of the beater. There are various arrangement of feeding systems used by the machine makers as pedal and roller arrangement or three roller arrangement.

The feeding arrangement used as a means of regulation is based on the theory underlying the use of the cone drums as means of varying the speed to counteract any irregularity in that feed. The regulator pedals – usually numbering about sixteen – across the width of the scutcher are connected to the cone drum strap-fork in such a way that to give proportional movement to the strap fork to increase or decrease the feed roller speed to compensate for the thick and thin places in the feed. If the sheet of the cotton being fed were perfectly regular, all the pedals would occupy the same position with which they co-operate but, if a thick place occurs, the particular pedal or pedals over which the thick place is passing would be depressed

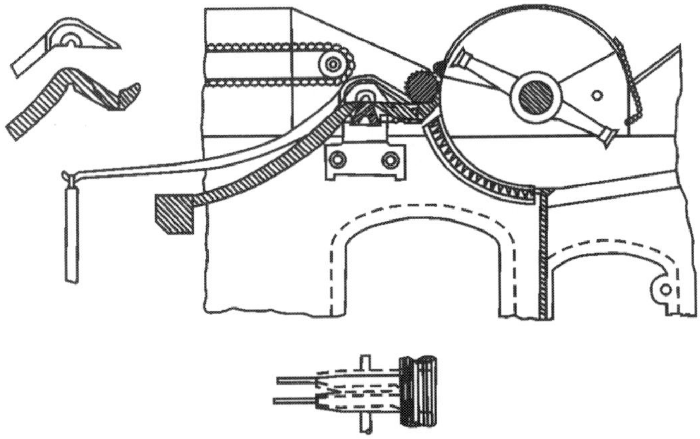

Figure 20.15

A thin portion of the feed material would have the opposite effect, and cause the pedals to rise. The primary objective is to deliver to the succeeding beater equal quantities of cotton in equal intervals of time. This will be achieved if the thick and thin place s were equally distributed throughout the sheet of cotton being fed. In such a case, one half of the pedal noses would be depressed and the other half would rise, thus neutralizing each other. The condition is to have a minimum variation, sometimes the variation is very slight and at other times excessive.

20.8 Cone drum construction (Fig. 20.16)

Referring to the diagram (Fig. 20.16) suppose that a 1/8 in. thickness of cotton is fed between the feed roller and pedal represented by *A*1 and that a 6 inch length *AB* is passed forward in one second. To obtain regularity, an equal quantity of cotton must be fed in every second, and it will readily be seen that if the thickness of the cotton is increased to ¼ in., as shown at *A*2, then to maintain the same amount of cotton passing forward per second, the length must be halved as at *AC*. In order to achieve this, the speed of the feed roller must be reduced to half.

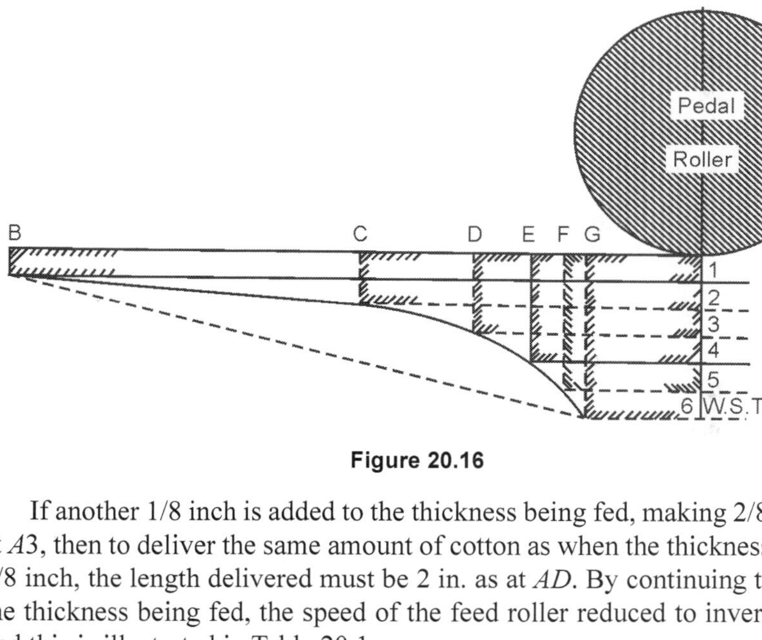

Figure 20.16

If another 1/8 inch is added to the thickness being fed, making 2/8 inch as at *A*3, then to deliver the same amount of cotton as when the thickness is only 1/8 inch, the length delivered must be 2 in. as at *AD*. By continuing to add to the thickness being fed, the speed of the feed roller reduced to inverse ratio, and this is illustrated in Table 20.1.

Table 20.1 Thickness of the material vs feed roller speed

Thickness of the material	Feed roller speed
1	1
2	1/2
3	1/3
4	¼
5	1/5
6	1/6

The above Table 20.1 is constructed assuming 1/8 in. by 1 and letting the speed of the feed roller as 1, the speed of the feed roller for different thicknesses.

As illustrated in Fig. 20.16, it will be seen that on joining together all the thickness plots, a hyperbolic cone drum profile is produced, indicating the curved form of the cone drums which are used to give necessary variable speed to the feed roller.

20.9 Design of cone drum profile (Fig. 20.17)

To design a pair of cone drums which will produce the requisite variation in feed roller speed to maintain a given quantity of cotton being fed in a succession of any given limits of time, certain limits must be assumed which, for example, can be a minimum thickness of ¼ in. and a maximum thickness of 1 in. with the bottom or driving cone drum rotating at a constant speed of say, 100 rev per min.

The speeds of the top cone drum necessary to give the required correction to the speed of the feed roller to compensate for variations in thickness can be obtained as shown in Table 20.2.

Table 20.2 Feed material thickness vs top cone drum speed

Thickness of the feed in inch.	Top cone drum speed in rpm
¼	X = 200 rev/min.
½	x/2 = 100 rev/min
¾	x/3 = 66.75 rev/min
1	x/4 = 50 rev/min.

This indicates a range in speed of the top cone drum of one to four.

Assuming convenient extreme diameters of the two cones to be 8 in. and 4 in. as shown in fig. The speeds resulting from these diameters are obtained as follows

$$8 \times 100/4 = 200 \text{ rev/min.}$$
$$4 \times 100/8 = 50 \text{ rev/min.}$$

Intermediate speeds at equal intervals along the cones are obtained thus: Since $x = 200$ rev/min, $x/2 = 100$ rev/min and $x/3 = 66.75$ rev/min.

As the driving belt drives the top or driving cone must always have uniform tension when driving at any point along the length of the cones, the sum of the cone diameters at every point along their length must conform to the sum of the minimum and maximum diameters, i.e.

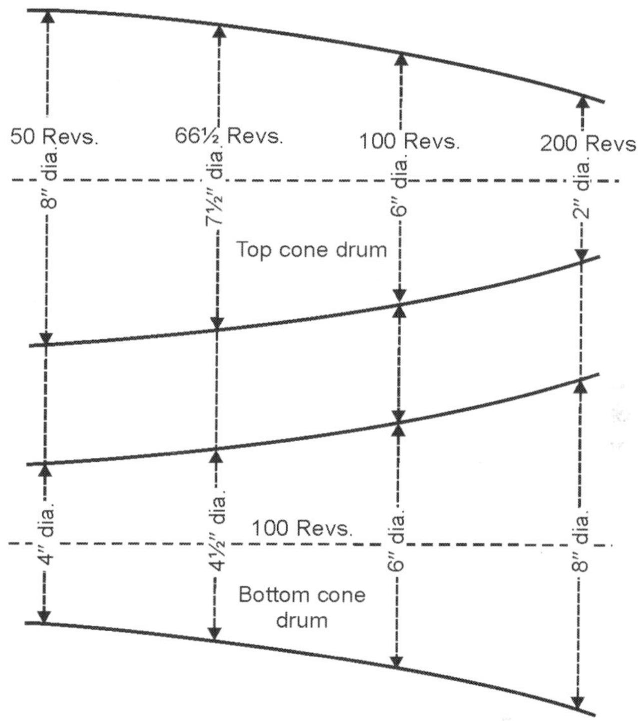

Figure 20.17

8 in. + 4 in. = 12 in.

Referring to Fig. 20.17, it will be seen that when the cone belt is driving from the large diameters of the bottom cone, the top cone has a speed of 200 rev/min, and when it is driving from the small diameter of the bottom cone, the top cone speed will be 50 rev/min.

The diameter of the top drum for 66.75 rev/min will be

$(8 + 4) \times 100/\ 100 + 66.75 = 7.2$ in.

Likewise, for 100 rev /min. the diameter of the top cone will be

$(6 + 6) \times 100\ /\ 100 + 100 = 6$ in.

Further, intermediate diameters at uniform intervals can be obtained, and when they are plotted and the points are joined, the correct curvature is obtained.

Uniform diameters in the thickness of cotton passing between the feed roller and the pedal lever must give uniform increments of movement to the belt along the cone drum. The examples given above include thickness of the driving belt connecting the two cone drums, since the centre line of the belt is the effective diameters and is allowed for when the cones are manufactured.

The cone-drum belt-shifting mechanism is connected to, and receives its motion from the pedal levers through the regulating motion. This device varies in appearance according to the maker, but in principle the action is always the same.

20.10 Design of cone drums in speed frame (Fig. 20.18)

The speed frame cone-drum can be constructed in the following way:

Let the front rollers deliver a length L per minute. Let the bobbin diameter at any instant be B. The top cone-drum diameter is D_T and bottom cone-drum diameter is D_B.

At any instant, the excess rev/min of the bobbin over the spindle is

$L/3.14 \times B$

Let the top cone-drum speed be R rev/min. Then, the bottom cone drum speed will be

$R \times D_T/D_B$

The bottom cone speed multiplied by the gearing constant of the drive from the bottom cone drum to the bobbin equals the excess rev/min. Let this constant be k1. We then have

$$R \times D_T/D_B \times k_1 = \mathrm{L}/3.14 \times \mathrm{B}$$

Since R, L, 3.14 and k_1 are all constants, we can rewrite the last line as

$$D_T/D_B = K/B \qquad \text{... (20.1)}$$

$$D_T/D_B \ \alpha \ 1/B$$

Thus the ratio of the cone drum diameters is inversely proportional to the bobbin diameter. Furthermore, since we have a belt drive, the sum of the cone drum diameters must remain constant, i.e.

$$D_T + D_B = k_2$$

Now we can determine the value of K for our example.

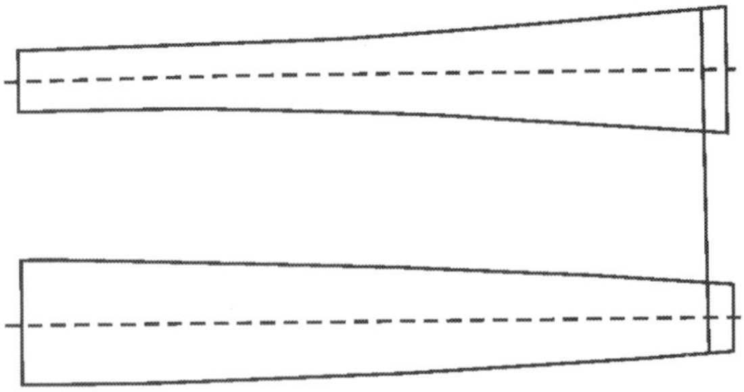

Figure 20.18 Cone drum profile of speed frame

When the bobbin diameter B is 3.35 cm, the ratio D_T/D_B was 1.845, D_T was 16.86 cm and D_B was 9.14 cm. Substituting this value in Eq. (20.1), we get

$$= K/\ 3.35$$

and hence $$K = 1.845 \times 3.35 = 6.18$$

Thus for any bobbin diameter, we can calculate the diameter ratio for the cone drums. We can also determine the values of the cone drum diameters.

Consider the ratio at a bobbin diameter B, i.e.

$$D_T / D_B = 6.18 / B$$

From which

$$D_B = D_T \times B / 6.18$$

Substituting for D_B in the Equation $D_T + D_B = 26$

i.e.

$$D_T (1 + B / 6.18) = 26$$

from which,

$$D_T = 26 \times 6.18 / 6.18 + B;\ D_T = 160.68 / 6.18 + B$$

From this, we can calculate directly the top cone drum diameter for any bobbin diameter and then obtain the bottom cone drum diameter by subtraction.

With the help of this constant of 160.68, a table can be drawn on the top and bottom cone drum diameters which correspond to bobbin size at any instant. This is shown in Table 20.3.

Table 20.3 Cone drum dimensions in relation to Bobbin diameter

Bobbin diameter B (cm)	Top cone Diameter D_T = 160.68/6.18 + B	Bottom cone Diameter $D_B = 26 - D_T$
3.00	17.5	8.5
3.35	16.86	9.14
4.00	15.79	10.21
5.00	14.37	11.63
6.00	13.19	12.81
7.00	12.19	13.81
8.00	11.33	14.67
9.00	10.59	15.41
10.00	9.93	16.09
11.00	9.35	16.65
12.00	8.83	17.17
13.00	8.38	17.62

In present day machines, the cones with hyperbolic profiles are replaced by straight- profile cone drums, and a specially designed cam is used to move the belt the required amount. The net effect is the same namely control of the bottom cone drum speed and hence control the excess revolutions of the bobbin over the spindle in order to wind on the roving delivered from the front rollers.

20.11 Tappets

The object of the shedding is to impart a vertical motion to warp ends for allowing passage to the shuttle so the healds must be moved at a varying speed to correspond with the warp tension. There are three distinct types of shedding mechanism.

1. Tappet shedding mechanism
2. Dobby shedding mechanism
3. Jacquard shedding mechanism

The object of all the different mechanisms is the same, but each of them works on different systems, and each kind has its own disadvantage. Tappet shedding mechanism is the simplest shedding mechanism used in the textile industry suitable for high speed looms. It is having the repeat of the pattern only on small number of ends and picks. The tappet shedding mechanism is suitable for simple weaves: but when the patterns consist of large number of ends for producing a particular design, then a dobby shedding mechanism is utilized. The dobby shedding mechanism is complex and the loom can be operated at slow speeds.

Jacquard shedding mechanism is still more complex but has the advantage of controlling large number of ends and suitable for weaving floral designs. Tappets, cams are the names given to those pieces of mechanism which are characterized as irregular in shape. A rotary motion is given to the tappet which produces reciprocating motion in the levers by sliding contact. The positive tappet controls healds in both the directions whereas a negative tappet is used either to pull down or lift up the heald shaft and an additional motion is required in the opposite direction.

20.12 Construction of shedding tappet

In order to design a shedding tappet, the following points and dimensions must be taken in to consideration:

1. Pattern to be produced in the fabric, i.e., the number of picks in a repeat of the design and the lifting order.
2. Lift or stroke of the tappet

3. Distance from the centre of the driving shaft to the nearest point of contact with the treadle bowl
4. Dwell of the tappet – Time during which the healds will remain stationary, and
5. Diameter of the treadle bowl

20.13 Determination of tappet lift

Consider the following dimensions for designing the tappet. All dimensions are in mm.

Sley sweep (S) = 140

Horizontal distance between cloth fell and from heald shaft (C) = 200

Horizontal distance between cloth fell and back heald shaft (d) = 250

Shuttle base width (W) = 45

Shuttle front wall height (h) = 34

Distance between treadle lever fulcrum and

(a) Bottom shaft centre (m) = 330

(b) Treadle bowl centre (n) = 315

(c) Front heald connection on treadle lever (p) = 630

(d) Back heald connection on treadle lever (q) = 580

Treadle bowl radius (R) = 34

Tappet shaft radius (r) = 19

Then,

(1) Shed height required at front wall of the shuttle when the loom is at back centre

= h + desired clearance between shuttle front top edge and the warp sheet + allowance for variation in dimensions and warp slackness

= (h + 3 + 3) = 40

(2) Shed required at front heald shaft

= $40 \times C / S - W$

= $40 \times 200 / 140 - 45$

= 84.3

(3) Shed height required at back heald shaft

= $40 \times d / S - W$

= $40 \times 250 / 140 - 45$

= 105.3

(4) Lift of the front heald tappet

= $97.2 \times n / p$

= $97.2 \times 315 / 830$

= 49

Lift of the back heald tappet
$= 118.3 \times n / q$
$= 118 \times 315 / 580$
$= 64$

20.14 Determination of NPC and FPC

From general engineering consideration, the minimum value of *NPC* should be equal to tappet shaft radius + 16 mm. Further, it is desirable that both the treadle levers are leveled when the healds are leveled. This means that the treadle levers should move equal distance above and below the line of leveling. Hence the *NPC* for the back tappet should be less than that of the front tappet; the sum *NPC* + *FPC* for the back tappet and for the front tappet will however be equal.

Thus,

(i) *NPC* for the back tappet = 19 + 16 = 35

(ii) *NPC* for the back tappet = NPC for back tappet + half the difference of lift of back and front tappets
$= 35 + 64 - 49 / 2$
$= 42.5$

(iii) FPC for back tappet = 35 + 64 = 99

(iv) FPC for front tappet = 42.5 + 49 = 91.5

20.15 Relation between dwell period and type of heald movement

For satisfactory flight and smooth lifting of heald shaft, a dwell period of 70° for looms up to 150 cm reed space and 80° for looms wider than 150 cm can be taken as standard guide line. Larger dwell periods are required for wider looms because of longer duration of crank cycle is necessary for the shuttle to pass from one box to another on these looms. The type of heald movement during change over follows simple harmonic.

20.16 Tappet design for plain weave

To design a shedding tappet for plain weave, the following particulars have been taken into consideration.

(i) Lift of tappet = 4 in.

(ii) Distance from the centre of the driving shaft to the nearest point of contact with the treadle bowl = 2 in.

(iii) Dwell of tappet = 1/3 of a pick

(iv) Diameter of the treadle bowl = 2 in.

Figure 20.19 shows the construction of the tappet for "Plain weave". The design is at *AA*. The tappet outline is shown at *BB*, and the picks to the round in a repeat of the design is at *CC*.

1. At a radius of 2 inches, draw a circle *A*. This circle represents the distance from the centre of the driving shaft (bottom shaft) to the nearest point of contact (*NPC*) with the treadle bowl. For plain weave tappet, the bottom shaft is used as the driving shaft whereas for twill and other weaves a counter shaft is used as tappet shaft.
2. Draw a circle of radius 3 inches which represents the circle *B*. One inch is added for the radius of the treadle bowl.
3. Draw a circle of radius 7 inches which represents the circle *C*. Four inches are added for the lift of the tappet.

   The circle *B* represents the centre of the treadle bowl when the inner circle of the tappet is acting upon the treadle bowl.

   The circle *C* represents the centre of the treadle bowl when pressed down by the tappet.

   The pattern being a plain one as shown at *AA*, the circle must be divided in to two equal parts, and each half circle will then represent one pick. By the line *DE* divide the circles in to two equal parts as shown in Fig. 20.19.

   Now as the healds must have a pause or dwell equal to 1/3 of a pick when at the top and the bottom of their stroke, divide each half circle in to three equal parts by lines *FK, GH.*
4. Divide *FH* and *GK* in to six equal parts and
5. Divide the space between the circles *B* and *C* in to six unequal parts, the largest being in the middle, gradually decreasing towards the circles *B* and *C*.

   To find out the six unequal parts draw a semi-circle *L* between *B* and *C* at a radius of half the lift of the tappet which is 2 inches. Now divide its circumference in to six equal parts and then draw perpendicular lines from them on the line *GH.* This gives six unequal divisions on the lift to obtain desired eccentric shape of the tappet.
6. From the corners of these unequal spaces, and with the radius of the treadle bowl, draw circles representing the positions of the treadle bowl at different parts of the movement.

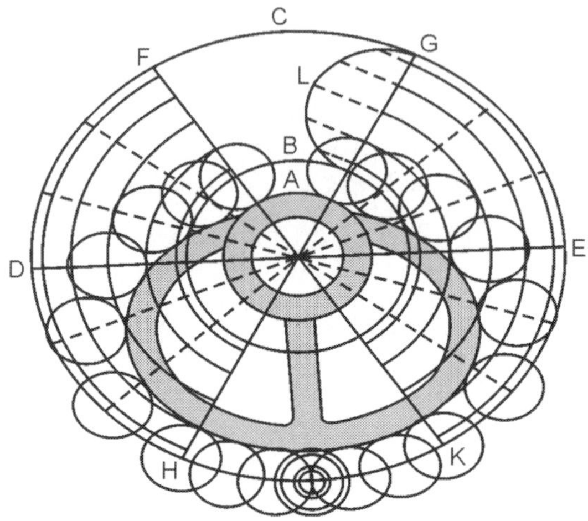

Figure 20.19(AA)

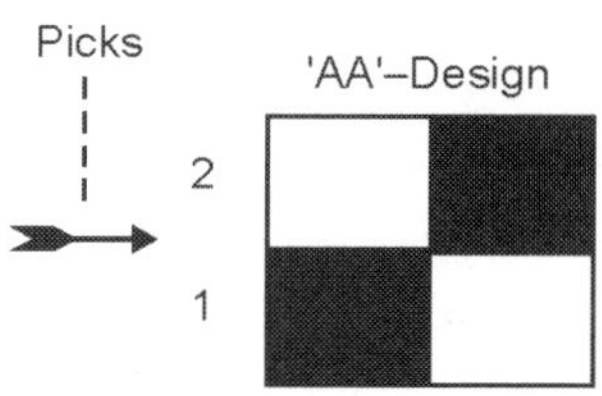

Figure 20.19(BB)

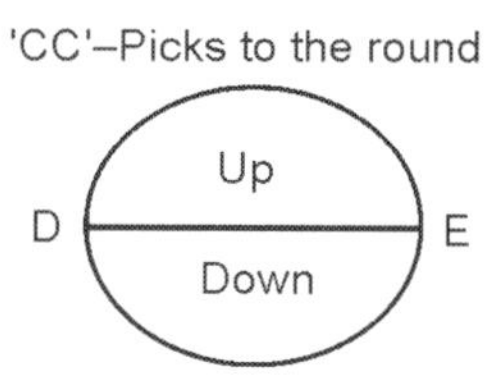

Figure 20.19(CC)

7. Draw curved lines touching the extremities of the treadle bowl. This gives the outline of the "Plain Weave Tappet."

In order to construct a plain weave, two tappets are required which must be secured at 180 degrees, the second being made fast by the side of number one.

It will be seen from the tappet construction that the treadle bowl is at rest from *F* to *G* and from *H* to *K*, or 1/3 of a pick at the top and bottom of the stroke. Therefore, the time allowed for change, or for moving the heald from top to bottom, or vice-versa, is equal to 2/3 of a pick.

If a dwell equal to half a pick is obtained by dividing the pick in to four equal parts and taking the middle two parts for dwell.

If two-thirds dwell is required divide the pick in to six equal parts and take four parts for dwell.

Tappets for twills and other simple weaves having more than two picks to the round are usually placed upon a 'counter shaft'.

20.17 Design of tappet for twill weave

Twill tappet is constructed with the following with the following particulars.

(i) Pattern or design – 3 up 1 down

(ii) Distance from centre of shaft to the nearest point of contact with treadle bowl – 3 inches

(iii) Lift of tappet – 3 inches

(iv) Diameter of treadle bowl – 2 inches

(v) Dwell – ½ a pick

Figure 20.20 shows the construction of the tappet of 3 up and 1 down twill weave. The design is shown at *AA*, the tappet construction at *BB* and the picks to the round in one repeat of the design at *CC*.

Construction of the tappet:

(i) Draw a circle of radius 3 inches and name it as *A*.

(ii) Draw a circle of radius 4 inches and name it as *B* (one inch added to treadle bowl).

(iii) Draw a circle of radius 7 inches in which 3 inches added for lift of tappet.

(iv) As there are four picks in the pattern, divide the circles in to four equal parts by the lines *DE*, *FG*. Then each quarter circle represents one pick and the tappets must be made to make one revolution for four revolutions of the crank shaft.

(v) As the dwell of the heald (when the shed is open) must be equal to half a pick or half a revolution of the crankshaft, divide the first pick in to four equal parts by the points *O*, *L*, *M*.

(vi) Make *DP* equal to *DO* and *FN* equal to *FM*, and draw line from *P*, *O*, *M*, and *N* to the centre. The distance *OM* represents the half pick dwell and the distance *OP* and *MN* represent the half pick which will be allowed for changing the heald from bottom to top of its stroke and vice-versa.

(vii) Divide *OP* and *MN* in to six equal parts and the lift of the tappet, or the distance between the circles *B* and *C* in to six unequal parts, the largest in the middle and gradually decreasing towards the two circles.

(viii) From the corners of the unequal spaces, draw small circles representing the treadle bowl at different parts of its stroke, and

(ix) Finally, draw the tappet outline which will touch the extremities of these circles. This gives the outline of the 3 up and 1 down Twill Weave Tappet.

A tappet designed in this way acts upon the treadle bowl 2 in. in diameter will take the heald down for one pick and allow it to go for three picks. The heald will be held stationary for exactly half a pick when at the bottom of its stroke, and will begin to rise slowly, and gradually increases in speed as it approaches the centre of its stroke, and will gradually decrease in speed as it approaches the top of its stroke.

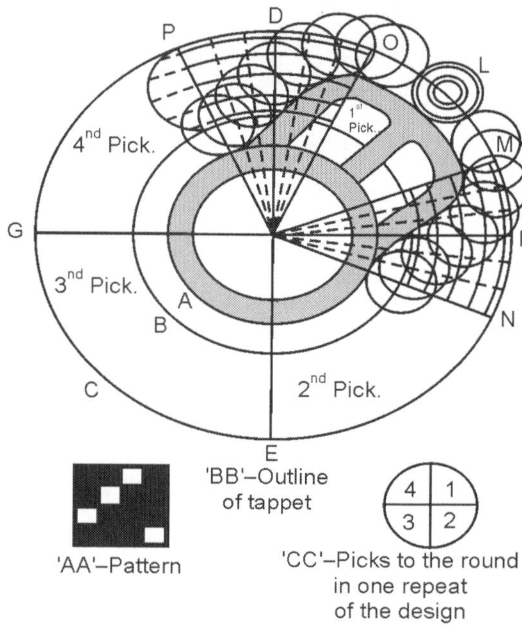

Figure 20.20 AA, BB, CC

For this twill weave there will be four treadle levers, each treadle being operated by a tappet of the same shape; but the tappet operating each succeeding treadle will be placed one quarter of a revolution than the previous one.

20.18 Construction of a shedding tappet for three-shaft weave

A shedding tappet is to be designed for a three-shaft weave. The shaft has to de down for one pick and up for two. The treadle lever is connected to the shaft, and the 'down' position of the bowl represents the 'down' position of the shaft. The lifting is carried by means of a spring. The following are the particulars of the tappet design.

NPC between bowl and tappet canters = 9 cm

Lift of tappet = 5 cm

Diameter of anti-friction bowl = 7.5 cm

Duration of dwell = 1/3 pick

Solution:

If the movement of the shaft is S.H.M. and the lift line is assumed to be straight and passing through the tappet shaft centre, design the tappet by geometrical construction.

Figure 20.22 shows the details of the construction used in tappet design procedure.

Since, the weave is three-shaft, the loom crank shaft will make three revolutions for each revolution of the tappet.

From a common point *A*, assumed as tappet centre, three straight lines are drawn such that they trisect any circle of centre *A*. The angle between any two these lines represents the angular rotation of the tappet during one revolution of the crank shaft. Also, we know that the dwell time is 1/3 of a pick, so that for one-third of each revolution of the crank shaft, there is no heald-shaft movement.

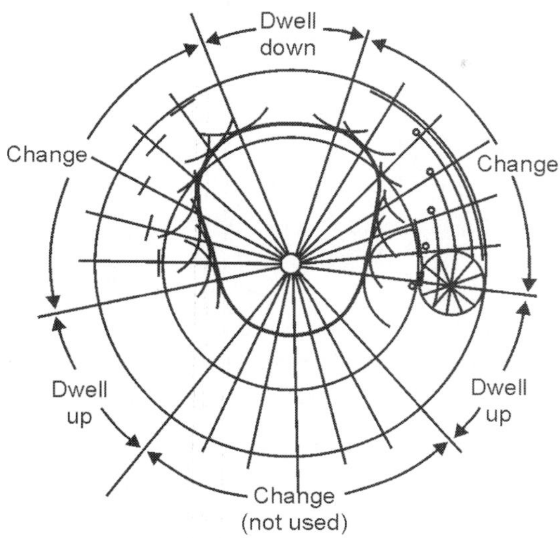

Figure 20.21 Shedding tappet design

Each of our three angles is therefore trisected as shown in Fig. 20.14, so that any circle of centre A is sub-divided into two angles of $2\frac{\pi}{9}$ and $4\frac{\pi}{9}$ repeated three times in sequence around the circle. Each angle of $2\pi/9$ then represents a dwell period and each angle of $4\frac{\pi}{9}$ a change period as shown in Fig. 20.21.

With centre *A* circles of 9 cm and 14 cm respectively are then drawn. At some positions on these circles, the minimum and maximum separation, respectively of tappet and bowl canters will occur. Each of the six $2\pi/9$ segments present in one or other change period is then trisected again, so that each change period angle of $4\pi/9$ has been divided in to six equal angles of $2\pi/27$.

Select as a reference line, the line *AB*, which thus separates a dwell period from a change period. On this line, construct a circle of diameter equal to the lift, located at the lift position of the line, which is just touching the two circles already drawn.

Divide the circle in to 12 equal segments, six on each side of the reference line. With centre *A*, draw a series of arcs that each arc cuts this circle at the point where a segment line meets it, as shown, and extend each arc to meet one of the lines marking the $2\pi/27$. segments of circles with centre A. With this construction, we get a series of points *C, D, E, F, G, H, I,* and since their distance from the start of the lift has been derived by projection of a circle on a straight line, their distance from the start of the lift must divide the lift in proportions equivalent to S.H.M.

From point *C*, a circle of diameter 7.5 cm (equal to that of anti-friction bowl) is then drawn, and a similar circle is drawn from each of the other six points *D* – I. The tappet out line required at this particular change period can then be obtained merely by drawing the line touching these circles.

For one of the other two change periods, the whole operation has to be repeated in the relevant area. For the third one, since no change takes place, when the heald is up, the tappet outline consists of a circle on which the bowl rests at its distance of minimum separation from the tappet shaft. The calculation shows this to be a circle of radius 5 – 25 cm obtained by subtracting the radius of the bowl from that of the circle at the start of the lift.

Similarly, at the dwell zone between the two active change periods, the tappet outline is a circle which allows the bowl to rest at its maximum distance from the tappet shaft centre and that has a radius of (14 – 3.75 cm) = 10.25 cm.

Now, the tappet outline can be drawn as shown in Fig. 20.21.

20.19 Cranks and eccentrics

Cranks and eccentrics are mere mechanical devices with a practical function in machine operation.

A crank is a device designed to convert a rotational movement in to a linear- oscillating one; the reverse operation is also possible if desired. The most familiar example is the bicycle pedal as the rider's leg moves up and down. Its reciprocating motion is converted to a rotational one by means of a crank connecting the pedal to the chain sprocket wheel. In general, the crank is a rigid rod fixed at one end to the centre of rotary movement and with a joint at the other end to provide a flexible connection to the oscillating component. Figure 20.22 shows the operation of the device.

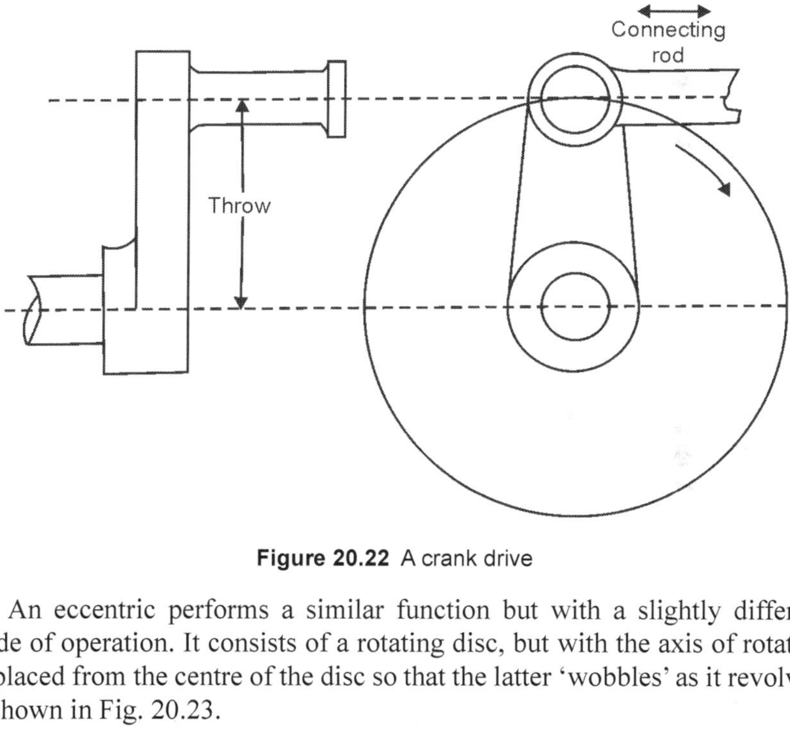

Figure 20.22 A crank drive

An eccentric performs a similar function but with a slightly different mode of operation. It consists of a rotating disc, but with the axis of rotation displaced from the centre of the disc so that the latter 'wobbles' as it revolves, as shown in Fig. 20.23.

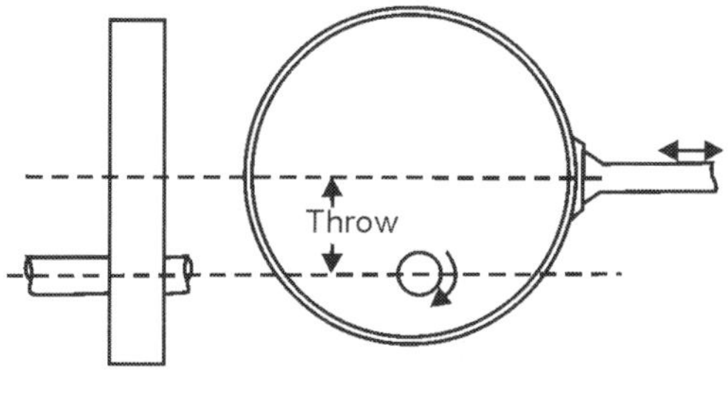

Figure 20.23 Operation of an eccentric

It is thus obvious that the rim of the disc moves in an oscillatory manner as rotation takes place, so that a reciprocating motion may be obtained by suitably harnessing this rim movement. An example, we can say, is the movement of the shedding griffe in a jacquard loom as shown in Fig. 20.24.

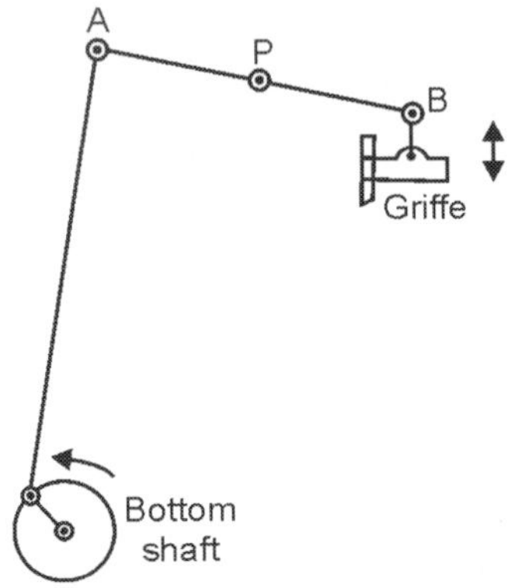

Figure 20.24 Shedding griffe in a jacquard loom

In the shedding mechanism of jacquard loom shown in Fig. 20.24, as the bottom shaft rotates, the crank attached to it moves the long shaft connected to the lever *AB*, pivoted at *P*, which controls the vertical movement of griffe. Since *P* is at the mid-point of *AB*, griffe movement is equal in magnitude but opposite in direction to the movement of *A*. In practice, A moves in a slightly curved path, because the length of the lever is much larger than the distance moved by *A*.

Index

A

B

C

K

L

M

N

O

P

Q

R

S

T

V

W

Y